STERLING
Education

Homeschool Physics

Questions & Explanations

3rd edition

Customer Satisfaction Guarantee

Your feedback is important because we strive to provide the highest quality prep materials. Email us comments or suggestions.

info@sterling–prep.com

We reply to emails – check your spam folder

3 2 1

ISBN-13: 979-8-8855711-7-3

Sterling Education materials are available at quantity discounts.

Contact info@sterling–prep.com

Sterling Education
6 Liberty Square #11
Boston, MA 02109

Published by Sterling Test Prep

Printed in the U.S.A.

STERLING
Education

Thousands of students use our educational products to achieve academic success!

Many students find physics interesting but a challenging discipline. To pass their state educational requirements, homeschool students are often required to take standardized tests. This book helps students develop and apply knowledge to solve physics problems. To do well in high school level physics, students need to understand core material, know how to extract and analyze information, and be able to distinguish between similar answer choices. With this book, students master essential physics content and develop the ability to apply their knowledge on quizzes and tests.

This book provides over one thousand practice questions covering physics topics for comprehensive preparation. The detailed explanations describe why an answer is correct and why another attractive choice is wrong. They provide step-by-step solutions for quantitative problems and teach the scientific foundations and details of essential physics needed to answer conceptual questions. Reading the explanations carefully is critical for understanding how they apply to the question and learning important physics principles and the relationships between them.

Experienced physics instructors prepared this practice material to build knowledge and skills crucial for success in physics class. Our editorial team reviewed and systematized the content for targeted and effective learning to significantly improve understanding of the subject.

We wish you great success in mastering physics!

251025akp

Homeschool Physics Comprehensive Content provides thorough coverage of physics, teaching the foundational ideas and theories necessary to master the core content and develop the ability to apply this knowledge on quizzes and tests.

Visit our Amazon store

For online practice resources visit
https://www.sterling-prep.com

If you benefited from this book, please leave a review on Amazon so others can learn from your input. Reviews help us understand our customers' needs and experiences while keeping our commitment to quality.

Table of Contents

Table of Contents (continued)

Topical Practice Questions

Translational Motion

1. Starting from rest, how long does a car take to reach 60 mi/h with an average acceleration of 13.1 mi/h·s?

A. 6.6 s
B. 3.1 s
C. 4.5 s
D. 4.6 s

2. A cannonball is fired with an initial 20 m/s at a 30° angle with the horizontal. Ignoring air resistance, how long does it take the cannonball to reach the top of its trajectory? (Use acceleration due to gravity $g = 10\ \text{m/s}^2$)

A. 0.5 s
B. 1 s
C. 1.5 s
D. 2 s

3. Darlene starts her car from rest and accelerates at a constant 2.5 m/s^2 for 9 s to get to her cruising speed. She then drives for 15 minutes at a constant speed. She arrives at her destination, a straight-line distance of 31.5 km, exactly 1.25 hours later. What is Darlene's average velocity during 1.25 hours?

A. 3 m/s
B. 7 m/s
C. 18 m/s
D. 22.5 m/s

4. Which of the following cannot be negative?

A. Instantaneous speed
B. Instantaneous acceleration
C. Acceleration of gravity
D. Displacement

5. How far does a car travel while accelerating from 5 m/s to 21 m/s at a rate of 3 m/s^2?

A. 15 m
B. 21 m
C. 69 m
D. 105 m

6. Acceleration is sometimes expressed in multiples of g, where g is the acceleration due to gravity. How many g are experienced, on average, by the driver in a car crash if the car's velocity changes from 30 m/s to 0 m/s in 0.15 s? (Use the acceleration due to gravity $g = 9.8\ \text{m/s}^2$)

A. 22 g
B. 28 g
C. 20 g
D. 14 g

7. Ignoring air resistance, how many forces act on a bullet fired horizontally after leaving the rifle?

A. Two (one from the gunpowder explosion and one from gravity)
B. One (from the motion of the bullet)
C. One (from the gunpowder explosion)
D. One (from the pull of gravity)

8. Suppose that a car traveling to the East begins to slow down as it approaches a traffic light. Which of the following statements about its acceleration is correct?

A. The acceleration is towards the East
B. The acceleration is towards the West
C. Since the car is slowing, its acceleration is positive
D. The acceleration is zero

9. On a planet where the acceleration due to gravity is 20 m/s^2, a freely falling object increases its speed each second by about:

A. 20 m/s
B. 10 m/s
C. 30 m/s
D. 40 m/s

10. What is a car's acceleration if it accelerates uniformly in one direction from 15 m/s to 40 m/s in 10 s?

A. 1.75 m/s^2
B. 2.5 m/s^2
C. 3.5 m/s^2
D. 7.6 m/s^2

11. If the fastest a person can drive is 65 mi/h, what is the longest time she can stop for lunch if she wants to travel 540 mi in 9.8 h?

A. 1 h
B. 2.4 h
C. 1.5 h
D. 2 h

12. What is a racecar's average velocity if it completes one lap around a 500 m track in 10 s?

A. 10 m/s
B. 0 m/s
C. 5 m/s
D. 20 m/s

13. What is a ball's net displacement after 5 s if it initially rolls up a slight incline at 0.2 m/s and decelerates uniformly at 0.05 m/s^2?

A. 0.38 m
B. 0.6 m
C. 0.9 m
D. 1.2 m

14. What does the slope of a line connecting two points on a velocity *vs.* time graph represent?

A. Change in acceleration
B. Instantaneous acceleration
C. Average acceleration
D. Instantaneous velocity

15. An airplane needs to reach a speed of 210.0 km/h to take off. On a 1,800.0 m runway, what minimum acceleration is necessary for the plane to reach this speed, assuming acceleration is constant?

A. 0.78 m/s^2
B. 0.95 m/s^2
C. 1.47 m/s^2
D. 1.1 m/s^2

16. A test rocket is fired straight up from rest with a net acceleration of 22 m/s^2. What maximum elevation does the rocket reach if the motor turns off after 4 s, but the rocket continues to coast upward? (Use the acceleration due to gravity $g = 10\ m/s^2$)

A. 408 m
B. 320 m
C. 357 m
D. 563 m

17. Without reference to direction, how fast an object moves refers to its:

A. speed
B. impulse
C. momentum
D. velocity

18. Ignoring air resistance, a 10 kg rock and a 20 kg rock are dropped simultaneously. If the 10 kg rock falls with acceleration a, what is the acceleration of the 20 kg rock?

A. $a / 2$
B. a
C. $2a$
D. $4a$

19. As an object falls freely, its magnitude of:

I. velocity increases II. acceleration increases III. displacement increases

A. I only
B. I and II only
C. II and III only
D. I and III only

20. A man stands in an elevator that is ascending at a constant velocity. What forces are being exerted on the man, and in which direction does the net force point?

A. Gravity pointing downward, normal force from the floor pointing upward, and tension force from the elevator cable pointing upward; net force points upward
B. Gravity pointing downward and the normal force from the floor pointing upward; net force points upward
C. Gravity pointing downward and normal force from the floor pointing upward; net force is 0
D. Gravity pointing downward; net force is 0

21. A football kicker is attempting a field goal from 44 m, and the ball just clears the lower bar with a time of flight of 2.9 s. What was the ball's initial speed if the angle of the kick was 45° with the horizontal?

A. 37 m/s
B. 2.5 m/s
C. 18.3 m/s
D. 21.4 m/s

22. Ignoring air resistance, if a rock, starting at rest, is dropped from a cliff and strikes the ground with an impact velocity of 14 m/s, from what height was it dropped? (Use acceleration due to gravity $g = 10$ m/s^2)

A. 10 m
B. 30 m
C. 45 m
D. 70 m

23. An SUV is traveling at 20 m/s. Then Joseph steps on the accelerator pedal, accelerating at a constant 1.4 m/s^2 for 7 s. How far does he travel during these 7 s?

A. 205 m
B. 174 m
C. 143 m
D. 158 m

24. Which of the following is NOT a scalar?

A. temperature
B. distance
C. mass
D. force

25. Two identical balls (A and B) fall from rest from different heights to the ground. Ignoring air resistance, what is the ratio of the heights from which A and B fall if ball B takes twice as long as ball A to reach the ground?

A. 1 : √2
B. 1 : 4
C. 1 : 2
D. 1 : 8

26. How far does a car travel in 10 s when it accelerates uniformly in one direction from 5 m/s to 30 m/s?

A. 175 m
B. 25 m
C. 250 m
D. 650 m

27. Which graph represents an acceleration of zero?

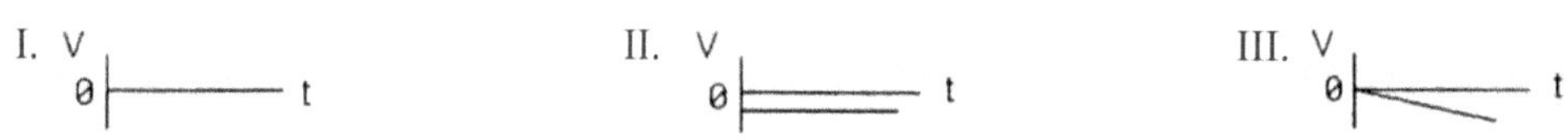

A. I only
B. II only
C. I and II only
D. II and III only

28. Doubling the distance between an orbiting satellite and the Earth results in what change in the gravitational attraction between the two?

A. Twice as much
B. Four times as much
C. One half as much
D. One fourth as much

29. An object is moving in a straight line. Consider its motion during some interval of time: under what conditions is it possible for the object's instantaneous velocity at some point during the interval to be equal to the average velocity over the interval?

I. When velocity is constant during the interval
II. When velocity is increasing at a constant rate during the interval
III. When velocity is increasing at an irregular rate during the interval

A. II only
B. I and III only
C. II and III only
D. I, II and III

30. A freely falling object on Earth, 10 s after starting from rest, has a speed of about: (Use the acceleration due to gravity $g = 10$ m/s^2)

A. 10 m/s
B. 20 m/s
C. 100 m/s
D. 150 m/s

31. A truck travels a certain distance at a constant velocity v for time t. If the truck travels three times as fast, covering the same distance, then by what factor does the time of travel in relation to t change?

A. Increases by 3
B. Decreases by 3
C. Decreases by $\sqrt{3}$
D. Increases by 9

32. Assuming equal acceleration rates, how much farther would Steve travel if he braked from 59 mi/h to rest than from 29 mi/h to rest?

A. 2 times farther
B. 16 times farther
C. 4 times farther
D. 3.2 times farther

33. What is the average speed if a horse does one lap around a 400 m track in 20 s?

A. 0 m/s
B. 7.5 m/s
C. 15 m/s
D. 20 m/s

34. What was a car's initial velocity if the car traveled up a slight slope while decelerating at 0.1 m/s2 and stopped after 5 s?

A. 0.5 m/s
B. 0.25 m/s
C. 2 m/s
D. 1.5 m/s

35. Average velocity equals the average of an object's initial and final velocity when acceleration is:

A. constantly decreasing
B. constantly increasing
C. constant
D. equal to zero

36. Ignoring air resistance, compared to a rock dropped from the same point, how much earlier does a thrown rock strike the ground if thrown downward with an initial velocity of 10 m/s from the top of a 300 m building? (Use acceleration due to gravity $g = 9.8 \text{ m/s}^2$)

A. 0.75 s
B. 0.33 s
C. 0.66 s
D. 0.95 s

37. With other factors equal, what happens to the acceleration of on an object with a given mass, if the unbalanced force is doubled?

A. Increased by one-fourth
B. Increased by one-half
C. Increased fourfold
D. Doubled

38. How fast an object is changing speed or direction of travel is a property of motion known as:

A. velocity
B. acceleration
C. speed
D. flow

39. Which statement concerning a car's acceleration must be correct if a car traveling to the North (+y direction) begins to slow down as it approaches a stop sign?

A. Acceleration is positive
B. Acceleration is zero
C. Acceleration is negative
D. Acceleration decreases in magnitude as the car slows

40. For the velocity *vs.* time graph of a basketball player traveling up and down the court in a straight-line path, what is the total distance run by the player in the 10 s?

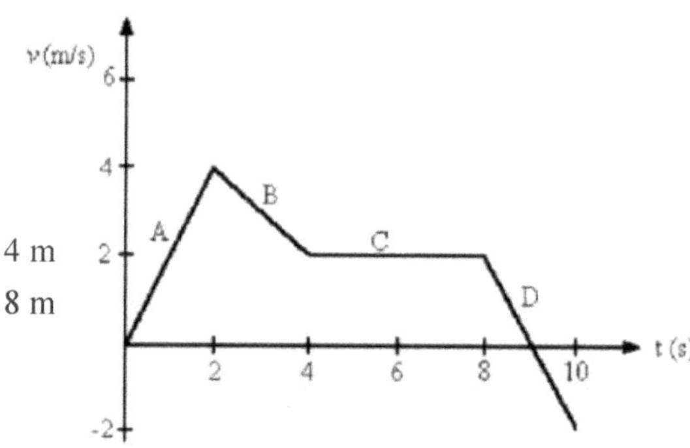

A. 20 m
B. 22 m
C. 14 m
D. 18 m

41. At the same time that a bullet is dropped into a river from a high bridge, another bullet is fired from a gun, straight down towards the water. Ignoring air resistance, the acceleration just before striking the water:

A. is greater for the dropped bullet
B. is greater for the fired bullet
C. is the same for each bullet
D. depends on how high the bullets started

42. Sarah starts her car from rest and accelerates at a constant 2.5 m/s^2 for 9 s to get to her cruising speed. What was her final velocity?

A. 22.5 m/s
B. 12.3 m/s
C. 4.6 m/s
D. 8.5 m/s

43. A bat hits a baseball, and the baseball's direction is completely reversed, and its speed is doubled. If the actual time of contact with the bat is 0.45 s, what is the ratio of the acceleration to the original velocity?

A. $-2.5\ s^{-1} : 1$
B. $-0.15\ s^{-1} : 1$
C. $-9.8\ s^{-1} : 1$
D. $-6.7\ s^{-1} : 1$

44. A 2 kg weight is thrown vertically upward from the surface of the Moon at a speed of 3.2 m/s, and it returns to its starting point in 4 s. What is the magnitude of the acceleration due to gravity on the Moon?

A. 0.8 m/s^2
B. 1.6 m/s^2
C. 3.7 m/s^2
D. 8.4 m/s^2

45. What is the change in velocity for a bird cruising at 1.5 m/s and then accelerating at a constant 0.3 m/s^2 for 3 s?

A. 0.9 m/s
B. 0.6 m/s
C. 1.6 m/s
D. 0.3 m/s

46. All of the following are vectors, except:

A. velocity
B. displacement
C. acceleration
D. mass

Questions **47-49** are based on the following:

A toy rocket is launched vertically from ground level where $y = 0$ m, at time $t = 0$ s. The rocket engine provides constant upward acceleration during the burn phase. At the instant of engine burnout, the rocket has risen to 64 m and acquired a velocity of 60 m/s. The rocket rises in unpowered flight, reaches the maximum height, and then falls back to the ground. (Use the acceleration due to gravity $g = 9.8\ m/s^2$)

47. What is the maximum height reached by rocket?

A. 274 m
B. 248 m
C. 223 m
D. 120 m

48. What is the upward acceleration of the rocket during the burn phase?

A. 9.9 m/s^2
B. 4.8 m/s^2
C. 28 m/s^2
D. 11.8 m/s^2

49. What is the time interval during which the rocket engine provides upward acceleration?

A. 1.5 s
B. 1.9 s
C. 2.3 s
D. 2.1 s

50. A car accelerates uniformly from rest along a straight track with markers spaced at equal distances. As it passes Marker 2, the car reaches a speed of 140 km/h. Where on the track is the car when it traveled at 70 km/h?

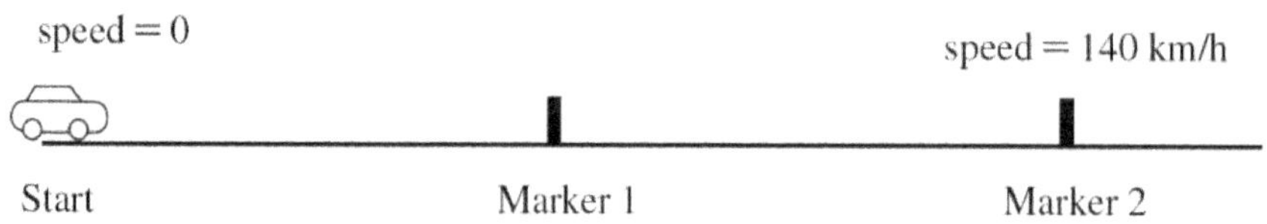

A. Close to Marker 2
B. Between Marker 1 and Marker 2
C. Before Marker 1
D. Close to the starting point

51. What are the two measurements necessary for calculating average speed?

A. Distance and time
B. Distance and acceleration
C. Velocity and time
D. Velocity and acceleration

52. A pedestrian traveling at speed v covers a distance x during a time interval t. If a bicycle travels at speed $3v$, how much time does it take to travel the same distance?

A. $t / 3$
B. $t - 3$
C. $t + 3^2$
D. $3t$

53. Ignoring air resistance, how much time passes before a ball strikes the ground if it is thrown straight upward with a velocity of 39 m/s? (Use acceleration due to gravity $g = 9.8$ m/s^2)

A. 2.2 s
B. 8 s
C. 12 s
D. 4 s

54. A particle travels to the right along a horizontal axis with a constantly decreasing speed. Which one of the following describes the direction of the particle's acceleration?

A. ↑
B. ↓
C. →
D. ←

55. Larry carries a 25-kg package at a constant velocity of 1.8 m/s across a room for 12 s. What is the work done by Larry on the package during the 12 s? (Use acceleration due to gravity $g = 10$ m/s^2)

A. 0 J
B. 280 J
C. 860 J
D. 2,200 J

56. What does the slope of a tangent line at a time on a velocity *vs.* time graph represent?

A. Instantaneous acceleration
B. Average acceleration
C. Instantaneous velocity
D. Position

57. A car is traveling North at 17.7 m/s. After 12 s, its velocity is 14.1 m/s in the same direction. What is the magnitude and direction of the car's average acceleration?

A. 0.3 m/s^2, North
B. 2.7 m/s^2, North
C. 0.3 m/s^2, South
D. 3.6 m/s^2, South

58. The graph shows the position of an object as a function of time. The letters A –E represent moments in time. At which moment in time is the speed of the object the highest?

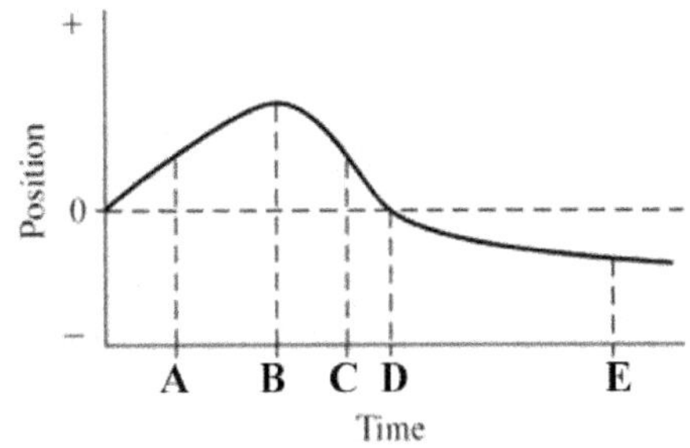

A. A
B. B
C. C
D. D

59. At speed less than terminal velocity, what is happening to the speed of an object falling toward the surface of the Earth?

A. Decreasing at a decreasing rate
B. Increasing at a decreasing rate
C. Decreasing
D. Constant

60. How far does the car travel if it starts from rest and accelerates at a constant 2 m/s^2 for 10 s, then travels with the constant speed it has achieved for another 10 s, and finally slows to a stop with a constant deceleration of magnitude 2 m/s^2?

A. 150 m
B. 200 m
C. 350 m
D. 400 m

Notes for active learning

Force and Motion

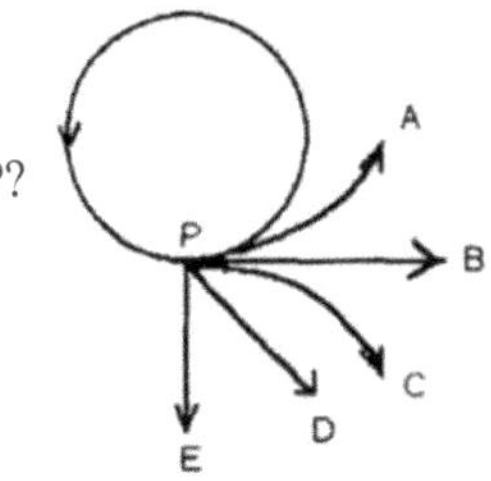

1. A boy attaches a weight to a string, which he swings counterclockwise in a horizontal circle. Which path does the weight follow when the string breaks at point P?

A. path A
B. path B
C. path C
D. path D

2. A garment bag hangs from a clothesline. The tension in the clothesline is 10 N on the right side of the garment bag and 10 N on the left side of the garment bag. The clothesline makes an angle of 60° from vertical. What is the mass of the garment bag? (Use the acceleration due to gravity g = 10.0 m/s^2)

A. 0.5 kg
B. 8 kg
C. 4 kg
D. 1 kg

3. A sheet of paper can be withdrawn from under a milk carton without toppling the carton if the paper is jerked away quickly. This demonstrates:

A. the inertia of the milk carton
B. that gravity tends to hold the milk carton secure
C. there is an action-reaction pair of forces
D. that the milk carton has no acceleration

4. A car of mass m goes up a shallow slope with an angle θ to the horizontal when the driver suddenly applies the brakes. The car skids as it comes to a stop. The coefficient of static friction between the tires and the road is μ_s, and the coefficient of kinetic friction is μ_k. Which expression represents the normal force on the car?

A. $mg \tan \theta$
B. $mg \sin \theta$
C. $mg \cos \theta$
D. mg

5. A 27 kg object is accelerated at a rate of 1.7 m/s^2. How much force does the object experience?

A. 62 N
B. 46 N
C. 7 N
D. 18 N

6. How are two identical masses moving if they are attached by a light string that passes over a small pulley? Assume that the table and the pulley are frictionless.

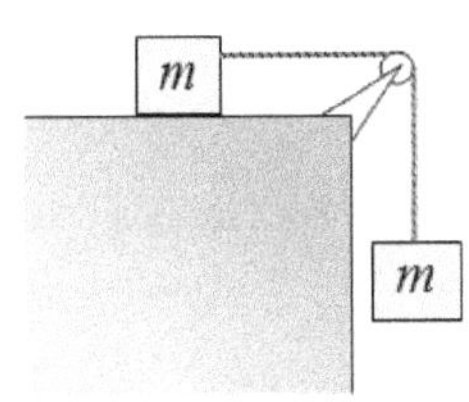

A. With an acceleration equal to g
B. With an acceleration greater than g
C. At a constant speed
D. With an acceleration less than g

7. An object is moving to the right in a straight line. The net force acting on the object is directed to the right, but the magnitude of the force decreases with time. What happens to the object?

A. Continues to move to the right with its speed increasing with time
B. Continues to move to the right with a constant speed
C. Continues to move to the right with its speed decreasing with time
D. Continues to move to the right, slowing quickly to a stop

8. A crate is sliding down an inclined ramp at a constant speed of 0.55 m/s. Where does the vector sum of the forces acting on this crate point?

A. Perpendicular to the ramp
B. Vertically downward
C. Vertically upward
D. None of the above

9. Consider an inclined plane that makes an angle θ with the horizontal. What is the relationship between the length of the ramp L and the vertical height of the ramp h?

A. $h = L \sin \theta$
B. $h = L \tan \theta$
C. $L = h \sin \theta$
D. $h = L \cos \theta$

10. What is the force exerted by the table on a 2 kg book on it? (Use acceleration due to gravity $g = 10\ \text{m/s}^2$)

A. 100 N
B. 20 N
C. 10 N
D. 0 N

11. Sean is pulling his son in a toy wagon. His son and the wagon are 60 kg. For 3s, Sean exerts a force that uniformly accelerates the wagon from 1.5 m/s to 3.5 m/s. What is the acceleration of the wagon with his son?

A. $0.67\ \text{m/s}^2$
B. $0.84\ \text{m/s}^2$
C. $1.66\ \text{m/s}^2$
D. $15.32\ \text{m/s}^2$

12. When an object moves in uniform circular motion, the direction of its acceleration is:

A. directed away from the center of its circular path
B. dependent on its speed
C. directed toward the center of its circular path
D. in the same direction as its velocity vector

13. What happens to a moving object in the absence of an external force?

A. Gradually accelerates until it reaches terminal velocity, at which point it continues at a constant velocity
B. Moves with constant velocity
C. Stops immediately
D. Slows and eventually stops

14. A force of 1 N causes a 1 kg mass to accelerate 1 m/s^2. A force of 9 N applied to a 9 kg mass would have what magnitude of acceleration?

A. 18 m/s^2
B. 9 m/s^2
C. 1 m/s^2
D. 3 m/s^2

15. Which statement is true about an object in two-dimensional projectile motion with no air resistance?

A. The acceleration of the object is zero at its highest point
B. The horizontal acceleration is zero, and the vertical acceleration is a nonzero constant downward
C. The velocity is always in the same direction as the acceleration
D. The acceleration of the object is $+g$ when the object is rising and $-g$ when it is falling

16. A can of paint with a mass of 10 kg hangs from a rope. If the can is to be pulled up to a rooftop with a constant velocity of 0.5 m/s, what must the tension on the rope be? (Use acceleration due to gravity $g = 10$ m/s^2)

A. 100 N
B. 40 N
C. 0 N
D. 120 N

17. What is the magnitude of the force exerted on a 1,000 kg object that accelerates at 2 m/s^2?

A. 500 N
B. 1,000 N
C. 1,200 N
D. 2,000 N

18. A 1,300 kg car is driven at a constant speed of 4 m/s and turns to the right on a curve on the road, an effective radius of 4 m. What is the acceleration of the car?

A. 0 m/s^2
B. 3 m/s^2
C. 4 m/s^2
D. 9.8 m/s^2

19. A block of mass m is resting on a 20° slope. The block has coefficients of friction $\mu_s = 0.55$ and $\mu_k = 0.45$ with the surface.

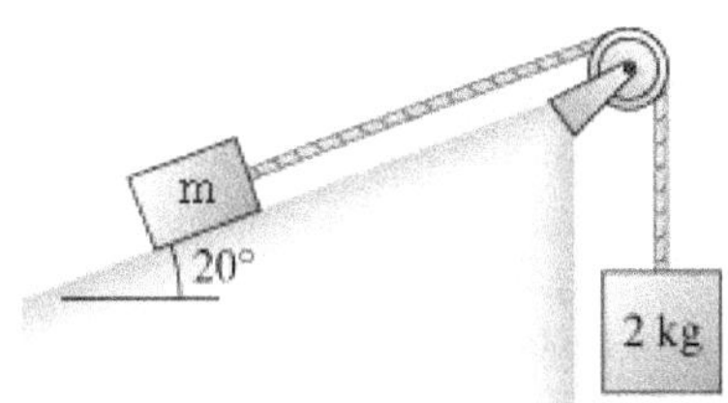

Block m is connected via a massless string over a massless, frictionless pulley to a hanging 2 kg block. What is the minimum mass of block m so that it does not slip? (Use the acceleration due to gravity $g = 9.8$ m/s^2)

A. 2.3 kg
B. 1.3 kg
C. 3.7 kg
D. 4.1 kg

20. As shown in the figure to the right, two identical masses, attached by a light cord passing over a massless, frictionless pulley on an Atwood's machine, are hanging at different heights. If the two masses are suddenly released, then the:

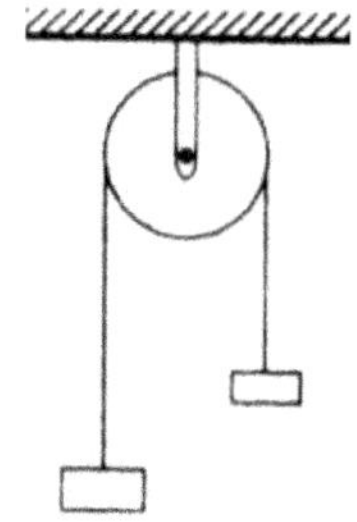

A. lower mass moves down
B. masses remain stationary
C. higher mass moves down
D. motion is unpredictable

21. When Victoria jumps up in the air, which of the following statements is the most accurate?

A. The ground cannot exert the upward force necessary to lift her into the air because the ground is stationary. Instead, Victoria is propelled into the air by the internal force of her muscles acting on her body
B. When Victoria pushes down on the Earth with force greater than her weight, the Earth pushes back with the same magnitude force and propels her into the air
C. Victoria is propelled up by the upward force exerted by the ground, but this force cannot be greater than her weight
D. The Earth exerts an upward force on Victoria that is stronger than the downward force she exerts on the Earth; therefore, Victoria is able to spring up

22. If a feather is pounded with a hammer, which experiences a greater force?

A. The magnitude of the force is always the same on both
B. If the feather moves, then it feels the greater force
C. Depends on the force with which the hammer strikes the feather
D. Always the hammer

23. A block is moving down a slope of a frictionless inclined plane. Compared to the weight of the block, what is the force parallel to the surface of the plane experienced by the block?

A. Greater
B. Unrelated
C. Less than
D. Equal

24. A package falls off a truck that is moving at 30 m/s. Ignoring air resistance, the horizontal speed of the package just before it hits the ground is:

A. 0 m/s
B. 30 m/s
C. $\sqrt{60}$ m/s
D. $\sqrt{30}$ m/s

25. A carousel with the radius *r* is turning counterclockwise at a frequency *f*. How does the velocity of a seat on the carousel change when *f* is doubled?

A. Increases by a factor of $2r$
B. Increases by a factor of r
C. Remains unchanged
D. Doubles

26. What is the mass of a car if it takes 4,500 N to accelerate it at a rate of 5 m/s^2?

A. 900 kg
B. 1,320 kg
C. 620 kg
D. 460 kg

27. Steve is standing facing forward in a moving bus. What force causes Steve to suddenly move forward when the bus comes to an abrupt stop?

A. Force due to the air pressure inside the previously moving bus
B. Force due to kinetic friction between Steve and the floor of the bus
C. Force due to stored kinetic energy
D. No forces were responsible for Steve's movement

28. A plastic ball in a liquid is acted upon by its weight and a buoyant force. The weight of the ball is 4.4 N. The buoyant force of 8.4 N acts vertically upward. An external force acting on the ball maintains it in a state of rest. What is the magnitude and direction of the external force?

A. 4 N, upward
B. 8.4 N, downward
C. 4.4 N, upward
D. 4 N, downward

29. A passenger on a train traveling in the forward direction notices that a piece of luggage starts to slide directly toward the front of the train. From this, it can be concluded that the train is:

A. slowing down
B. speeding up
C. moving at a constant velocity forward
D. changing direction

30. An object has a mass of 36 kg and weighs 360 N at the surface of the Earth. If this object is transported to an altitude twice the Earth's radius, what is the object's mass and weight, respectively?

A. 9 kg and 90 N
B. 36 kg and 90 N
C. 4 kg and 90 N
D. 36 kg and 40 N

31. A truck is moving at constant velocity. Inside the storage compartment, a rock is dropped from the midpoint of the ceiling and strikes the floor below. The rock hits the floor:

A. just behind the midpoint of the ceiling
B. exactly halfway between the midpoint and the front of the truck
C. exactly below the midpoint of the ceiling
D. just ahead of the midpoint of the ceiling

32. Jason takes off across level water on his jet-powered skis. The combined mass of Jason and his skis is 75 kg (the mass of the fuel is negligible). The skis have a thrust of 200 N and a coefficient of kinetic friction on the water of 0.1. If the skis run out of fuel after only 67 s, how far has Jason traveled before he stops?

A. 10,331 m
B. 3,793 m
C. 8,224 m
D. 7,642 m

33. A 200 g hockey puck is launched up a metal ramp inclined at a 30° angle. The puck's initial speed is 63 m/s. What vertical height does the puck reach above its starting point? (Use acceleration due to gravity $g = 9.8$ m/s^2, coefficient of static friction $\mu_s = 0.40$ and kinetic friction $\mu_k = 0.30$ between the puck and metal ramp)

A. 66 m
B. 200 m
C. 170 m
D. 130 m

34. When a 4 kg mass and a 10 kg mass are pushed from rest with equal force:

A. 4 kg mass accelerates 2.5 times faster than the 10 kg mass
B. 10 kg mass accelerates 10 times faster than the 4 kg mass
C. 4 kg mass accelerates at the same rate as the 10 kg mass
D. 10 kg mass accelerates 2.5 times faster than the 4 kg mass

35. An object at rest on an inclined plane starts to slide when the incline is increased to 17°. What is the coefficient of static friction between the object and the plane? (Use the acceleration due to gravity $g = 9.8$ m/s^2)

A. 0.37
B. 0.43
C. 0.24
D. 0.31

36. Which of the following statements must be true when a 20-ton truck collides with a 1,500 lb car?

A. During the collision, the force on the truck is equal to the force on the car
B. The truck did not slow down during the collision, but the car did
C. During the collision, the force on the truck is greater than the force on the car
D. During the collision, the force on the truck is smaller than the force on the car

37. A block is on a frictionless table on Earth. The block accelerates at 3 m/s^2 when a 20 N horizontal force is applied to it. The block and table are transported to the Moon. What is the weight of the block on the Moon? (Use the acceleration due to gravity at the surface of the Moon = 1.62 m/s^2)

A. 5.8 N
B. 14.2 N
C. 8.5 N
D. 11 N

38. Two forces of equal magnitude are acting on an object, as shown. If the magnitude of each force is 2.3 N and the angle between them is 40°, which third force causes the object to be in equilibrium?

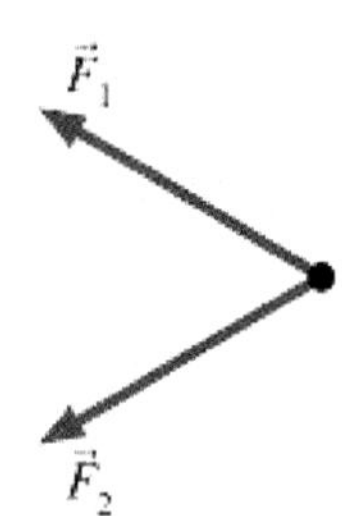

A. 4.3 N pointing to the right
B. 2.2 N pointing to the right
C. 3.5 N pointing to the right
D. 6.6 N pointing to the right

39. Car A starts from rest and accelerates uniformly for time t to travel a distance of d. Car B, which has four times the mass of car A, starts from rest and accelerates uniformly. If the magnitudes of the forces accelerating car A and car B are the same, how long does it take car B to travel the same distance d?

A. t
B. $2t$
C. $t / 2$
D. $16t$

40. A 1,100 kg vehicle is traveling at 27 m/s when it starts to decelerate. What is the average braking force acting on the vehicle, if after 578 m, it comes to a complete stop?

A. –440 N
B. –740 N
C. –690 N
D. –540 N

41. An ornament of mass M is suspended by a string from the ceiling inside an elevator. What is the tension in the string holding the ornament when the elevator travels upward at a constant speed?

A. Equal to Mg
B. Less than Mg
C. Greater than Mg
D. Equal to M / g

42. An object that weighs 75 N is pulled on a horizontal surface by a force of 50 N to the right. The friction force on this object is 30 N to the left. What is the acceleration of the object? (Use the acceleration due to gravity $g = 9.8$ m/s^2)

A. 0.46 m/s^2
B. 1.7 m/s^2
C. 2.6 m/s^2
D. 10.3 m/s^2

43. While flying horizontally in an airplane, a string attached from the overhead luggage compartment hangs at rest 15° from the vertical toward the front of the plane. It can be concluded that the airplane is:

A. accelerating forward
B. accelerating backward
C. accelerating upward at 15° from horizontal
D. moving backward

44. An object slides down an inclined ramp at a constant speed. If the ramp's incline angle is θ, what is the coefficient of kinetic friction (μ_k) between the object and the ramp?

A. $\mu_k = 1$
B. $\mu_k = \cos \theta / \sin \theta$
C. $\mu_k = \sin \theta / \cos \theta$
D. $\mu_k = \sin \theta$

45. What are the readings on the spring scales when a 17 kg fish is weighed with two spring scales if each scale has negligible weight?

A. The top scale reads 17 kg, and the bottom scale reads 0 kg
B. Each scale reads greater than 0 kg and less than 17 kg, but the sum of the scales is 17 kg
C. The bottom scale reads 17 kg, and the top scale reads 0 kg
D. The sum of the two scales is 34 kg

46. What is the acceleration of a 105 kg tiger that accelerates uniformly from rest to 20 m/s in 10 s?

A. 4.7 m/s^2
B. 1.5 m/s^2
C. 2 m/s^2
D. 3.4 m/s^2

47. Yana tries to pull an object by tugging on a rope attached to the object with a force of F. If the object does not move, what does this imply?

A. The object has reached its natural state of rest and can no longer be set into motion
B. The rope is not transmitting the force to the object
C. No other forces are acting on the object
D. There are one or more other forces that act on the object with a sum of $-F$

48. If a force F is exerted on an object, the force which the object exerts back:

A. equals $-F$
B. depends on the density of the object
C. depends on if the object is moving
D. depends on if the object is stationary

49. Two forces acting on an object have magnitudes $F_1 = -6.6$ N and $F_2 = 2.2$ N. Which third force causes the object to be in equilibrium?

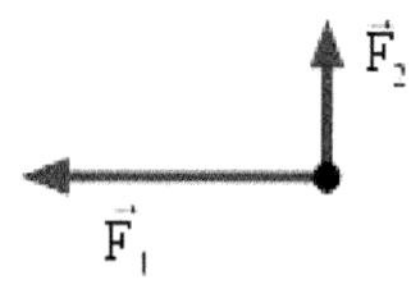

A. 4.4 N at 162° counterclockwise from F_1
B. 4.4 N at 108° counterclockwise from F_1
C. 7 N at 162° counterclockwise from F_1
D. 7 N at 108° counterclockwise from F_1

50. Sarah and her father Bob (who weighs four times as much) are standing on identical skateboards (with frictionless ball bearings), both initially at rest. For a brief time, Bob pushes Sarah on the skateboard. When Bob stops pushing:

A. Sarah and Bob move away from each other, and Sarah's speed is four times that of Bob's
B. Sarah and Bob move away from each other, and Sarah's speed is one-fourth of Bob's
C. Sarah and Bob move away from each other with equal speeds
D. Sarah moves away from Bob, and Bob is stationary

51. Considering the effects of friction, which statement best describes the motion of an object along a surface?

A. Less force is required to start than to keep the object in motion at a constant velocity
B. The same force is required to start to keep the object in motion at a constant velocity
C. More force is required to start than to keep the object in motion at a constant velocity
D. Once the object is set in motion, no force is required to keep it in motion at constant velocity

52. An object maintains its state of motion because it has:

A. mass
B. acceleration
C. speed
D. weight

53. Joe and Bill are playing tug-of-war. Joe is pulling with a force of 200 N, while Bill is simply holding onto the rope. What is the tension of the rope if neither person is moving?

A. 75 N
B. 0 N
C. 100 N
D. 200 N

54. A 4 kg wooden block A slides on a frictionless table pulled by a hanging 5 kg block B via a massless string and pulley system as shown. What is the acceleration of block A as it slides? (Use the acceleration due to gravity $g = 9.8$ m/s^2)

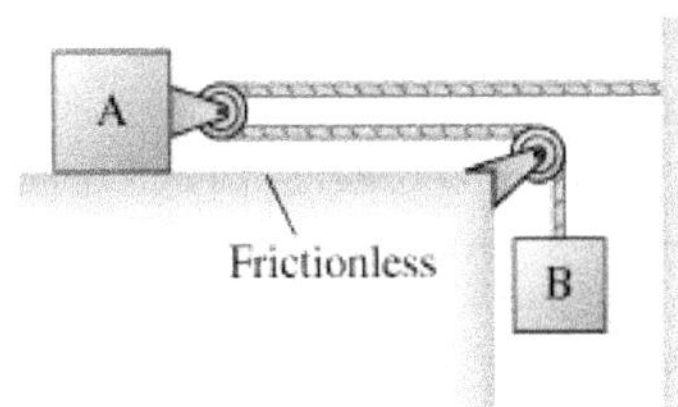

A. 2.8 m/s^2
B. 1.6 m/s^2
C. 4.1 m/s^2
D. 4.9 m/s^2

55. Which of the following best describes the direction in which the force of kinetic friction acts relative to the interface between the interacting bodies?

A. Parallel to the interface and in the same direction as the relative velocity
B. Parallel to the interface and in the opposite direction of the relative velocity
C. Perpendicular to the interface and in the same direction as the relative velocity
D. Perpendicular to the interface and in the opposite direction of the relative velocity

56. A person who normally weighs 600 N is standing on a scale in an elevator. The elevator is initially moving upwards at a constant speed of 8 m/s and starts to slow down at a rate of 6 m/s^2. What is the reading of the person's weight on the scale in the elevator during the slowdown? (Use acceleration due to gravity $g = 9.8$ m/s^2)

A. 600 N
B. 588 N
C. 98 N
D. 233 N

57. How large is the force of friction impeding the motion of a bureau when the 120 N bureau is pulled across the sidewalk at a constant speed by a force of 30 N?

A. 0 N
B. 30 N
C. 120 N
D. 3 N

58. What is the acceleration of a 40 kg crate pulled along a frictionless surface by a force of 140 N that makes an angle of 30° with the surface?

A. 1.5 m/s^2
B. 2 m/s^2
C. 2.5 m/s^2
D. 3 m/s^2

59. A force is a vector quantity because it has both:

I. action and reaction counterparts
II. mass and acceleration
III. magnitude and direction

A. I only
B. II only
C. III only
D. I and II only

Questions **60-61** are based on the following:

Alice pulls her daughter on a sled by a rope on level snow. Alice is 70 kg, and her daughter is 20 kg. The sled has a mass of 10 kg, which slides along the snow with a coefficient of kinetic friction of 0.09. The tension in the rope is 30 N, making an angle of 30° with the ground. They are moving at a constant of 2.5 m/s for 4 s. (Use the acceleration due to gravity $g = 10$ m/s^2)

60. What is the work done by the force of gravity on the sled?

A. –3,000 J
B. 0 J
C. 1,000 J
D. 3,000 J

61. What is the work done by the rope on the sled?

A. 260 J
B. 130 J
C. 65 J
D. 520 J

Notes for active learning

Notes for active learning

Equilibrium and Momentum

1. When is the angular momentum of a system constant?

A. When no net external torque acts on the system
B. When the linear momentum and the energy are constant
C. When no net external force acts on the system
D. When the total kinetic energy is positive

2. When a rock rolls down a mountainside at 7 m/s, the horizontal component of its velocity vector is 1.8 m/s. What was the angle of the mountain surface above the horizontal?

A. 15°
B. 63°
C. 40°
D. 75°

3. A 200 N sled slides down a frictionless hill at an angle of 37° to the horizontal. What is the magnitude of the force that the hill exerts on the sled parallel to the surface of the hill?

A. 170 N
B. 200 N
C. 74 N
D. 0 N

4. Water causes a water wheel to turn as it passes by. The force of the water is 300 N, and the radius of the wheel is 10 m. What is the torque around the center of the wheel?

A. 0 N·m
B. 300 N·m
C. 3,000 N·m
D. 3 N·m

5. Through what angle, in degrees, does a 33 rpm record turn in 0.32 s?

A. 44°
B. 94°
C. 113°
D. 63°

6. A freight train rolls along a track with considerable momentum. What is its momentum if it rolls at the same speed but has twice the mass?

A. Zero
B. Doubled
C. Quadrupled
D. Unchanged

Questions **7-9** are based on the following:

Three carts run along a level, frictionless one-dimensional track. Furthest to the left is a 1 kg cart I, moving at 0.5 m/s to the right. In the middle is a 1.5 kg cart II, moving at 0.3 m/s to the left. Furthest to the right is a 3.5 kg cart III moving at 0.5 m/s to the left. The carts collide in sequence, sticking together. (Assume the direction to the right is the positive direction)

7. What is the total momentum of the system before the collision?

A. –2.6 kg·m/s
B. 1.4 kg·m/s
C. 0.6 kg·m/s
D. –1.7 kg·m/s

8. Assuming cart I and cart II collide first, and cart III is still independent, what is the total momentum of the system just after cart I and cart II collide?

A. –1.7 kg·m/s
B. 0.1 kg·m/s
C. 0.9 kg·m/s
D. –0.9 kg·m/s

9. What is the final velocity of the three carts?

A. –0.35 m/s
B. –0.28 m/s
C. –0.87 m/s
D. 0.35 m/s

10. A 480 kg car is moving at 14.4 m/s when it collides with another car moving at 13.3 m/s in the same direction. If the second car has a mass of 570 kg and a new velocity of 17.9 m/s after the collision, what is the velocity of the first car after the collision?

A. 19 m/s
B. –9 m/s
C. 9 m/s
D. 14 m/s

11. An 8 g bullet is shot into a 4 kg block at rest on a frictionless horizontal surface. The bullet remains lodged in the block. The block moves into a spring and compresses it by 8.9 cm. After the block comes to a stop, the spring fully decompresses and sends the block in the opposite direction. What is the magnitude of the impulse of the block (including the bullet), due to the spring, during the entire time interval in which the block and spring are in contact? (Use the spring constant = 1,400 N/m)

A. 11 N·s
B. 8.3 N·s
C. 6.4 N·s
D. 13 N·s

12. An ice skater performs a fast spin by pulling in her outstretched arms close to her body. What happens to her rotational kinetic energy about the axis of rotation?

A. Decreases
B. Remains the same
C. Increases
D. It changes, but it depends on her body mass

13. A toy car is traveling in a circular path. The force required to maintain this motion is F. If the velocity of the object is doubled, what is the force required to maintain its motion?

A. $2F$
B. F
C. $\frac{1}{2}F$
D. $4F$

14. Which of the following are units of momentum?

A. $kg \cdot m/s^2$
B. $J \cdot s/m$
C. $N \cdot m$
D. $kg \cdot s$

15. The impulse on an apple hitting the ground depends on:

I. the speed of the apple just before it hits
II. whether the apple bounces
III. the time of impact with the ground

A. I only
B. II only
C. III only
D. I, II and III

16. A 55 kg girl throws a 0.8 kg ball against a wall. The ball strikes the wall horizontally with a speed of 25 m/s and bounces back at the same speed. The ball is in contact with the wall for 0.05 s. What is the average force exerted on the wall by the ball?

A. 27,500 N
B. 55,000 N
C. 400 N
D. 800 N

17. Three objects are moving along a straight line as shown. If the positive direction is to the right, what is the total momentum of this system?

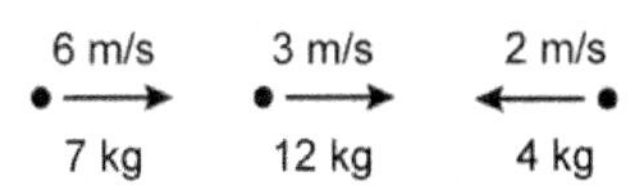

A. −70 kg·m/s
B. +70 kg·m/s
C. +86 kg·m/s
D. −86 kg·m/s

Questions **18-19** are based on the following:

Two ice skaters, Vladimir (60 kg) and Olga (40 kg), collide in midair. Before the collision, Vladimir was going North at 0.5 m/s, and Olga was going West at 1 m/s. Right after the collision and well before they land on the ground, they stick. Assume they have no vertical velocity.

18. What is the magnitude of their velocity just after the collision?

A. 0.1 m/s
B. 1.8 m/s
C. 0.9 m/s
D. 0.5 m/s

19. What is the magnitude of the total momentum just after the collision?

A. 25 kg·m/s
B. 50 kg·m/s
C. 65 kg·m/s
D. 80 kg·m/s

20. A horse is running in a straight line. If both the mass and the speed of the horse are doubled, by what factor does its momentum increase?

A. $\sqrt{2}$
B. 2
C. 4
D. 8

21. The mass of box P is greater than the mass of box Q. Both boxes are on a frictionless horizontal surface and connected by a light cord. A horizontal force F is applied to box Q, accelerating the boxes to the right. What is the magnitude of the force exerted by the connecting cord on box P?

A. equal to F
B. equal to $2F$
C. zero
D. less than F but > 0

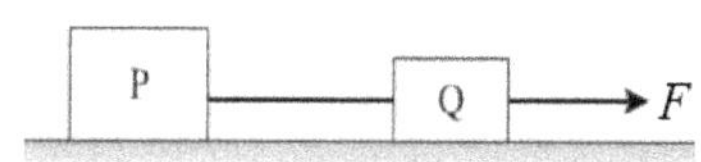

22. Which of the following is true when Melissa (M) and her friend Samantha (S) are riding on a merry-go-round, as viewed from above?

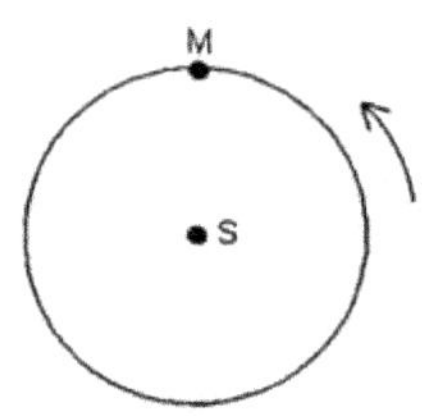

A. They have the same speed, but different angular velocity
B. They have different speeds, but the same angular velocity
C. They have the same speed and the same angular velocity
D. They have different speeds and different angular velocities

23. The relationship between impulse and impact force involves the:

A. time the force acts
B. distance the force acts
C. difference between acceleration and velocity
D. mass and its effect on resisting a change in velocity

24. Angular momentum cannot be conserved if the:

A. moment of inertia changes
B. system is experiencing a net force
C. angular velocity changes
D. system has a net torque

25. A 6.8-kg block moves on a frictionless surface with a speed of $v_i = 5.4$ m/s and makes a perfectly elastic collision with a 4.8-kg stationary block. After the collision, the 6.8-kg block recoils with a speed of $v_f = 3.2$ m/s. What is the magnitude of the average force on the 6.8-kg block while the two blocks are in contact for 2 s?

A. 4.4 N
B. 46.1 N
C. 32.6 N
D. 29.2 N

Questions **26-27** are based on the following:

A 4 kg rifle imparts a high velocity to a small 10 g bullet by exploding a charge that causes the bullet to leave the barrel at 300 m/s. Take the system as the combination of the rifle and bullet. Typically, the rifle is fired with the butt of the gun pressed against the shooter's shoulder and ignore the force of the shoulder on the rifle.

26. What is the momentum of the system just after the bullet leaves the barrel?

A. 0 kg·m/s
B. 3 kg·m/s
C. 9 kg·m/s
D. 30 kg·m/s

27. What is the recoil velocity of the rifle (i.e., the velocity of the rifle just after firing)?

A. 23 m/s
B. 1.5 m/s
C. 5.6 m/s
D. 0.75 m/s

28. A ball thrown horizontally from a point 24 m above the ground strikes the ground after traveling horizontally at a distance of 18 m. With what speed was it thrown, assuming negligible air resistance? (Use the acceleration due to gravity $g = 9.8\ \text{m/s}^2$)

A. 6.8 m/s
B. 7.5 m/s
C. 8.1 m/s
D. 8.6 m/s

29. An object is moving in a circle at a constant speed. Its acceleration vector is directed:

A. toward the center of the circle
B. away from the center of the circle
C. tangent to the circle and in the direction of the motion
D. behind the normal and toward the center of the circle

30. Impulse is equal to the:

I. force multiplied by the distance over which the force acts
II. change in momentum
III. momentum

A. I only
B. II only
C. III only
D. I and II only

31. A 4 kg object is at the height of 10 m above the Earth's surface. Ignoring air resistance, what is its kinetic energy immediately before impacting the ground if it is thrown straight downward with an initial speed of 20 m/s? (Use the acceleration due to gravity $g = 10$ m/s^2)

A. 150 J
B. 300 J
C. 1,200 J
D. 900 J

32. A car traveling along the highway needs a certain amount of force to stop. More stopping force may be required when the car has:

I. less stopping distance II. more momentum III. more mass

A. I only
B. II only
C. III only
D. I, II and III

33. A table tennis ball moving East at a speed of 4 m/s collides with a stationary bowling ball. The table tennis ball bounces back to the West, and the bowling ball moves very slowly to the East. Which ball experiences the greater magnitude of impulse during the collision?

A. Bowling ball
B. Table tennis ball
C. Neither because both experience the same magnitude of the impulse
D. It is not possible to determine since the velocities after the collision are unknown

34. Assume that a massless bar of 5 m is suspended from a rope attached to the bar at a distance of x from the bar's left end. If a 30 kg mass hangs from the right side of the bar and a 6 kg mass hangs from the left side, what value of x results in equilibrium? (Use acceleration due to gravity $g = 9.8$ m/s^2)

A. 2.8 m
B. 4.2 m
C. 3.2 m
D. 1.6 m

35. A block of mass m sits at rest on a rough inclined ramp that makes an angle θ with the horizontal. What must be true about the force of static friction (f) on the block?

A. $f > mg \sin \theta$
B. $f = mg \cos \theta$
C. $f = mg$
D. $f = mg \sin \theta$

36. A 30 kg block is pushed in a straight line across a horizontal surface. What is the coefficient of kinetic friction μ_k between the block and the surface if a constant force of 45 N must be applied to the block to maintain a constant velocity of 3 m/s? (Use the acceleration due to gravity $g = 10$ m/s^2)

A. 0.1
B. 0.33
C. 0.15
D. 0.5

37. The impulse-momentum relationship is a direct result of:

I. Newton's First Law II. Newton's Second Law III. Newton's Third Law

A. I only
B. II only
C. III only
D. I and II only

Questions **38-40** are based on the following:

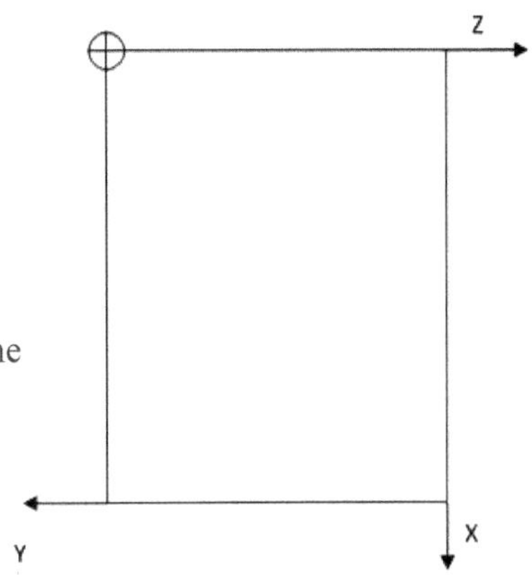

A 0.5 m by 0.6 m rectangular piece of metal is hinged (⊗) (as shown) in the upper left corner, hanging so that the long edge is vertical. A 25 N force (Y) acts to the left at the lower-left corner. A 15 N force (X) acts down at the lower right corner, and a 30 N force (Z) acts to the right at the upper right corner. Each force vector is in the plane of the metal. Use counterclockwise as the positive direction.

38. What is the torque of force X about the pivot?

A. 5 N·m
B. 3 N·m
C. –7.5 N·m
D. 0 N·m

39. What is the torque of force Z about the pivot?

A. –10 N·m
B. –4.5 N·m
C. 4.5 N·m
D. 0 N·m

40. What is the torque of force Y about the pivot?

A. –15 N·m
B. –3 N·m
C. 0 N·m
D. 3 N·m

41. A 50 g weight is tied to the end of a string and whirled at 20 m/s in a horizontal circle with a radius of 2 m. Ignoring the force of gravity, what is the tension in the string?

A. 5 N
B. 10 N
C. 50 N
D. 150 N

42. A small car collides with a large truck in a head-on collision. Which of the following statements concerning the magnitude of the average force during the collision is correct?

A. The small car and the truck experience the same average force
B. The force experienced by each one is inversely proportional to its velocity
C. The truck experiences the greater average force
D. The small car experiences the greater average force

43. A 10 kg bar that is 2 m long extends perpendicularly from a vertical wall. The free end of the bar is attached to a point on the wall by a light cable, which makes an angle of 30° with the bar. What is the tension in the cable? (Use the acceleration due to gravity $g = 10\ \text{m/s}^2$)

A. 75 N
B. 150 N
C. 100 N
D. 125 N

44. Object A has the same size and shape as object B but is twice as heavy. When objects A and B are dropped simultaneously from a tower, they reach the ground simultaneously. Object A has greater:

I. speed II. momentum III. acceleration

A. I only
B. II only
C. III only
D. I and II only

45. Two vehicles approach a right-angle intersection and then collide. After the collision, they become entangled. If their mass ratio was 1 : 4 and their respective speeds as they approached at 12 m/s, what is the magnitude of the velocity immediately following the collision?

A. 16.4 m/s
B. 11.9 m/s
C. 13.4 m/s
D. 9.9 m/s

46. A skater stands stationary on frictionless ice, and she throws a heavy ball to the right at an angle of 5° above the horizontal. With respect to the ice, if the ball weighs one-third as much as the skater and she is measured to be moving with a speed of 2.9 m/s to the left after the throw, how fast did she throw the ball?

A. 10.2 m/s
B. 7.2 m/s
C. 8.73 m/s
D. 9.8 m/s

47. Ignoring the forces of friction, what horizontal force must be applied to an object with a weight of 98 N to give it a horizontal acceleration of $10\ \text{m/s}^2$? (Use the acceleration due to gravity $g = 9.8\ \text{m/s}^2$)

A. 9.8 N
B. 100 N
C. 79 N
D. 125 N

48. Consider a winch that pulls a cart at constant speed up an incline. Point A is at the bottom of the incline, and point B is at the top. Which of the following statements is/are true from point A to B?

I. The KE of the cart is constant
II. The PE of the cart is constant
III. The sum of the KE and PE of the cart is constant

A. I only
B. II only
C. III only
D. I and II only

49. A high-speed dart is shot from ground level with a speed of 140 m/s at an angle of 35° above the horizontal. What is the vertical component of its velocity after 4 s if air resistance is ignored? (Use the acceleration due to gravity g = 9.8 m/s^2)

A. 59 m/s
B. 75 m/s
C. 34 m/s
D. 41 m/s

50. What does the area under the curve of a force *vs.* time graph represent for a diver as she leaves the platform during her approach to the water below?

A. Work
B. Momentum
C. Impulse
D. Displacement

51. A rifle of mass 2 kg is suspended by strings. The rifle fires a bullet of mass 0.01 kg at a speed of 220 m/s. What is the recoil velocity of the rifle?

A. 0.001 m/s
B. 0.01 m/s
C. 0.1 m/s
D. 1.1 m/s

52. How do automobile airbags reduce injury during a collision?

A. They reduce the kinetic energy transferred to the passenger
B. They reduce the momentum transferred to the passenger
C. They reduce the acceleration of the automobile
D. They reduce the forces exerted upon the passenger

Questions **53-55** are based on the following:

Tim nails a meter stick to a board at the meter stick's 0 m mark. Force I acts at the 0.5 m mark perpendicular to the meter stick with a force of 10 N, as shown in the figure. Force II acts at the end of the meter stick with a force of 5 N, making a 35° angle. Force III acts at the same point with a force of 20 N, providing tension but no shear stress. Use counterclockwise as the positive direction.

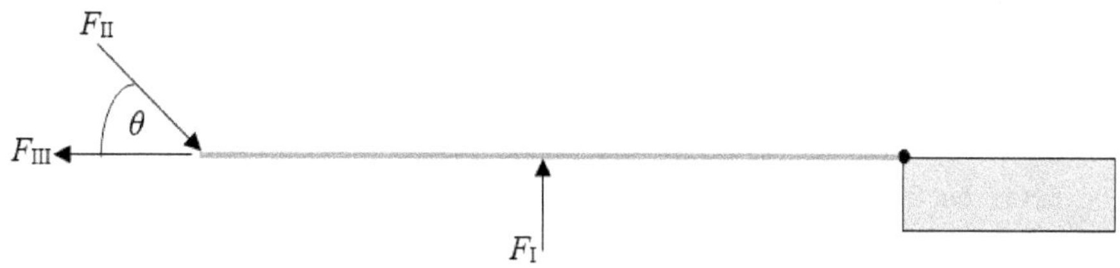

53. What is the torque of Force I about the fixed point?

A. –5 N·m
B. 0 N·m
C. 5 N·m
D. 10 N·m

54. What is the torque of Force II about the fixed point?

A. –4.8 N·m
B. –2.9 N·m
C. 4.8 N·m
D. 2.9 N·m

55. What is the torque of Force III about the fixed point?

A. –20 N·m
B. 0 N·m
C. 10 N·m
D. 20 N·m

56. Two equal mass balls (one yellow and the other red) are dropped from the same height and rebound off the floor. The yellow ball rebounds to a higher position. Which ball is subjected to the greater magnitude of impulse during its collision with the floor?

A. Both balls were subjected to the same magnitude of impulse
B. Red ball
C. Yellow ball
D. Requires the time intervals and forces

57. Calculate the impulse associated with a force of 4.5 N that lasts for 1.4 s:

A. 5.4 kg·m/s
B. 6.8 kg·m/s
C. 4.6 kg·m/s
D. 6.3 kg·m/s

58. A heavy truck and a small truck roll down a hill. Ignoring friction, at the bottom of the hill, the heavy truck has greater:

I. momentum II. acceleration III. speed

A. I only
B. II only
C. III only
D. I and II only

59. A 78 g steel ball is released from rest and falls vertically onto a rigid surface. The ball strikes the surface and is in contact with it for 0.5 ms. The ball rebounds elastically and returns to its original height during a round trip of 4 s. Assume that the surface does not deform during contact. What is the maximum elastic energy stored by the ball? (Use the acceleration due to gravity $g = 9.8\ m/s^2$)

A. 23 J
B. 43 J
C. 11 J
D. 15 J

60. Cart A is 5 kg, and cart B is 10 kg, and they are initially stationary on a frictionless horizontal surface. A force of 3 N to the right acts on cart A for 2 s. Subsequently, it hits cart B, and the two carts stick. What is the final velocity of the two carts?

A. 1 m/s
B. 0.4 m/s
C. 1.2 m/s
D. 1.8 m/s

Notes for active learning

Rotational Motion

1. Suppose a uniform solid sphere of mass M and radius R rolls without slipping down an inclined plane starting from rest. The linear velocity of the sphere at the bottom of the incline depends on:

A. the radius of the sphere
B. the mass of the sphere
C. both the mass and the radius of the sphere
D. neither the mass nor the radius of the sphere

2. A solid, uniform sphere of mass 2.0 kg and radius 1.7 m rolls from rest without slipping down an inclined plane of height 5.3 m. What is the angular velocity of the sphere at the bottom of the inclined plane?

A. 3.7 rad/s
B. 5.1 rad/s
C. 6.7 rad/s
D. 8.3 rad/s

3. A solid uniform ball with a mass of 125.0 g is rolling without slipping along the horizontal surface of a table with a speed of 4.5 m/s when it rolls off the edge and falls towards the floor, 1.1 m below. What is the rotational kinetic energy of the ball just before it hits the floor?

A. 0.51 J
B. 0.87 J
C. 1.03 J
D. 2.26 J

4. David swings a 0.38 kg ball in a circle on a string that is 1.3 m long. What is the magnitude of the ball's angular momentum if the ball makes 1.2 rev/s?

A. 0.6 kg·m^2/s
B. 2.2 kg·m^2/s
C. 3.6 kg·m^2/s
D. 4.8 kg·m^2/s

5. An ice skater has a moment of inertia of 5.0 kg·m^2 when her arms are outstretched, and at this time, she is spinning at 3.0 rev/s. If she pulls in her arms and decreases her moment of inertia to 2.0 kg·m^2, how fast will she be spinning?

A. 1.8 rev/s
B. 4.5 rev/s
C. 7.5 rev/s
D. 10.5 rev/s

6. The angular momentum of a system remains constant when:

A. its total kinetic energy is constant
B. the moment of inertia is constant
C. no net external torque acts on the system
D. no net external force acts on the system

7. A bicycle has wheels 60.0 cm in diameter. What is the angular speed of these wheels moving at 4.0 m/s?

A. 0.28 rad/s
B. 1.6 rad/s
C. 3.4 rad/s
D. 13.3 rad/s

8. What is the kinetic energy of a thin uniform rod of length 120.0 cm with a mass of 450.0 g that rotates about its center along the short axis at 3.60 rad/s? (The short axis is perpendicular to the axis of the rod. Imagine spinning the rod like an airplane propeller.)

A. 0.350 J
B. 1.30 J
C. 2.70 J
D. 4.96 J

9. A rope is wrapped around a wheel of radius R = 2.0 meters. The wheel is mounted with frictionless bearings on an axle through its center. A block of mass 14.0 kg is suspended from the end of the rope. When the system is released from rest, it is observed that the block descends 10.0 meters in 2.0 seconds. What is the moment of inertia of the wheel?

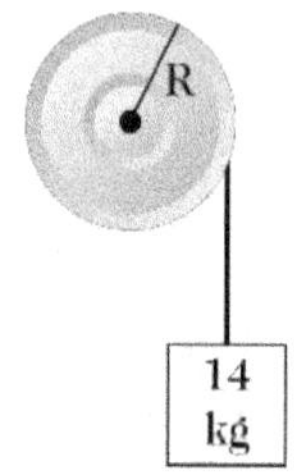

A. 300.0 kg·m^2
B. 185.0 kg·m^2
C. 53.8 kg·m^2
D. 521.0 kg·m^2

10. A string is wrapped tightly around a fixed frictionless pulley with a moment of inertia of 0.0352 kg·m^2 and a radius of 12.5 cm. The string is pulled away from the pulley with a constant force of 5.00 N, causing the pulley to rotate. If the string does not slip on the pulley, what is the speed of the string after it has unwound 1.25 m? Consider the string to be massless.

A. 0.69 m/s
B. 2.36 m/s
C. 3.62 m/s
D. 4.90 m/s

11. When a rigid object rotates about a fixed axis, what is true about all the points in the object?

I. They have the same angular acceleration
II. They have the same tangential acceleration
III. They have the same radial acceleration

A. I only
B. II only
C. III only
D. I and II only

12. A small mass is placed on a record turntable that is rotating at 33.33 rpm. The linear velocity of the mass is:

A. zero
B. directed parallel to the line joining the mass and the center of rotation
C. independent (in magnitude) of the position of the mass on the turntable
D. greater the farther the mass is from the center

13. To drive a midsize car at 40.0 mph on a level road for one hour requires about 3.2×10^7 J of energy. Suppose this much energy was attempted to be stored in a spinning, solid, uniform, cylindrical flywheel. If a flywheel with a diameter of 1.2 m and mass of 400.0 kg were used, what angular speed would be required to store 3.2×10^7 J?

A. 380 rad/s
B. 620 rad/s
C. 940 rad/s
D. 1,450 rad/s

14. A wheel having a moment of inertia of 5.0 kg·m² starts from rest and accelerates for 8.0 s under a constant torque of 3.0 N·m. What is the wheel's rotational kinetic energy at the end of 8.0 s?

A. 29 J
B. 58 J
C. 83 J
D. 112 J

15. When a rigid object rotates about a fixed axis, what is true about all the points in the object?

I. They have the same angular speed
II. They have the same tangential speed
III. They have the same angular acceleration

A. I only
B. II only
C. III only
D. I and III only

16. A uniform, solid cylindrical flywheel of radius 1.4 m and mass 15.0 kg rotates at 2.4 rad/s. What is the magnitude of the flywheel's angular momentum?

A. 11 kg·m²/s
B. 18 kg·m²/s
C. 25 kg·m²/s
D. 35 kg·m²/s

17. A uniform solid disk is released from rest and rolls without slipping down an inclined plane that makes an angle of 25° with the horizontal. What is the forward speed of the disk after it has rolled 3.0 m, measured along the plane?

A. 0.8 m/s
B. 1.8 m/s
C. 2.9 m/s
D. 4.1 m/s

18. A tire is rolling along a road, without slipping, with a center-of-mass velocity v. A piece of tape is attached to the tire. When the tape is opposite the road (top of the tire), what is its velocity with respect to the road?

A. $2v$
B. v
C. $1.5v$
D. $\sqrt{v}$

19. A string is wound tightly around a fixed pulley with a radius of 5.0 cm. As the string is pulled, the pulley rotates without any slipping of the string. What is the angular speed of the pulley when the string is moving at 5.0 m/s?

A. 10.0 rad/s
B. 25.0 rad/s
C. 75.0 rad/s
D. 100.0 rad/s

20. A 1.4 kg object at $x = 2.00$ m, $y = 3.10$ m moves at 4.62 m/s at an angle 45° north of east. What is the magnitude of the object's angular momentum about the origin?

A. 1.2 kg·m²/s
B. 2.6 kg·m²/s
C. 3.8 kg·m²/s
D. 5.0 kg·m²/s

21. When a fan is turned off, its angular speed decreases from 10.0 rad/s to 6.3 rad/s in 5.0 s. What is the magnitude of the average angular acceleration of the fan?

A. 0.46 rad/s²
B. 0.74 rad/s²
C. 1.86 rad/s²
D. 2.80 rad/s²

22. At time $t = 0$ s, a wheel has an angular displacement of 0 radians and an angular velocity of +26.0 rad/s. The wheel has a constant acceleration of –0.43 rad/s². In this situation, what is the time t (after $t = 0$ s), at which the kinetic energy of the wheel is twice the initial value?

A. 48 s
B. 86 s
C. 115 s
D. 146 s

23. A solid uniform disk of diameter 3.20 m and mass 42.0 kg rolls without slipping to the bottom of a hill, starting from rest. If the angular speed of the disk is 4.27 rad/s at the bottom, how high vertically did it start on the hill above the bottom?

A. 2.46 m
B. 3.57 m
C. 4.85 m
D. 6.24 m

24. When Steve rides a bicycle, in what direction is the angular velocity of the wheels?

A. to his left
B. to his right
C. forward
D. backward

25. A rolling wheel of a diameter of 68.0 cm slows down uniformly from 8.4 m/s to rest over a distance of 115.0 m. What is the magnitude of its angular acceleration if there was no slipping?

A. 0.90 rad/s²
B. 1.6 rad/s²
C. 4.2 rad/s²
D. 7.8 rad/s²

26. A uniform solid cylinder with a radius of 10.0 cm and a mass of 3.0 kg is rotating about its center axis with an angular speed of 33.4 rpm. What is the kinetic energy of the uniform solid cylinder?

A. 0.091 J
B. 0.19 J
C. 0.66 J
D. 1.14 J

27. A uniform 135.0-g meter stick rotates about an axis perpendicular to the stick, passing through its center with an angular speed of 3.50 rad/s. What is the magnitude of the angular momentum of the stick?

A. 0.0394 kg·m^2/s
B. 0.0848 kg·m^2/s
C. 0.286 kg·m^2/s
D. 0.458 kg·m^2/s

28. A 23.0 kg mass is connected to a nail on a frictionless table by a massless string of length 1.3 m. If the tension in the string is 51.0 N while the mass moves in a uniform circle on the table, how long does it take for the mass to make one complete revolution?

A. 2.8 s
B. 3.6 s
C. 4.8 s
D. 5.4 s

29. A machinist turns on the power to a grinding wheel at time $t = 0$ s. The wheel accelerates uniformly from rest for 10.0 s and reaches the operating angular speed of 38.0 rad/s. The wheel is run at that angular speed for 30.0 s, and then power is shut off. The wheel slows down uniformly at 2.1 rad/s^2 until the wheel stops. What is the angular acceleration of the wheel between $t = 0$ s and $t = 10.0$ s?

A. 1.21 rad/s^2
B. 2.63 rad/s^2
C. 3.80 rad/s^2
D. 5.40 rad/s^2

30. A force of 17.0 N is applied to the end of a 0.63 m long torque wrench at an angle 45° from a line joining the pivot point to the handle. What is the magnitude of the torque generated about the pivot point?

A. 4.3 N·m
B. 8.2 N·m
C. 7.6 N·m
D. 11.8 N·m

31. A solid disk of radius 1.60 m and mass 2.30 kg rolls from rest without slipping to the bottom of an inclined plane. If the angular velocity of the disk is 4.27 rad/s at the bottom, what is the height of the inclined plane?

A. 0.57 m
B. 1.08 m
C. 2.84 m
D. 3.57 m

32. A merry-go-round spins freely when Paul moves quickly to the center along a radius of the merry-go-round. As he does this, the moment of inertia of the system:

A. increases, and the angular speed increases
B. decreases, and the angular speed remains the same
C. decreases, and the angular speed decreases
D. decreases, and the angular speed increases

33. What is the angular speed of a compact disc that, at a specific instant, is rotating at 210.0 rpm?

A. 8.5 rad/s
B. 22.0 rad/s
C. 36.4 rad/s
D. 52.6 rad/s

34. A solid uniform sphere is rolling without slipping along a horizontal surface with a speed of 5.5 m/s when it starts up a ramp that makes an angle of 25° with the horizontal. What is the speed of the sphere after it has rolled 3.0 m up as measured along the surface of the ramp?

A. 0.8 m/s
B. 1.6 m/s
C. 3.5 m/s
D. 4.8 m/s

35. A force of 16.88 N is applied tangentially to a wheel of radius 0.340 m and gives rise to angular acceleration of 1.20 rad/s^2. What is the rotational inertia of the wheel?

A. 1.48 kg·m^2
B. 2.26 kg·m^2
C. 3.48 kg·m^2
D. 4.78 kg·m^2

36. A machine does 3.9 kJ of work on a spinning flywheel to bring it from 500.0 rpm to rest. This flywheel is in the shape of a solid uniform disk of a radius of 1.2 m. What is the mass of this flywheel?

A. 2.6 kg
B. 4.0 kg
C. 5.2 kg
D. 6.4 kg

37. A rectangular billboard with h = 20.0 cm high and w = 11.0 cm wide loses three of its four support bolts and rotates into the position as shown, with P_1 directly over P_3. It is supported by P_2, which is so tight that it holds the billboard from further rotation. What is the gravitational torque about P_2 if the mass of the billboard is 5.0 kg?

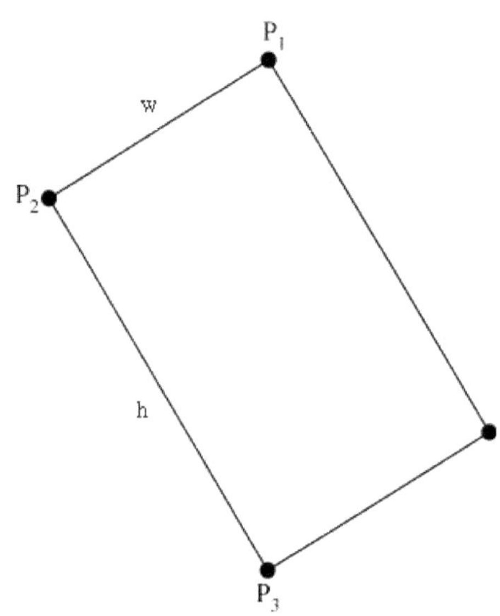

A. 1.2 Nm
B. 2.5 Nm
C. 4.7 Nm
D. 6.8 Nm

38. A disk, a hoop, and a solid sphere are released simultaneously at the top of an inclined plane. In which order do they reach the bottom if each is uniform and rolls without slipping?

A. sphere, hoop, disk
B. sphere, disk, hoop
C. hoop, sphere, disk
D. disk, hoop, sphere

39. A uniform disk is attached at the rim to a vertical shaft and is used as a cam. A side view and top view of the disk and shaft are shown.

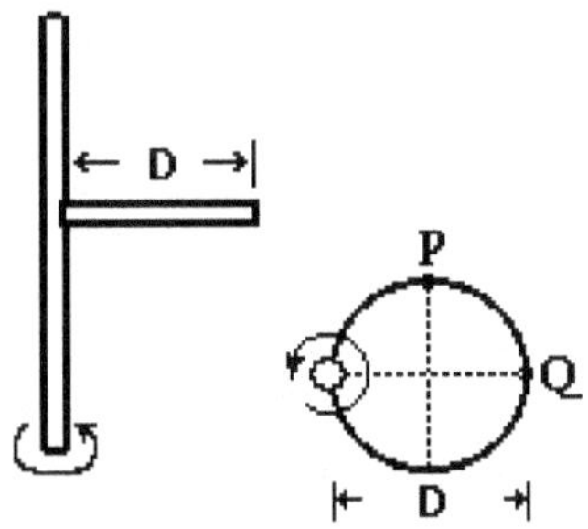

The disk has a diameter of 80.0 cm. The moment of inertia of the disk about the axis of the shaft is 6.0×10^{-3} kg·m^2. What is the kinetic energy of the disk as the shaft rotates uniformly about its axis at 96.0 rpm?

A. 0.18 J
B. 0.30 J
C. 0.49 J
D. 0.57 J

40. A scooter has wheels with a diameter of 240.0 mm. What is the angular speed of the wheels when the scooter is moving forward at 6.00 m/s?

A. 128.6 rpm
B. 248.2 rpm
C. 472.0 rpm
D. 478.0 rpm

41. A spinning ice skater on frictionless ice can control the rate at which she rotates by pulling in her arms. Which of the following statements are true about the skater during this process?

I. Her kinetic energy remains constant
II. Her moment of inertia remains constant
III. Her angular momentum remains constant

A. I only
B. II only
C. III only
D. I and II only

42. Tanya is riding a merry-go-round with an instantaneous angular speed of 1.25 rad/s and angular acceleration of 0.745 rad/s^2. Tanya is standing 4.65 m from the center of the merry-go-round. What is the magnitude of the linear acceleration of Tanya?

A. 2.45 m/s^2
B. 4.20 m/s^2
C. 6.82 m/s^2
D. 8.05 m/s^2

43. Through how many degrees does a 33.0 rpm turntable rotate in 0.32 s?

A. 31°
B. 42°
C. 63°
D. 76°

44. A 50.0 kg uniform ladder, length $L = 5.00$ m long, is placed against a smooth wall at the height of $h = 3.70$ m. The ladder's base rests on a rough horizontal surface whose coefficient of static friction $\mu = 0.750$. An 80.0 kg block is suspended from the top rung of the ladder, just at the wall. What is the approximate magnitude of the force exerted on the base of the ladder due to contact with the rough horizontal surface?

A. 1,370 N
B. 1,460 N
C. 1,580 N
D. 1,640 N

45. At time $t = 0$ s, a wheel has an angular displacement of zero radians and an angular velocity of +29.0 rad/s. The wheel has a constant acceleration of –0.52 rad/s^2. In this situation, what is the maximum value of the angular displacement?

A. +467 rad
B. +809 rad
C. +1,110 rad
D. +1,460 rad

46. A disk lies in the xz-plane with its center at the origin. When viewed from the positive y-axis (i.e., above the disk), the direction of rotation appears clockwise. In what direction does the angular velocity of the disk point?

A. to her right
B. to her left
C. down
D. up

47. How long does it take for a rotating object to speed up from 15.0 rad/s to 33.3 rad/s if it has a uniform angular acceleration of 3.45 rad/s^2?

A. 3.45 s
B. 5.30 s
C. 8.35 s
D. 14.60 s

48. A solid uniform sphere of mass 120.0 kg and radius 1.7 m starts from rest and rolls without slipping down an inclined plane of vertical height 5.3 m; the sphere starts at the top of the ramp. What is the angular speed of the sphere at the bottom of the inclined plane? The moment of inertia of a solid sphere is $(2/5)mR^2$.

A. 0.81 rad/s
B. 1.7 rad/s
C. 2.9 rad/s
D. 5.1 rad/s

49. Three solid, uniform, cylindrically shaped flywheels, each of mass 65.0 kg and radius 1.47 m, rotate independently around a common axis. Two flywheels rotate in one direction at 3.83 rad/s; the other rotates in the opposite direction at 3.42 rad/s. What is the magnitude of the net angular momentum of the system?

A. 168.0 kg·m²/s
B. 298.0 kg·m²/s
C. 456.0 kg·m²/s
D. 622.0 kg·m²/s

50. A cylinder and a sphere are released simultaneously at the top of an inclined plane. Which reaches the bottom first if they roll down the inclined plane without slipping?

A. The one of smallest diameter
B. The one of greatest mass
C. The disk
D. The sphere

51. What is the angular speed of a flywheel turning at 813.0 rpm?

A. 8.33 rad/s
B. 56.23 rad/s
C. 33.84 rad/s
D. 85.14 rad/s

52. A machinist turns the power on to a grinding wheel at time $t = 0$ s. The wheel accelerates uniformly from rest for 10.0 s and reaches the operating angular speed of 96.0 rad/s. The wheel is run at that angular velocity for 40.0 s, and then power is shut off. The wheel slows down uniformly at 1.5 rad/s^2 until the wheel stops. For how long after the power is shut off does it take the wheel to stop?

A. 56.0 s
B. 64.0 s
C. 72.0 s
D. 82.0 s

53. A futuristic design for a car is to have a large disk-like flywheel within the car storing kinetic energy. The flywheel has a mass 370.0 kg with a radius of 0.500 m and can rotate up to 200.0 rev/s. Assuming this stored kinetic energy could be transferred to the linear velocity of the 1500.0-kg car, what is the maximum attainable speed of the car?

A. 29.6 m/s
B. 88.4 m/s
C. 162 m/s
D. 221 m/s

54. An electrical motor spins at a constant 2,695.0 rpm. If the rotor radius is 7.165 cm, what is the linear acceleration of the edge of the rotor?

A. 707.0 m/s^2
B. 1,280 m/s^2
C. 3,272 m/s^2
D. 5,707 m/s^2

55. To drive a typical car at 40.0 mph on a level road for one hour requires about 3.2×10^7 J of energy. Suppose one tried to store this much energy in a spinning solid cylindrical flywheel which was then coupled to the wheels of the car. What angular speed would be required to store 3.2×10^7 J if the flywheel has a radius of 0.60 m and mass 400.0 kg?

A. 943.0 rad/s
B. 1,384.2 rad/s
C. 1,822.4 rad/s
D. 2,584.5 rad/s

56. A uniform, solid, cylindrical flywheel of radius 1.4 m and mass 15.0 kg rotates at 2.7 rad/s about an axis through its circular faces. What is the magnitude of the flywheel's angular momentum?

A. 22 kg·m^2/s
B. 40 kg·m^2/s
C. 64 kg·m^2/s
D. 80 kg·m^2/s

57. A particular motor can provide a maximum torque of 110.0 N·m. Assuming that this torque is used to accelerate a solid, uniform, cylindrical flywheel of mass 10.0 kg and radius 3.00 m, how long will it take for the flywheel to accelerate from rest to 8.13 rad/s?

A. 2.13 s
B. 3.33 s
C. 4.65 s
D. 5.46 s

58. A satellite is in a circular orbit around a planet. What is the satellite's orbital speed if the orbital radius is 34.0 km and the gravitational acceleration at that height is 2.3 m/s^2?

A. 26 m/s
B. 150 m/s
C. 280 m/s
D. 310 m/s

Notes for active learning

Notes for active learning

Work and Energy

1. Consider the following ways that a girl might throw a stone from a bridge. The speed of the stone as it leaves her hand is the same in each of the three cases.

I. Thrown straight up
II. Thrown straight down
III. Thrown straight out horizontally

Ignoring air resistance, in which case is the vertical speed of the stone the greatest when it hits the water below?

A. I only
B. II only
C. III only
D. I and II only

2. A package is pulled along the ground by a 5 N force *F* directed 45° above the horizontal. How much work is done by the force when it pulls the package 10 m?

A. 14 J
B. 35 J
C. 70 J
D. 46 J

3. Which quantity has the greatest influence on the amount of kinetic energy that a large truck has while moving down the highway?

A. Velocity
B. Mass
C. Density
D. Direction

4. No work is done by gravity on a bowling ball that rolls along the floor of a bowling alley because:

A. no potential energy is converted to kinetic energy
B. the force on the ball is at a right angle to the ball's motion
C. its velocity is constant
D. the total force on the ball is zero

5. A 5 kg toy car is moving along the level ground. At a given time, it travels at a speed of 2 m/s and accelerates at 3 m/s^2. What is the cart's kinetic energy at this time?

A. 20 J
B. 8 J
C. 10 J
D. 4 J

6. A treehouse is 8 m above the ground. If Peter does 360 J of work while pulling a box from the ground up to his treehouse with a rope, what is the mass of a box? (Use acceleration due to gravity g = 10 m/s^2)

A. 4.5 kg
B. 3.5 kg
C. 5.8 kg
D. 2.5 kg

7. For an ideal elastic spring, what does the slope of the curve represent for a displacement (x) *vs.* applied force (F) graph?

A. The acceleration of gravity
B. The square root of the spring constant
C. The spring constant
D. The reciprocal of the spring constant

8. A spring with a spring constant of 22 N/m is stretched from equilibrium to 3 m. How much work is done in the process?

A. 33 J
B. 66 J
C. 99 J
D. 198 J

9. A baseball is thrown straight up. Compare the sign of the work done by gravity while the ball goes up with the sign of the work done by gravity while it goes down:

A. negative on the way up and positive on the way down
B. negative on the way up and negative on the way down
C. positive on the way up and positive on the way down
D. positive on the way up and negative on the way down

10. Let A_1 represent the magnitude of the work done by gravity as mass A's gravitational energy increases by 400 J. Let B_1 represent the total amount of work necessary to increase mass B's kinetic energy by 400 J. How do A_1 and B_1 compare?

A. $A_1 > B_1$
B. $A_1 = B_1$
C. $A_1 < B_1$
D. $A_1 = 400\ B_1$

11. According to the definition of work, pushing on a rock accomplishes no work unless there is:

A. an applied force equal to the rock's weight
B. movement perpendicular to the force
C. an applied force greater than the rock's weight
D. movement parallel to the force

12. A job is done slowly, while an identical job is done quickly. Both jobs require the same amount of work, but different amounts of:

I. energy II. power III. torque

A. I only
B. II only
C. I and II only
D. I and III only

13. On a force (F) *vs.* distance (d) graph, what represents the work done by the force F?

A. The area under the curve
B. A line connecting two points on the curve
C. The slope of the curve
D. The length of the curve

14. A 3 kg cat leaps from a tree to the ground, a distance of 4 m. What is its kinetic energy just before the cat reaches the ground? (Use acceleration due to gravity $g = 10\ m/s^2$)

A. 0 J
B. 9 J
C. 120 J
D. 60 J

15. A book is resting on a plank of wood. Jackie pushes the plank and accelerates it so that the book is stationary with respect to the plank. The work done by static friction is:

A. zero
B. positive
C. negative
D. parallel to the surface

16. 350 J of work is required to drive a stake into the ground thoroughly. If the average resistive force on the stake by the ground is 900 N, how long is the stake?

A. 2.3 m
B. 0.23 m
C. 3 m
D. 0.39 m

17. A lightweight and a very heavy object are sliding with equal speeds along a level, frictionless surface. They both slide up the same frictionless hill with no air resistance. Which object rises to a greater height?

A. They both slide to the same height
B. The heavy object because it has more kinetic energy to carry it up the hill
C. The heavy object because it has greater potential energy
D. The lightweight object because it has more kinetic energy to carry it up the hill

18. If Investigator II does 3 times the work of Investigator I in one third the time, the power output of Investigator II is:

A. 9 times greater
B. 3 times greater
C. 1/3 times greater
D. the same

19. A diver who weighs 450 N steps off a diving board that is 9 m above the water. What is the kinetic energy when the diver strikes the water?

A. 160 J
B. 540 J
C. 45 J
D. 4,050 J

20. A vertical, hanging spring stretches by 23 cm when a 160 N object is attached. What is the weight of a hanging plant that stretches the spring by 34 cm?

A. 237 N
B. 167 N
C. 158 N
D. 309 N

21. A mule pulls with a horizontal force F on a covered wagon of mass M. The mule and covered wagon travel at a constant speed v on level ground. How much work is done by the mule on the covered wagon during time Δt? (Use acceleration due to gravity $g = 10\ \text{m/s}^2$)

A. $-Fv\Delta t$
B. $Fv\Delta t$
C. 0 J
D. $-F\sqrt{v}\Delta t$

22. Jane pulls on the strap of a sled at an angle of 32° above the horizontal. If 540 J of work is done by the strap while moving the sled at a horizontal distance of 18 m, what is the tension in the strap?

A. 86 N
B. 112 N
C. 24 N
D. 35 N

23. A vertical spring stretches 6 cm from equilibrium when a 120 g mass is attached to the bottom. If an additional 120 g mass is added to the spring, how does the potential energy of the spring change?

A. the same
B. 4 times greater
C. 2 times greater
D. $\sqrt{2}$ times greater

24. A Ferrari, Maserati, and Lamborghini move at the same speed, and each driver slams on his brakes and brings the car to a stop. The most massive is the Ferrari, and the least massive is the Lamborghini. If the tires of the three cars have identical coefficients of friction with the road surface, which car experiences the greatest amount of work done by friction?

A. Maserati
B. Lamborghini
C. Ferrari
D. The amount is the same

25. A hammer does the work of driving a nail into a wooden board. Compared to the moment before the hammer strikes the nail after it impacts the nail, the hammer's mechanical energy is:

A. the same
B. less, because work has been done on the hammer
C. greater, because the hammer has done work
D. less, because the hammer has done work

26. A 1,500 kg car travels at 25 m/s on a level road, and the driver slams on the brakes. The skid marks are 10 m long. What is the work done by the road on the car?

A. -4.7×10^5 J
B. 0 J
C. 2×10^5 J
D. 3.5×10^5 J

27. A 1,000 kg car is traveling at 4.72 m/s. How fast is the truck traveling if a 2,000 kg truck has 20 times the car's kinetic energy?

A. 23.6 m/s
B. 47.2 m/s
C. 94.4 m/s
D. 14.9 m/s

28. A 1,500 kg car travels at 25 m/s on a level road, and the driver slams on the brakes. The skid marks are 30 m long. What forces are acting on the car while it is coming to a stop?

A. Gravity down, normal force up, and a frictional force forward
B. Gravity down, normal force up, and the engine force forward
C. Gravity down, normal force up, and a frictional force backward
D. Gravity down, normal force forward, and the engine force backward

29. A 6,000 N piano is raised via a pulley. For every 1 m that the rope is pulled down, the piano rises 0.15 m. In this pulley system, what is the force needed to lift the piano?

A. 60 N
B. 900 N
C. 600 N
D. 300 N

30. What does the area under the curve on a force *vs.* position graph represent?

A. Kinetic energy
B. Momentum
C. Work
D. Displacement

31. What is the form in which most energy comes to and leaves the Earth?

A. Kinetic
B. Radiant
C. Chemical
D. Light

32. A driver abruptly slams on the brakes in her car, and the car skids a certain distance on a straight level road. If she had been traveling twice as fast, what distance would the car have skid under the same conditions?

A. 1.4 times farther
B. ½ as far
C. 4 times farther
D. 2 times farther

33. A crane hoists an object weighing 2,000 N to the top of a building. The crane raises the object straight upward at a constant rate. Ignoring the forces of friction, at what rate is energy consumed by the electric motor of the crane if it takes 60 s to lift the mass 320 m?

A. 2.5 kW
B. 6.9 kW
C. 3.50 kW
D. 10.7 kW

34. A barbell with a mass of 25 kg is raised 3.0 m in 3.0 s before it reaches constant velocity. What is the net power expended by forces in raising the barbell? (Use acceleration due to gravity $g = 9.8 \text{ m/s}^2$ and the acceleration of the barbell is constant)

A. 17 W
B. 34 W
C. 67 W
D. 98 W

35. Susan carried a 6.5 kg bag of groceries 1.4 m above the ground at a constant velocity for 2.4 m across the kitchen. How much work did Susan do on the bag in the process? (Use acceleration due to gravity $g = 10 \text{ m/s}^2$)

A. 52 J
B. 0 J
C. 164 J
D. 138 J

36. A 1,000 kg car experiences a net force of 9,600 N while decelerating from 30 m/s to 22 m/s. How far does it travel while slowing down?

A. 17 m
B. 22 m
C. 12 m
D. 34 m

37. What is the power output in relation to the work W if a person exerts 100 J in 50 s?

A. ¼ W
B. ½ W
C. 2 W
D. 4 W

38. If a ball is released from a cliff ledge 58 m above the ground, how fast is the ball traveling when it reaches the ground? (Use the acceleration due to gravity $g = 10 \text{ m/s}^2$)

A. 68 m/s
B. 16 m/s
C. 44 m/s
D. 34 m/s

39. A stone is held at a height h above the ground, and a second stone with four times the mass is held at the same height. What is the gravitational potential energy of the second stone compared to that of the first stone?

A. Four times as much
B. The same
C. One-fourth as much
D. One-half as much

40. A 1.3 kg coconut falls off a coconut tree, landing on the ground 600 cm below. How much work is done on the coconut by the gravitational force? (Use the acceleration due to gravity $g = 10\ \text{m/s}^2$)

A. 6 J
B. 78 J
C. 168 J
D. 340 J

41. The potential energy of a pair of interacting objects is related to their:

A. relative position
B. momentum
C. acceleration
D. kinetic energy

42. A spring has a spring constant of 65 N/m. One end of the spring is fixed at point P, while the other is connected to a 7 kg mass m. The fixed end and the mass sit on a horizontal, frictionless surface so that the mass and the spring can rotate about P. The mass moves in a circle of radius $r = 4$ m, and the centripetal force of the mass is 15 N. What is the potential energy stored in the spring?

A. 1.7 J
B. 2.8 J
C. 3.7 J
D. 7.5 J

43. If electricity costs 8.16 cents/kW·h, how much would it cost to run a 120 W stereo system 3.5 hours per day for five weeks?

A. $1.11
B. $1.46
C. $1.20
D. $0.34

44. A boy does 120 J of work to pull his sister back on a swing with a 5.1 m chain until the swing makes an angle of 32° with the vertical. What is the mass of his sister? (Use the acceleration due to gravity $g = 9.8\ \text{m/s}^2$)

A. 18 kg
B. 15.8 kg
C. 13.6 kg
D. 11.8 kg

45. What is the value of the spring constant if 111 J of work is needed to stretch a spring from 1.4 m to 2.9 m if the spring's equilibrium position is at 0.0 m?

A. 58 N/m
B. 53 N/m
C. 67 N/m
D. 34 N/m

46. The metric unit of a joule (J) is a unit of:

I. potential energy II. kinetic energy III. work

A. I only
B. II only
C. III only
D. I, II and III

47. A horizontal spring-mass system oscillates on a frictionless table. Find the maximum extension of the spring if the ratio of the mass to the spring constant is 0.038 kg·m/N, and the maximum speed of the mass is 18 m/s?

A. 3.5 m
B. 0.67 m
C. 3.4 cm
D. 67 cm

48. A truck weighs twice as much as a car and is moving at twice the speed of the car. Which statement is true about the truck's kinetic energy compared to that of the car?

A. The truck has 8 times the KE
B. The truck has twice the KE
C. The truck has $\sqrt{2}$ times the KE
D. The truck has 4 times the KE

49. When a car brakes to a stop, its kinetic energy is transformed into:

A. energy of rest
B. energy of momentum
C. heat
D. stopping energy

50. A 30 kg block hangs from a spring with a spring constant of 900 N/m. How far does the spring stretch from its equilibrium position? (Use acceleration due to gravity $g = 10\ \text{m/s}^2$)

A. 12 cm
B. 33 cm
C. 50 cm
D. 0.5 cm

51. What is the kinetic energy of a 0.33 kg baseball thrown at a velocity of 40 m/s?

A. 426 J
B. 574 J
C. 318 J
D. 264 J

52. An object is acted upon by force represented by the force *vs.* position graph below. What is the work done as the object moves from 0 m to 4 m?

A. 10 J
B. 50 J
C. 20 J
D. 30 J

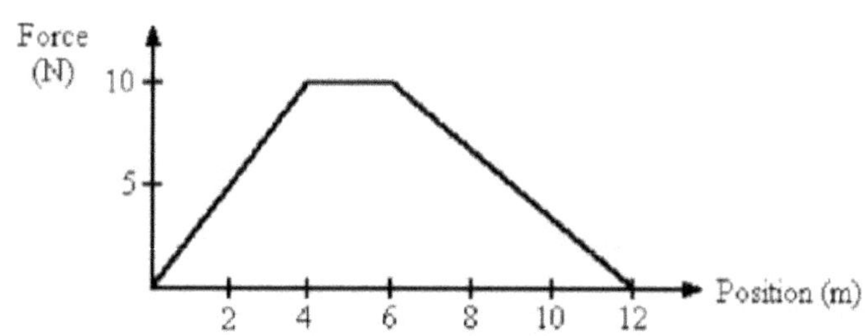

53. James and Bob throw identical balls vertically upward. James throws his ball with an initial speed twice that of Bob's. Assuming no air resistance, what is the maximum height of James's ball compared with that of Bob's ball?

A. Equal
B. Eight times
C. Four times
D. Two times

54. The graphs show the magnitude of the force (F) exerted by a spring as a function of the distance (x) the spring has been stretched. Which graph shows a spring that obeys Hooke's Law?

A.

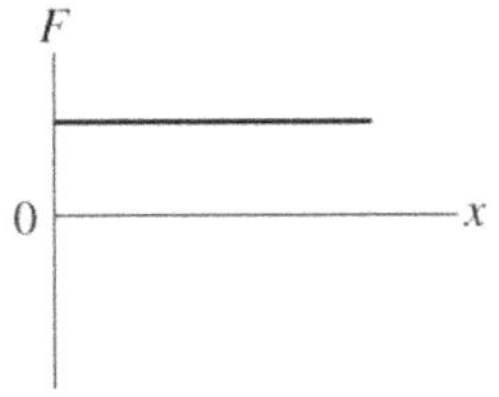

C.

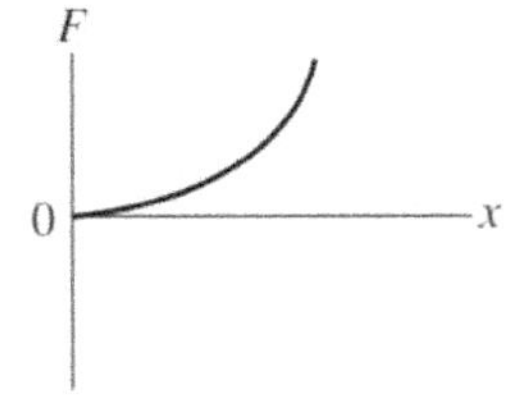

B.

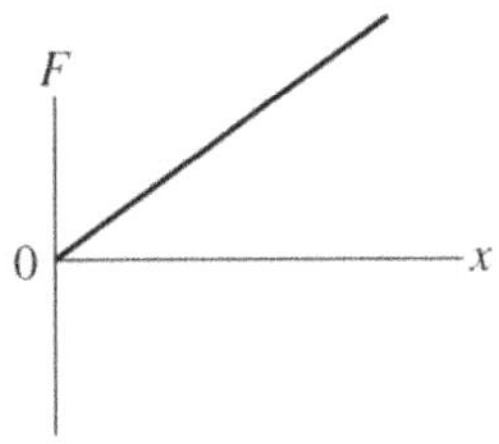

D.

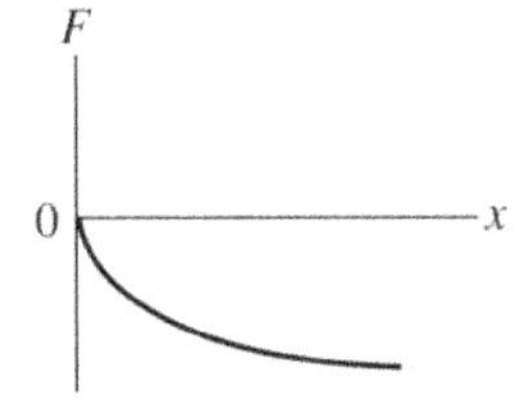

55. If a rocket travels through the air, it loses kinetic energy due to air resistance. Some transferred energy:

A. decreases the temperature of the air around the rocket
B. is found in increased KE of the rocket
C. is found in increased KE of the air molecules
D. decreases the temperature of the rocket

56. A car moves four times as fast as an identical car. Compared to the slower car, the faster car has how much more kinetic energy?

A. 4 times
B. 8 times
C. $\sqrt{2}$ times
D. 16 times

57. A massless, ideal spring with spring constant k is connected to a wall and a massless plate on the other end. A mass m is sitting on a frictionless floor. The mass m is slid against the plate and pushed back a distance x. After release, it achieves a maximum speed v_1. In a second experiment, the same mass is pushed back a distance $4x$. After its release, it reaches a maximum speed v_2. How does v_2 compare with v_1?

A. $v_2 = v_1$
B. $v_2 = 2v_1$
C. $v_2 = 4v_1$
D. $v_2 = 16v_1$

58. A N·m/s is a unit of:

I. work II. force III. power

A. I only
B. II only
C. III only
D. I and II only

59. For the work-energy theorem, which statement is accurate regarding the net work done?

A. The net work done plus the initial KE is the final KE
B. Final KE plus the net work done is the initial KE
C. The net work done minus the final KE is the initial KE
D. The net work done is equal to the initial KE plus the final KE

60. A 1,320 kg car climbs a 5° slope at a constant velocity of 70 km/h. Ignoring air resistance, at what rate must the engine deliver energy to drive the car? (Use the acceleration due to gravity $g = 9.8\ \text{m/s}^2$)

A. 45.1 kW
B. 12.7 kW
C. 6.3 kW
D. 22.6 kW

Notes for active learning

Notes for active learning

Periodic Motion

1. A simple harmonic oscillator oscillates with frequency f when its amplitude is A. What is the new frequency if the amplitude is doubled to 2A?

A. $f/2$
B. f
C. $4f$
D. $2f$

2. Springs A and B are attached in series, with the free end of spring B attached to a wall. The free end of spring A is pulled, and both springs expand from their equilibrium lengths. The length of spring A increases by L_A, and the length of spring B increases by L_B. What is the expression for the spring constant k_B of spring B?

A. L_B/k_A
B. k_A^2
C. $k_A L_B$
D. $(k_A L_A) / L_B$

3. Particles of material moving back and forth in the same direction the wave moves are in what type of wave?

A. Standing
B. Torsional
C. Transverse
D. Longitudinal

4. If a wave has a wavelength of 25 cm and a frequency of 1.68 kHz, what is its speed?

A. 44 m/s
B. 160 m/s
C. 420 m/s
D. 314 m/s

5. The total stored energy in a system undergoing simple harmonic motion (SHM) is proportional to the:

A. $(\text{amplitude})^2$
B. wavelength
C. $(\text{spring constant})^2$
D. amplitude

6. An 11 kg mass m is attached to a spring and allowed to hang in the Earth's gravitational field. The spring stretches 3 cm before reaching its equilibrium position. If the spring were allowed to oscillate, what would be its frequency? (Use the acceleration due to gravity $g = 9.8$ m/s^2)

A. 0.7 Hz
B. 1.8 Hz
C. 4.1 Hz
D. 2.9 Hz

7. The time required for one cycle of a repeating event is the:

A. amplitude
B. frequency
C. period
D. rotation

8. A pendulum of length L is suspended from the ceiling of an elevator. When the elevator is at rest, the period of the pendulum is T. How does T change when the elevator moves upward with a constant velocity?

A. Decreases only if the upward acceleration is less than ½g
B. Decreases
C. Increases
D. Remains the same

9. What is the period of a transverse wave with a frequency of 100 Hz?

A. 0.01 s
B. 0.05 s
C. 0.2 s
D. 20 s

10. Two radio antennae are located on a seacoast 10 km apart on a North-South axis. The antennas broadcast identical in-phase AM radio waves at a frequency of 4.7 MHz 200 km offshore, a steamship travels North at 15 km/h, passing East of the antennae with a radio tuned to the broadcast frequency. From the moment of the maximum reception of the radio signal on the ship, what is the time interval until the next occurrence of maximum reception? (Use the speed of radio waves equals the speed of light $c = 3 \times 10^8$ m/s and the path difference = 1 λ)

A. 7.7 min
B. 5.1 min
C. 3.8 min
D. 8.9 min

11. A 2.31 kg rope is stretched between supports 10.4 m apart. If one end of the rope is tweaked, how long will it take for the resulting disturbance to reach the other end? Assume that the tension in the rope is 74.4 N.

A. 0.33 s
B. 0.74 s
C. 0.65 s
D. 0.57 s

12. Simple pendulum A swings back and forth twice the frequency of simple pendulum B. Which statement is correct?

A. Pendulum A is ¼ as long as B
B. Pendulum A is twice as massive as B
C. Pendulum A is ½ as long as B
D. Pendulum B is twice as massive as A

13. A weight attached to the free end of an anchored spring is allowed to slide back and forth in simple harmonic motion on a frictionless table. How many times greater is the spring's restoring force at $x = 5$ cm compared to $x = 1$ cm (measured from equilibrium)?

A. 2.5
B. 5
C. 7.5
D. 15

14. A massless, ideal spring projects horizontally from a wall and is connected to a 1 kg mass. The mass is oscillating in one dimension, such that it moves 0.5 m from one end of its oscillation to the other. It undergoes ten complete oscillations in 60 s. What is the period of the oscillation?

A. 9 s
B. 3 s
C. 6 s
D. 12 s

15. The total mechanical energy of a simple harmonic oscillating system is:

A. a nonzero constant
B. maximum when it reaches the maximum displacement
C. zero when it reaches the maximum displacement
D. zero as it passes the equilibrium point

16. What is the frequency of the oscillations when a vibrating spring moves from its maximum elongation position to its maximum compression position in 1 s?

A. 0.75 Hz
B. 0.5 Hz
C. 1 Hz
D. 2.5 Hz

17. Which of the following is not a transverse wave?

I. Radio II. Light III. Sound

A. I only
B. II only
C. III only
D. I and II only

18. If a wave has a speed of 362 m/s and a period of 4 ms, its wavelength is closest to:

A. 8.6 m
B. 1.5 m
C. 0.86 m
D. 15 m

19. Simple harmonic motion is characterized by:

A. acceleration that is proportional to the negative displacement
B. acceleration that is proportional to the velocity
C. constant positive acceleration
D. acceleration that is inversely proportional to the negative displacement

20. If the frequency of a harmonic oscillator doubles, by what factor does the maximum value of acceleration change?

A. $2/\pi$
B. $\sqrt{2}$
C. 2
D. 4

21. An object that hangs from the ceiling of a stationary elevator by an ideal spring oscillates with a period T. If the elevator accelerates upwards with an acceleration of $2g$, what is the period of oscillation of the object?

A. T/2
B. T
C. 2T
D. 4T

22. Which of the following changes made to a transverse wave must increase wavelength?

A. An increase in frequency and a decrease in speed
B. The wavelength is only affected by a change in amplitude
C. A decrease in frequency and an increase in speed
D. A decrease in frequency and a decrease in speed

23. If a wave travels 30 m in 1 s, making 60 vibrations per second, what are its frequency and speed, respectively?

A. 30 Hz and 60 m/s
B. 60 Hz and 30 m/s
C. 30 Hz and 30 m/s
D. 60 Hz and 15 m/s

24. Transverse waves propagate at 40 m/s in a string subjected to a tension of 60 N. If the string is 16 m long, what is its mass?

A. 0.6 kg
B. 0.9 kg
C. 0.2 kg
D. 9 kg

25. Doubling only the amplitude of a vibrating mass-on-spring system changes the system frequency by what factor?

A. Increases by 3
B. Increases by 2
C. Increases by 5
D. Remains the same

26. A leaky faucet drips 60 times in 40 s. What is the frequency of the dripping?

A. 0.75 Hz
B. 0.67 Hz
C. 1.5 Hz
D. 12 Hz

27. Particles of material move up and down perpendicular to the direction that the wave is moving. What is the type of wave?

A. Torsional
B. Mechanical
C. Longitudinal
D. Transverse

28. The figure shows velocity v as a function of time t for a system undergoing simple harmonic motion. Which one of the following graphs represents the acceleration of this system as a function of time?

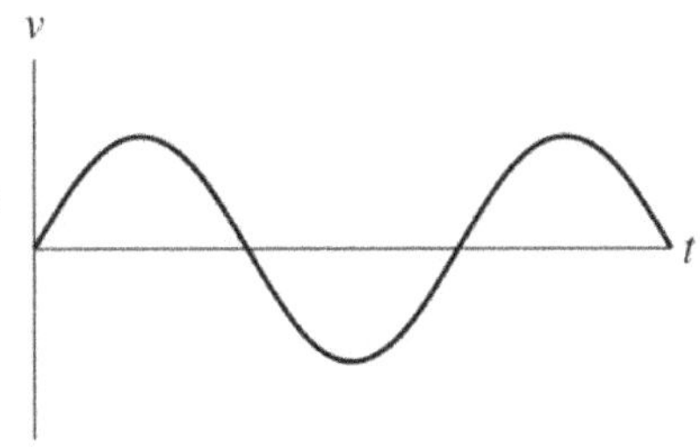

A. 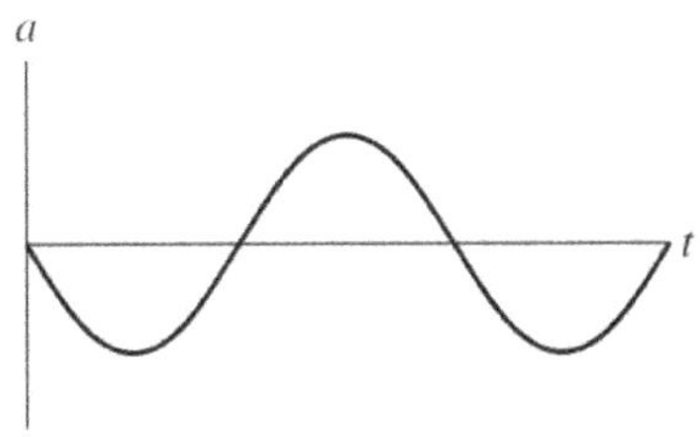

C. 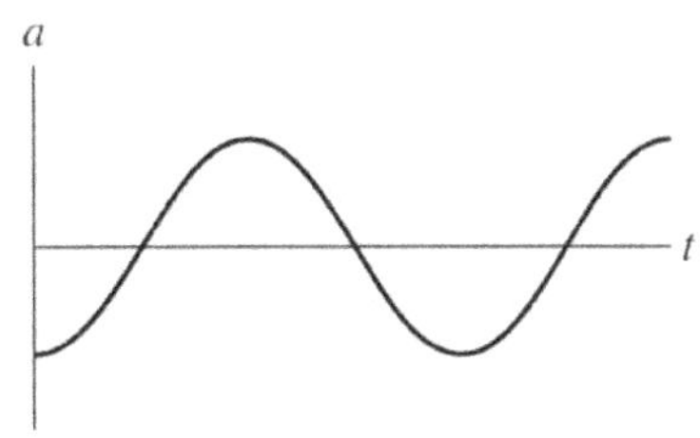

B. 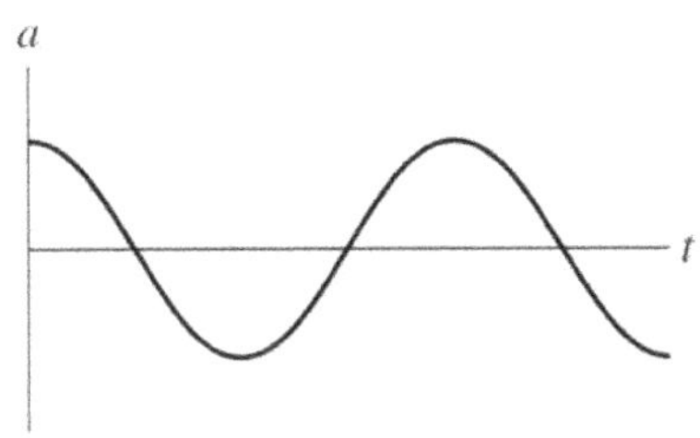

D. 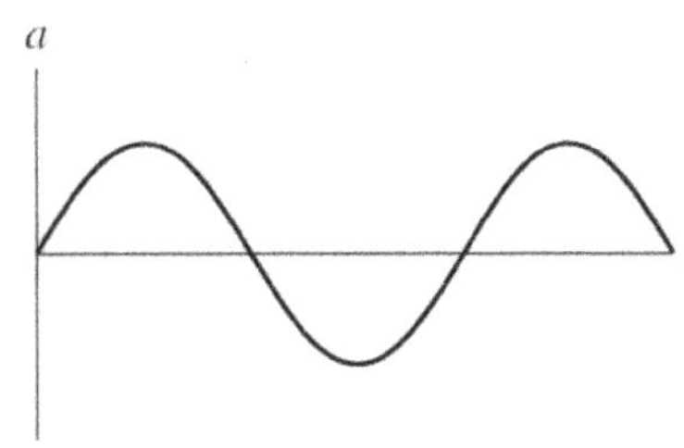

29. When compared, a transverse wave and a longitudinal wave have amplitudes of equal magnitude. Which statement is true about their speeds?

A. The waves have the same speeds
B. The transverse wave has exactly twice the speed of the longitudinal wave
C. The speeds of the two waves are unrelated to their amplitudes
D. The longitudinal wave has a slower speed

30. What is the frequency when the weight on the end of a spring bobs up and down and completes one cycle every 2 s?

A. 0.5 Hz
B. 1 Hz
C. 2 Hz
D. 2.5 Hz

31. The velocity of a given longitudinal sound wave in an ideal gas is $v = 340$ m/s at constant pressure and constant volume. Assuming an ideal gas, what is the wavelength for a 2,100 Hz sound wave?

A. 0.08 m
B. 0.16 m
C. 1.6 m
D. 7.3 m

32. When the mass of a simple pendulum is quadrupled, how does the time t required for one complete oscillation change?

A. Decreases to ¼t
B. Decreases to ¾t
C. Increases to 4t
D. Remains the same

33. An object undergoing simple harmonic motion has an amplitude of 2.5 m. If the maximum velocity of the object is 15 m/s, what is the object's angular frequency (ω)?

A. 6.0 rad/s
B. 3.6 rad/s
C. 37.5 rad/s
D. 8.8 rad/s

34. Unpolarized light is incident upon two polarization filters that do not have their transmission axes aligned. If 14% of the light passes through, what is the angle between the transmission axes of the filters?

A. 73°
B. 81°
C. 43°
D. 58°

35. A mass on a spring undergoes simple harmonic motion. Which of the statements is true when the mass is at its maximum distance from the equilibrium position?

A. KE is nonzero
B. Acceleration is at a minimum
C. Speed is zero
D. Speed is maximum

36. What is the frequency if the speed of a sound wave is 240 m/s and its wavelength is 10 cm?

A. 2.4 Hz
B. 24 Hz
C. 240 Hz
D. 2,400 Hz

37. Unlike a transverse wave, a longitudinal wave has no:

A. wavelength
B. crests or troughs
C. amplitude
D. frequency

38. The density of aluminum is 2,700 kg/m^3. If transverse waves propagate at 36 m/s in a 9.2 mm diameter aluminum wire, what is the tension in the wire?

A. 43 N
B. 68 N
C. 233 N
D. 350 N

39. When a wave obliquely crosses a boundary into another medium, it is:

A. always slowed down
B. reflected
C. diffracted
D. refracted

40. A floating leaf oscillates up and down two complete cycles each second as a water wave passes. What is the wave's frequency?

A. 0.5 Hz
B. 1 Hz
C. 2 Hz
D. 3 Hz

41. A higher pitch for a sound wave means the wave has a greater:

A. frequency
B. wavelength
C. amplitude
D. period

42. An object is attached to a vertical spring and bobs up and down between points A and B. Where is the object located when its kinetic energy is at a maximum?

A. One-fourth of the way between A and B
B. One-third of the way between A and B
C. Midway between A and B
D. At A or B

43. A pendulum consists of a 0.5 kg mass attached to the end of a 1 m rod of negligible mass. What is the magnitude of the torque τ about the pivot when the rod makes an angle θ of 60° with the vertical? (Use the acceleration due to gravity $g = 10\ m/s^2$)

A. 2.7 N·m
B. 4.4 N·m
C. 5.2 N·m
D. 10.6 N·m

44. The Doppler effect is characteristic of:

I. light waves II. sound waves III. water waves

A. I only
B. II only
C. III only
D. I, II and III

45. A crane lifts a 2,500 kg cement block using a steel cable with a mass per unit length of 0.65 kg/m. What is the speed of the transverse waves on this cable? (Use acceleration due to gravity $g = 10\ m/s^2$)

A. 196 m/s
B. 1,162 m/s
C. 322 m/s
D. 558 m/s

46. A simple pendulum consists of a mass M attached to a weightless string of length L. Which statement about the frequency f is accurate for this system when it experiences small oscillations?

A. The f is directly proportional to period
B. The f is independent of the mass M
C. The f is inversely proportional to the amplitude
D. The f is independent of the length L

47. A child on a swing set swings back and forth. If the length of the supporting cables for the swing is 3.3 m, what is the period of oscillation? (Use the acceleration due to gravity $g = 10$ m/s^2)

A. 3.6 s
B. 5.9 s
C. 4.3 s
D. 2.7 s

48. A massless, ideal spring projects horizontally from a wall and is connected to a 0.3 kg mass. The mass is oscillating in one dimension, such that it moves 0.4 m from one end of its oscillation to the other. It undergoes 15 complete oscillations in 60 s. How does the frequency change if the spring constant increases by a factor of 2?

A. Increases by 200%
B. Decreases by 59%
C. Increases by 41%
D. Decreases by 41%

49. A ball swinging at the end of a massless string undergoes simple harmonic motion (SHM). At what point(s) is the instantaneous acceleration of the ball the greatest?

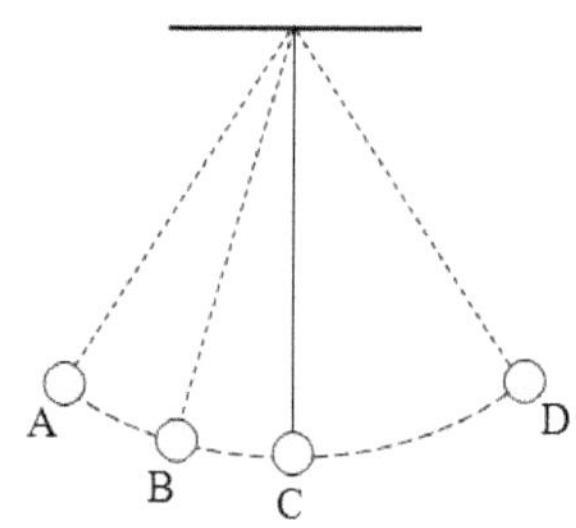

A. A
B. B
C. C
D. A and D

50. A simple pendulum, consisting of a 2 kg weight connected to a 10 m massless rod, is brought to an angle of 90° from the vertical and then released. What is the speed of the weight at its lowest point? (Use the acceleration due to gravity $g = 10$ m/s^2)

A. 14 m/s
B. 10 m/s
C. 20 m/s
D. 25 m/s

51. A sound source of high pitch emits a wave with a high:

I. frequency II. amplitude III. speed

A. I only
B. II only
C. III only
D. I, II and III

52. Find the wavelength of a train whistle that a fixed observer hears as the train moves toward him with a velocity of 50 m/s. The wind blows at 5 m/s from the observer to the train. The whistle has a natural frequency of 500 Hz. (Use the velocity v of sound = 340 m/s)

A. 0.75 m
B. 0.43 m
C. 0.58 m
D. 7.5 m

53. Considering a vibrating mass on a spring, what effect on the system's mechanical energy is caused by only doubling the amplitude?

A. Increases by a factor of two
B. Increases by a factor of four
C. Increases by a factor of three
D. Produces no change

54. Which of the following is an accurate statement?

A. Tensile stress is measured in N·m
B. Stress is a measure of external forces on a body
C. The ratio stress/strain is the elastic modulus
D. Tensile strain is measured in meters

55. The efficient transfer of energy taking place at a natural frequency occurs in a phenomenon called:

A. reverberation
B. the Doppler effect
C. beats
D. resonance

56. A simple pendulum and a mass oscillating on an ideal spring both have period T in an elevator at rest. If the elevator now accelerates downward uniformly at 2 m/s^2, what is true about the periods of these two systems?

A. The period of the pendulum increases, but the period of the spring remains the same
B. The period of the pendulum increases, and the period of the spring decreases
C. The period of the pendulum decreases, but the period of the spring remains the same
D. The periods of the pendulum and the spring both increase

57. All of the following is true of a pendulum that has swung to the top of its arc and has not yet reversed its direction, EXCEPT:

A. The PE of the pendulum is at a maximum
B. The acceleration of the pendulum equals zero
C. The KE of the pendulum equals zero
D. The velocity of the pendulum equals zero

58. The Doppler effect occurs when a source of sound moves:

I. toward the observer
II. away from the observer
III. with the observer

A. I only
B. II only
C. III only
D. I and II only

59. Consider the wave shown in the figure. The amplitude is:

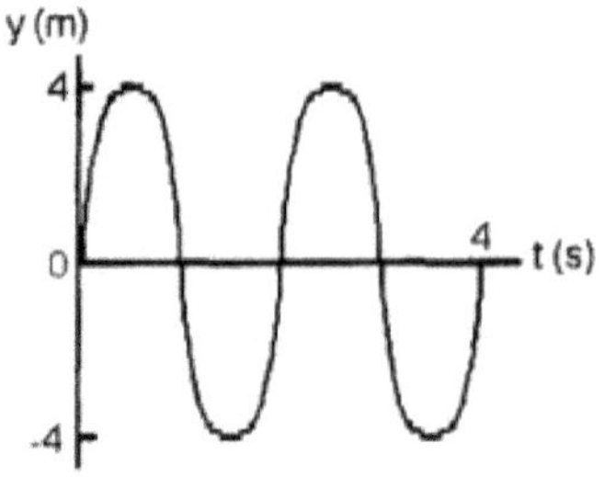

A. 1 m
B. 2 m
C. 4 m
D. 8 m

60. Increasing the mass *m* of a mass-and-spring system causes what kind of change on the resonant frequency *f* of the system?

A. The *f* decreases
B. There is no change in *f*
C. The *f* decreases only if the ratio k / m is < 1
D. The *f* increases

Notes for active learning

Notes for active learning

Fluids and Solids

Questions **1-3** are based on the following:

A container has a vertical tube with an inner radius of 20 mm connected to the container at its side. An unknown liquid reaches level A in the container and level B in the tube. Level A is 5 cm higher than level B. The liquid supports a 20 cm high column of oil between levels B and C with a density of 850 kg/m^3. (Use the acceleration due to gravity $g = 9.8$ m/s^2)

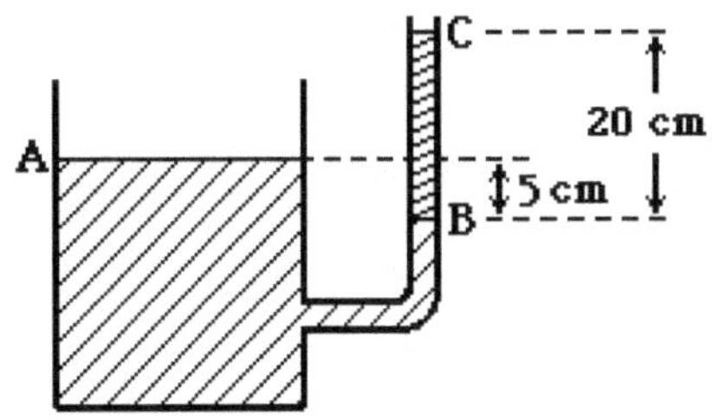

1. What is the density of the unknown liquid?

A. 2,800 kg/m^3
B. 2,100 kg/m^3
C. 3,400 kg/m^3
D. 3,850 kg/m^3

2. The gauge pressure at level B is closest to:

A. 1,250 Pa
B. 1,830 Pa
C. 340 Pa
D. 1,666 Pa

3. What is the mass of the oil?

A. 210 g
B. 453 g
C. 620 g
D. 847 g

4. A cubical block of stone is lowered at a steady rate into the ocean by a crane, keeping the top and bottom faces horizontal. Which of the following graphs best describes the gauge pressure P on the bottom of this block as a function of time t if the block just enters the water at time $t = 0$ s?

A.

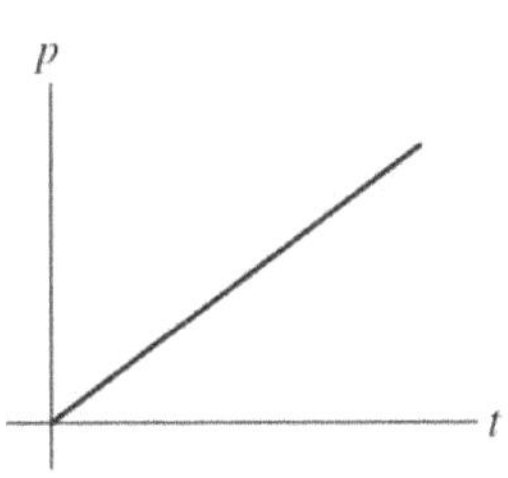

C.

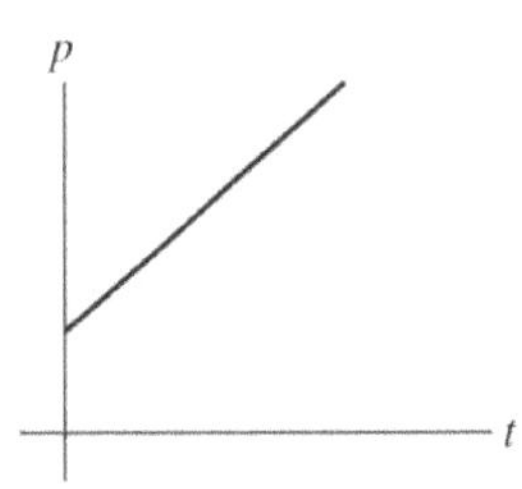

B.

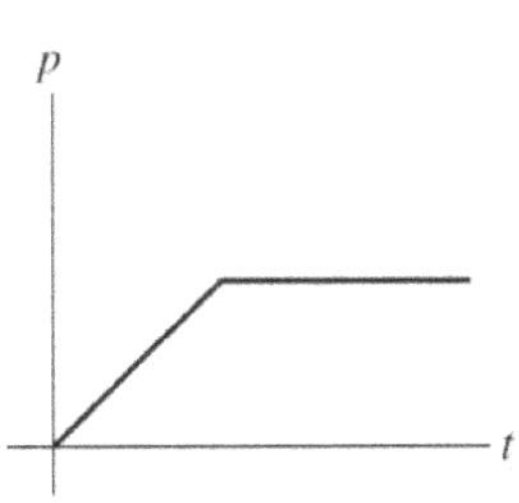

D.

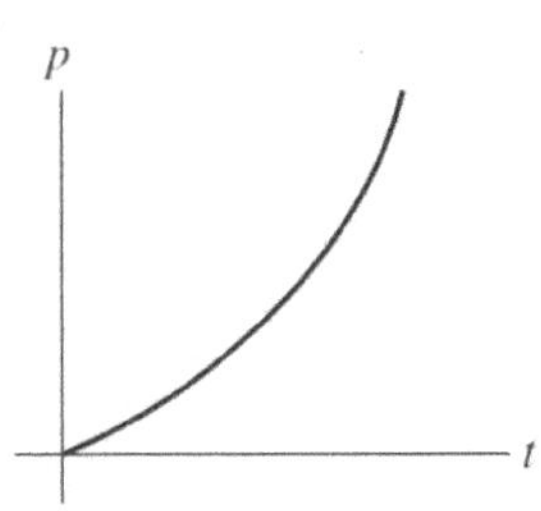

5. Consider a very small hole in the bottom of a tank that is 19 cm in diameter and is filled with water to a height of 80 cm. What is the speed at which the water exits the tank through the hole? (Use the acceleration due to gravity $g = 9.8\ m/s^2$)

A. 8.6 m/s
B. 12 m/s
C. 14.8 m/s
D. 4 m/s

6. An ideal gas at standard temperature and pressure is compressed until its volume is half the initial volume, and then it is allowed to expand until its pressure is half the initial pressure. This is achieved while holding the temperature constant. If the initial internal energy of the gas is U, the final internal energy of the gas is:

A. U/2
B. U/3
C. U
D. 2U

7. When 9.5 kg mass is suspended from a 4.5 m long wire with 1.2 mm diameter, the wire stretches by 4 mm. What is the effective spring constant for the wire? (Use the acceleration due to gravity $g = 9.8\ m/s^2$)

A. 5.6×10^4 N/m
B. 2.3×10^4 N/m
C. 3.7×10^5 N/m
D. 7.2×10^3 N/m

8. An object is sinking in a fluid. What is the weight of the fluid displaced by the sinking object when the object is completely submerged?

A. Dependent on the viscosity of the liquid
B. Equal to the weight of the object
C. Greater than the weight of the object
D. Less than the weight of the object

9. A submarine in neutral buoyancy is 100 m below the surface of the water. How much air pressure must be supplied to remove water from the ballast tanks for the submarine to the surface? (Use the acceleration due to gravity $g = 9.8\ m/s^2$ and density of water $\rho = 10^3\ kg/m^3$)

A. $9.8 \times 10^5\ N/m^2$
B. $4.7 \times 10^5\ N/m^2$
C. $7.6 \times 10^5\ N/m^2$
D. $5.6 \times 10^5\ N/m^2$

10. When atmospheric pressure increases, what happens to the absolute pressure at the bottom of a pool?

A. It does not change
B. It increases by double the amount
C. It increases by the same amount
D. It increases by half the amount

11. When soup gets cold, it often tastes greasy because oil spreads out on the surface of the soup instead of staying in small globules. This is explained in terms of the:

A. increase in the surface tension of water with a decreasing temperature
B. Archimedes' principle
C. decrease in the surface tension of water with a decreasing temperature
D. Joule-Thomson effect

12. For a wire of diameter d and fixed length L, the weight that causes the wire to stretch a given distance is:

A. directly proportional to d^2 of the wire
B. inversely proportional to d^2 of the wire
C. proportional to d of the wire
D. independent of d of the wire

13. An object whose weight is 60 N is floating at the surface of a container of water. How much of the object's volume is submerged? (Use acceleration due to gravity $g = 10\ m/s^2$)

A. 0.006 m^3
B. 0.06 m^3
C. 0.6 m^3
D. 6%

14. What is the volume flow rate of fluid if it flows at 2.5 m/s through a pipe of diameter 3 cm?

A. 0.9 m^3/s
B. 4.7 m^3/s
C. $5.7 \times 10^{-4}\ m^3/s$
D. $1.8 \times 10^{-3}\ m^3/s$

15. What is the specific gravity of a cork that floats with three-quarters of its volume in and one-quarter of its volume out of the water?

A. 0.25
B. 0.5
C. 0.75
D. 2

16. What volume does 600 g of cottonseed oil occupy if the density of cottonseed oil is 0.93 g/cm^3?

A. 255 cm^3
B. 360 cm^3
C. 470 cm^3
D. 645 cm^3

17. The kinetic theory of a monatomic gas suggests the average kinetic energy per molecule is:

A. 1/3 k_BT
B. 2 k_BT
C. 3/2 k_BT
D. 2/3 k_BT

18. An object is weighed in air and while totally submerged in water. If it weighs 150 N less when submerged, find the volume of the object. (Use acceleration due to gravity $g = 10\ m/s^2$ and the density of water $\rho = 1{,}000\ kg/m^3$)

A. 0.0015 m^3
B. 0.015 m^3
C. 0.15 m^3
D. 1 m^3

19. When a container of water is on a laboratory scale, the scale reads 140 g. Now a 30 g piece of copper is suspended from a thread and lowered into the water without contacting the bottom of the container. What does the scale read? (Use acceleration due to gravity $g = 9.8\ m/s^2$, density of water $\rho = 1\ g/cm^3$ and density of copper $\rho = 8.9\ g/cm^3$)

A. 122 g
B. 168 g
C. 143 g
D. 110 g

20. An ideal, incompressible fluid flows through a 6 cm diameter pipe at 1 m/s. There is a 3 cm decrease in diameter within the pipe. What is the speed of the fluid in this constriction?

A. 3 m/s
B. 1.5 m/s
C. 2.5 m/s
D. 4 m/s

21. Two blocks are submerged in a fluid. Block A has dimensions 2 cm high × 3 cm wide × 4 cm long, and Block B is 2 cm × 3 cm × 8 cm. Both blocks are submerged with their large faces pointing up and down (i.e., the blocks are horizontal), and they are submerged to the same depth. Compared to the fluid pressure on the bottom of Block A, the bottom of Block B experiences:

A. pressure that depends on the density of the objects
B. greater fluid pressure
C. exactly double the fluid pressure
D. less fluid pressure

22. A piece of thread of diameter d is in the shape of a rectangle (length l, width w) and is lying on the surface of the water in a beaker. If A is the surface tension of the water, what is the maximum weight that the thread can have without sinking?

A. $\mathrm{A}d(l + w) / \pi$
B. $\mathrm{A}(l + w)$
C. $4\mathrm{A}(l + w)$
D. $\mathrm{A}(l + w) / 2$

23. What is the difference between the pressure inside and outside a tire called?

A. Absolute pressure
B. Fluid pressure
C. Atmospheric pressure
D. Gauge pressure

24. Which of the following is NOT a unit of pressure?

A. atm
B. $\mathrm{N \cdot m^2}$
C. inches of mercury
D. Pascal

25. Which of the following is a dimensionless number?

I. Reynolds number II. specific gravity III. shear stress

A. I only
B. II only
C. III only
D. I and II only

26. Water flows out of a large reservoir through a 5 cm diameter pipe. The pipe connects to a 3 cm diameter pipe that is open to the atmosphere, as shown. What is the speed of the water in the 5 cm pipe? Treat the water as an ideal incompressible fluid. (Use the acceleration due to gravity $g = 9.8$ m/s^2)

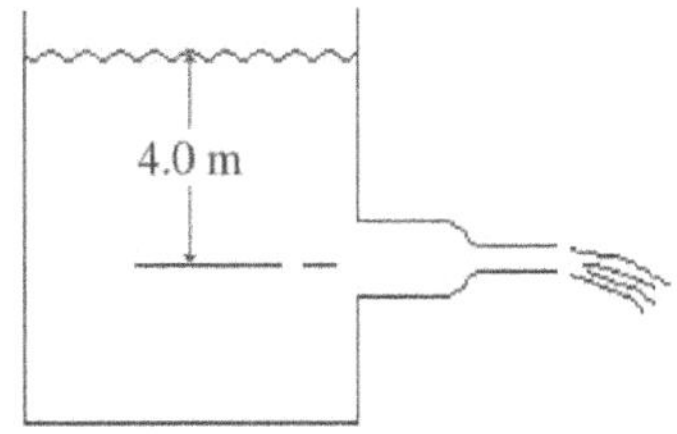

A. 2.6 m/s
B. 3.2 m/s
C. 4.8 m/s
D. 8.9 m/s

27. If the pressure acting on an ideal gas at constant temperature is tripled, what is the resulting volume of the ideal gas?

A. Increased by a factor of two
B. Remains the same
C. Reduced to one-half
D. Reduced to one-third

28. A circular plate with an area of 1 m^2 covers a drain-hole at the bottom of a tank of water 1 m deep. Approximately how much force is required to lift the cover if it weighs 1,500 N? (Use the acceleration due to gravity $g = 10$ m/s^2)

A. 4,250 N
B. 9,550 N
C. 16,000 N
D. 11,500 N

29. A bowling ball that weighs 80 N is dropped into a swimming pool filled with water. If the buoyant force on the bowling ball is 20 N when the ball is 1 m below the surface (and sinking), what is the normal force exerted by the bottom of the pool on the ball when it comes to rest there 4 m below the surface?

A. 0 N
B. 60 N
C. 50 N
D. 70 N

30. A block of unknown material is floating in a fluid, half-submerged. If the specific gravity of the fluid is 1.6, what is the block's density? (Use the specific gravity = $\rho_{\text{fluid}} / \rho_{\text{water}}$ and the density of water $\rho = 1{,}000$ kg/m^3)

A. 350 kg/m^3
B. 800 kg/m^3
C. 900 kg/m^3
D. 1,250 kg/m^3

31. Three equal mass Styrofoam balls of radii R, 2R, and 3R, are released simultaneously from a tall tower. Which reaches the ground last?

A. The smallest Styrofoam ball
B. The middle size Styrofoam ball
C. The largest Styrofoam ball
D. All three reach the ground simultaneously

32. In a closed container of fluid, object A is submerged at 6 m from the bottom, and object B is submerged at 12 m from the bottom. Compared to object A, object B experiences:

A. less fluid pressure
B. double the fluid pressure
C. equal fluid pressure
D. triple the fluid pressure

33. Ideal, incompressible water flows at 14 m/s in a horizontal pipe with a pressure of 3.5×10^4 Pa. If the pipe widens to twice its original radius, what is the pressure in the wider section? (Use the density of water $\rho = 1{,}000$ kg/m^3)

A. 7.6×10^4 Pa
B. 12.7×10^4 Pa
C. 2×10^5 Pa
D. 11.1×10^3 Pa

34. Two kilometers above the surface of the Earth, the atmospheric pressure is:

A. unrelated to the atmospheric pressure at the surface
B. twice the atmospheric pressure at the surface
C. triple the atmospheric pressure at the surface
D. less than the atmospheric pressure at the surface

35. An 80 kg man would weigh 784 N if there were no atmosphere. By how much does the buoyancy due to air reduce the man's weight? (Use the density of the man = 1 g/cm^3, the density of the air = 1.2×10^{-3} g/cm^3, the mass $m = 80$ kg and the acceleration due to gravity $g = 9.8$ m/s^2)

A. 0.58 N
B. 0.32 N
C. 0.94 N
D. 2.8 N

36. Diffusion is described by which law?

A. Dulong's
B. Faraday's
C. Kepler's
D. Graham's

37. As a metal rod is stretched, which condition is reached first?

A. Elastic limit
B. Proportional limit
C. Breaking point
D. Fracture point

38. A pump uses a piston 12 cm in diameter that moves 3 cm/s. What is the fluid velocity in a tube that is 2 mm in diameter?

A. 218 cm/s
B. 88 cm/s
C. 136 cm/s
D. 108 m/s

39. What is the specific gravity of an object floating with one-tenth of its volume out of the water?

A. 0.3
B. 0.9
C. 1.3
D. 2.1

40. If each factor listed below were changed by 15%, which would have the greatest effect on the flow rate?

A. Fluid density
B. Pressure difference
C. Length of the pipe
D. Radius of the pipe

41. A 680 g steel hammer (m_h) is tied to a string hanging from a force meter. A 5 kg container of water (m_w) sits on a scale. The hammer is lowered completely into the water but above the bottom. What does the force meter read? (Use the density of steel $\rho = 7.9\ g/cm^3$, density of water $\rho = 1\ g/cm^3$ and acceleration due to gravity $g = 10\ m/s^2$)

A. 5.9 N
B. 8.4 N
C. 10.7 N
D. 5.2 N

42. An external pressure applied to an enclosed fluid is transmitted unchanged to every point within the fluid by:

A. Torricelli's law
B. Bernoulli's principle
C. Archimedes' principle
D. Pascal's principle

43. A submarine rests on the bottom of the sea. What is the normal force exerted upon the submarine by the seafloor equal to?

A. weight of the submarine
B. weight of the submarine minus the weight of the displaced water
C. buoyant force minus the atmospheric pressure acting on the submarine
D. weight of the submarine plus the weight of the displaced water

44. Consider a brick totally immersed in water, with the long edge of the brick vertical. Which statement describes the pressure on the brick?

A. Highest on the sides of the brick
B. Highest on the top of the brick
C. Highest on the bottom of the brick
D. Same on all surfaces of the brick

45. Water is flowing in a drainage channel of a rectangular cross-section. The width of the channel is 14 m, the depth of water is 7 m, and the flow speed is 3 m/s. What is the mass flow rate of the water? (Use the density of water $\rho = 1{,}000\ kg/m^3$)

A. 2.9×10^5 kg/s
B. 4.8×10^4 kg/s
C. 6.2×10^5 kg/s
D. 9.3×10^4 kg/s

46. What is the magnitude of the buoyant force if a 3 kg object floats motionlessly in a fluid of specific gravity 0.8? (Use acceleration due to gravity $g = 10\ \text{m/s}^2$)

A. 15 N
B. 7.5 N
C. 30 N
D. 45 N

47. What is the pressure 6 m below the surface of the ocean? (Use density of water $\rho = 10^3\ \text{kg/m}^3$, atmospheric pressure $P_{atm} = 1.01 \times 10^5$ Pa and acceleration due to gravity $g = 10\ \text{m/s}^2$)

A. 1.6×10^5 Pa
B. 0.8×10^5 Pa
C. 2.7×10^4 Pa
D. 3.3×10^4 Pa

48. Density is:

A. inversely proportional to mass and volume
B. proportional to mass and inversely proportional to the volume
C. inversely proportional to mass and proportional to the volume
D. proportional to mass and volume

49. Which of the following would be expected to have the smallest bulk modulus?

A. Helium vapor
B. Liquid water
C. Solid lead
D. Liquid mercury

50. When an 8 kg object is suspended from a steel wire that is 2.7 m long and 0.8 mm in diameter, how much does the wire stretch? (Use the acceleration due to gravity $g = 9.8\ \text{m/s}^2$ and Young's modulus = $20 \times 10^{10}\ \text{N/m}^2$)

A. 2.8 mm
B. 3.2 mm
C. 0.8 mm
D. 2.1 mm

51. For the wire and the object from the previous question, if the diameter of the wire is doubled, what would be the stretch? (Use the acceleration due to gravity $g = 9.8\ \text{m/s}^2$ and Young's modulus = $20 \times 10^{10}\ \text{N/m}^2$)

A. 0.5 mm
B. 2.1 mm
C. 1.2 mm
D. 0.2 mm

52. If atmospheric pressure increases by an amount ΔP, which of the following statements about the pressure in a large pond is true?

A. The gauge pressure increases by ΔP
B. The absolute pressure increases by ΔP
C. The absolute pressure increases, but by an amount less than ΔP
D. The absolute pressure does not change

53. Strain is defined as the:

A. stress per unit area
B. applied force per unit area
C. ratio of stress to the elastic modulus
D. ratio of the change in length to the original length

54. Fluid flows through a 19 cm long tube with a radius of 2.1 mm at an average speed of 1.8 m/s. What is the viscosity of the fluid if the pressure drop is 970 Pa?

A. 0.036 $N \cdot s/m^2$
B. 0.013 $N \cdot s/m^2$
C. 0.0044 $N \cdot s/m^2$
D. 0.0016 $N \cdot s/m^2$

55. The pressure differential across the cross-section of a condor's wing due to the difference in airflow is explained by:

A. Torricelli's law
B. Poiseuille's law
C. Bernoulli's equation
D. Newton's First Law

Notes for active learning

Electrostatics and Magnetism

1. How many excess electrons are present for an object that has a charge of –1 Coulomb? (Use Coulomb's constant $k = 9 \times 10^9$ N·m²/C² and charge of an electron $e = -1.6 \times 10^{-19}$ C)

A. 3.1×10^{19} electrons
B. 6.3×10^{19} electrons
C. 6.3×10^{18} electrons
D. 1.6×10^{19} electrons

2. A flat disk 1 m in diameter is oriented so that the area vector of the disk makes an angle of $\pi/6$ radians with a uniform electric field. What is the electric flux through the surface if the field strength is 740 N/C?

A. 196π N·m²/C
B. $250/\pi$ N·m²/C
C. 644π N·m²/C
D. 160π N·m²/C

3. A positive charge $Q = 1.3 \times 10^{-9}$ C is located along the x-axis at $x = -10^{-3}$ m, and a negative charge of the same magnitude is at the origin. What is the magnitude and direction of the electric field at the point along the x-axis where $x = 10^{-3}$ m? (Use Coulomb's constant $k = 9 \times 10^9$ N·m²/C² and to the right as the positive direction)

A. 8.8×10^6 N/C to the left
B. 3.25×10^7 N/C to the right
C. 5.5×10^7 N/C to the right
D. 2.75×10^6 N/C to the right

4. Two charges $Q_1 = 2.4 \times 10^{-10}$ C and $Q_2 = 9.2 \times 10^{-10}$ C, are near each other, and charge Q_1 exerts a force F_1 on Q_2. How does F_1 change if the distance between Q_1 and Q_2 is increased by a factor of 4?

A. Decreases by a factor of 4
B. Increases by a factor of 16
C. Decreases by a factor of 16
D. Increases by a factor of 4

5. A proton is located at ($x = 1$ nm, $y = 0$ nm) and an electron is located at ($x = 0$ nm, $y = 4$ nm). Find the attractive Coulomb force between them. (Coulomb's constant $k = 9 \times 10^9$ N·m²/C² and charge of electron $e = -1.6 \times 10^{-19}$ C)

A. 5.3×10^8 N
B. 1.4×10^{-11} N
C. 9.3×10^4 N
D. 2.6×10^{-18} N

6. Which form of electromagnetic radiation has photons with the highest energy?

A. Gamma rays
B. Visible light
C. Microwaves
D. Ultraviolet radiation

7. A 54,000 kg asteroid carrying a negative charge of 15 μC is 180 m from another 51,000 kg asteroid carrying a negative charge of 11 μC. What is the net force the asteroids exert upon each other? (Use the gravitational constant $G = 6.673 \times 10^{-11}$ N·m²/kg² and Coulomb's constant $k = 9 \times 10^9$ N·m²/C²)

A. 400,000 N
B. 5,700 N
C. -4.0×10^{-5} N
D. 4.0×10^{-5} N

8. Two small beads are 30 cm apart with no other charges or fields present. Bead A has 20 μC of charge, and bead B has 5 μC. Which of the following statements is true about the electric forces on these beads?

A. The force on A is 120 times the force on B
B. The force on A is precisely equal to the force on B
C. The force on B is 4 times the force on A
D. The force on A is 20 times the force on B

Questions **9-10** are based on the following:

Two parallel metal plates separated by 0.01 m are charged to create a uniform electric field of 3.5×10^4 N/C between them, which points down. A small, stationary 0.008 kg plastic ball *m* is located between the plates and has a small charge *Q*. The only forces acting on it are the force of gravity and the electric field. (Use Coulomb's constant $k = 9 \times 10^9$ N·m²/C², the charge of an electron = -1.6×10^{-19} C, the charge of a proton = 1.6×10^{-19} C, the mass of a proton = 1.67×10^{-27} kg, the mass of an electron = 9.11×10^{-31} kg and the acceleration due to gravity $g = 9.8$ m/s²)

9. What is the charge on the ball?

A. –250 C
B. 250 C
C. 3.8×10^{-6} C
D. -2.2×10^{-6} C

10. How would the acceleration of an electron between the plates compare to the acceleration of a proton between the plates?

A. One thousand eight hundred thirty times as large, and in the opposite direction
B. The square root times as large and in the opposite direction
C. Twice as large and in the opposite direction
D. The same magnitude, but in the opposite direction

11. A positive charge $Q = 2.3 \times 10^{-11}$ C is 10^{-2} m from a negative charge of equal magnitude. Point P is located equidistant between them. What is the magnitude of the electric field at point P? (Use Coulomb's constant $k = 9 \times 10^9$ N·m²/C²)

A. 9.0×10^3 N/C
B. 4.5×10^3 N/C
C. 3.0×10^4 N/C
D. 1.7×10^4 N/C

12. A point charge $Q = -10\ \mu C$. What is the number of excess electrons on charge Q? (Use the charge of an electron $e = -1.6 \times 10^{-19}$ C)

A. 6.3×10^{13} electrons
B. 1.6×10^{13} electrons
C. 9.0×10^{13} electrons
D. 8.5×10^{13} electrons

13. A distance of 3 m separates an electron and a proton. What happens to the magnitude of the force on the proton if the electron is moved 1.5 m closer to the proton?

A. It increases to twice its original value
B. It decreases to one-fourth its original value
C. It increases to four times its original value
D. It decreases to one-half its original value

14. How will the magnitude of the electrostatic force between two objects be affected if the distance between them and both of their charges are doubled?

A. It will increase by a factor of 4
B. It will increase by a factor of 2
C. It will decrease by a factor of 2
D. It will be unchanged

15. Which statement is true for an H nucleus, which has a charge $+e$ situated to the left of a C nucleus, which has a charge $+6e$? The electrical force experienced by the H nucleus is:

A. to the right, and the magnitude is equal to the force exerted on the C nucleus
B. to the right, and the magnitude is less than the force exerted on the C nucleus
C. to the left, and the magnitude is greater than the force exerted on the C nucleus
D. to the left, and the magnitude is equal to the force exerted on the C nucleus

16. Two oppositely charged particles are slowly separated from each other. What happens to the force as the particles are slowly moved apart?

A. attractive and decreasing
B. repulsive and decreasing
C. attractive and increasing
D. repulsive and increasing

17. Electrons move in an electrical circuit:

A. because the wires are so thin
B. by interacting with an established electric field
C. by colliding with each other
D. by being repelled by protons

18. If the number of turns on the secondary coil of a transformer is less than those on the primary, the result is:

A. 240 V transformer
B. 110 V transformer
C. step-up transformer
D. step-down transformer

19. Two charges $Q_1 = 3 \times 10^{-8}$ C and $Q_2 = 9 \times 10^{-8}$ C, are near each other, and charge Q_1 exerts a force F_1 on Q_2. What is F_2, the force that charge Q_2 exerts on charge Q_1?

A. $F_1 / 3$
B. F_1
C. $3F_1$
D. $2F_1$

20. Two electrons are passing 30 mm apart. What is the repulsive electric force that they exert on each other? (Use Coulomb's constant $k = 9 \times 10^9$ N·m²/C² and the charge of an electron $= -1.6 \times 10^{-19}$ C)

A. 1.3×10^{-25} N
B. 3.4×10^{-27} N
C. 2.56×10^{-25} N
D. 3.4×10^{10} N

21. A light bulb is connected to a circuit and has a wire leading to it in a loop. What happens when a strong magnet is quickly passed through the loop?

A. The brightness of the light bulb dims or gets brighter due to an induced emf produced by the magnet
B. The light bulb's brightness remains the same although current decreases
C. The light bulb gets brighter because more energy is added to the system by the magnet inside the coil
D. The light bulb gets brighter because there is an induced emf driving more current through the light bulb

22. Suppose a van de Graaff generator builds a negative static charge, and a grounded conductor is near enough to it so that an 8 μC of negative charge arcs to the conductor. What is the number of electrons transferred? (Use the charge of an electron $e = -1.6 \times 10^{-19}$ C)

A. 1.8×10^{14} electrons
B. 48 electrons
C. 5×10^{13} electrons
D. 74 electrons

23. Which statement must be true if two objects are electrically attracted to each other?

A. One of the objects could be electrically neutral
B. One object must be negatively charged, and the other must be positively charged
C. At least one of the objects must be positively charged
D. At least one of the objects must be negatively charged

24. An amp is a unit of electrical:

A. capacity
B. current
C. potential difference
D. charge

25. A loop of wire is rotated about a diameter (which is perpendicular to a given magnetic field). In one revolution, the induced current in the loop reverses direction how many times?

A. 2
B. 1
C. 0
D. 4

26. Two charges ($Q_1 = 2.3 \times 10^{-8}$ C and $Q_2 = 2.5 \times 10^{-9}$ C) are a distance 0.1 m apart. How much energy is required to bring them to a distance 0.01 m apart? (Use Coulomb's constant $k = 9 \times 10^9$ N·m²/C²)

A. 2.2×10^{-4} J
B. 8.9×10^{-5} J
C. 1.7×10^{-5} J
D. 4.7×10^{-5} J

27. A solid aluminum cube rests on a wooden table in a region where a uniform external electric field is directed straight upward. What can be concluded regarding the charge on the top surface of the cube?

A. The top surface is neutral
B. The top surface is charged negatively
C. The top surface is charged positively
D. The top surface's charge cannot be determined without further information

28. A point charge of $+Q$ is at the center of an equilateral triangle, as shown. When the second charge of $+Q$ is at one of the triangle's vertices, an electrostatic force of 5 N acts on it. What is the magnitude of the force that acts on the center charge when the third charge of $+Q$ is at one of the other vertices?

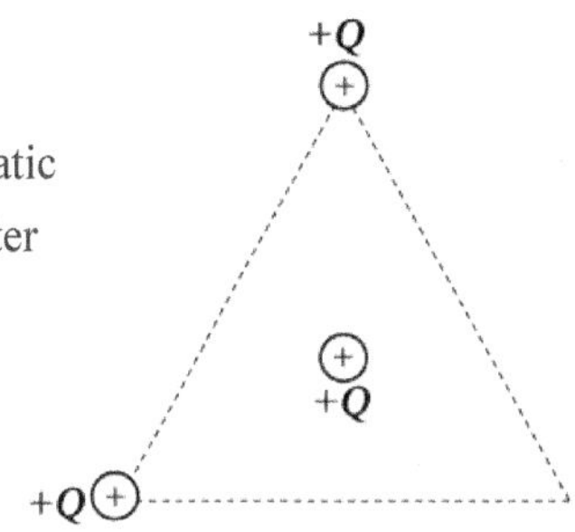

A. 0 N
B. 4 N
C. 5 N
D. 8 N

29. In the figure below, the charge in the middle is fixed and $Q = -7.5$ nC. For what fixed, positive charge q_1 will non-stationary, negative charge q_2 be in static equilibrium?

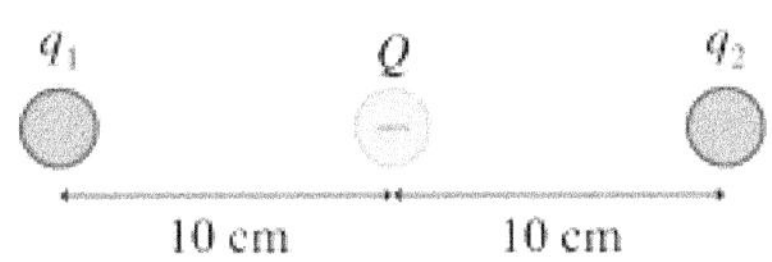

A. 53 nC
B. 7.5 nC
C. 15 nC
D. 30 nC

30. Which form of electromagnetic radiation has the highest frequency?

A. Gamma radiation
B. Ultraviolet radiation
C. Visible light
D. Radio waves

31. The following affect the electrostatic field strength at a point at a distance from a source charge, EXCEPT:

A. the sign of the source charge
B. the distance from the source charge
C. the magnitude of the source charge
D. the nature of the medium surrounding the source charge

32. A charged particle is observed traveling in a circular path in a uniform magnetic field. If the particle had been traveling twice as fast, the radius of the circular path would be:

A. three times the original radius
B. twice the original radius
C. one-half of the original radius
D. four times the original radius

33. Two charges separated by 1 m exert a 1 N force on each other. If the magnitude of each charge is doubled, the force on each charge is:

A. 1 N
B. 2 N
C. 4 N
D. 6 N

34. In a water solution of NaCl, the NaCl dissociates into ions surrounded by water molecules. Consider a water molecule near a Na^+ ion. What tends to be the orientation of the water molecule?

A. The hydrogen atoms are nearer the Na^+ ion because of their positive charge
B. The hydrogen atoms are nearer the Na^+ ion because of their negative charge
C. The oxygen atom is nearer the Na^+ ion because of the oxygen's positive charge
D. The oxygen atom is nearer the Na^+ ion because of the oxygen's negative charge

35. A metal sphere is insulated electrically and is given a charge. If 30 electrons are added to the sphere giving a charge, how many Coulombs are added? (Use Coulomb's constant $k = 9 \times 10^9$ N·m²/C² and the charge of an electron $e = -1.6 \times 10^{-19}$ C)

A. –2.4 C
B. –30 C
C. -4.8×10^{-18} C
D. -4.8×10^{-16} C

36. What happens to the cyclotron frequency of a charged particle if its speed doubles?

A. It remains the same
B. It is ½ as large
C. It doubles
D. It is $\sqrt{2}$ times as large

37. A positive test charge q is released near a positive fixed charge Q. As q moves away from Q, it experiences:

A. increasing acceleration
B. decreasing acceleration
C. constant velocity
D. decreasing velocity

38. If a value has SI units kg m²/s²/C, this value can be:

A. electric potential difference
B. resistance
C. electric field strength
D. Newton's forces

39. A Coulomb is a unit of electrical:

A. capacity
B. resistance
C. charge
D. potential difference

40. To say that electric charge is conserved means that no case has ever been found where:

A. charge has been created or destroyed
B. the total charge on an object has increased
C. the net negative charge on an object is unbalanced by a positive charge on another object
D. the total charge on an object has changed by a significant amount

41. Two charges, $Q_1 = 1.7 \times 10^{-10}$ C and $Q_2 = 6.8 \times 10^{-10}$ C, are near each other. How would F change if the charges were both doubled, but the distance between them remained the same?

A. F increases by a factor of 2
B. F increases by a factor of 4
C. F decreases by a factor of $\sqrt{2}$
D. F decreases by a factor of 4

42. Two like charges of the same magnitude are 10 mm apart. If the force of repulsion they exert upon each other is 4 N, what is the magnitude of each charge? (Use Coulomb's constant $k = 9 \times 10^9$ N·m²/C²)

A. 6×10^{-5} C
B. 6×10^{5} C
C. 2×10^{-7} C
D. 1.5×10^{-7} C

43. A circular loop of wire is rotated about an axis whose direction at constant angular speed can be varied. In a region where a uniform magnetic field points straight down, what orientation of the axis of the rotation guarantees that the emf will be zero (regardless of how the axis is aligned to the loop)?

A. It must be vertical
B. It must make an angle of 45° to the direction South
C. It could have any horizontal orientation
D. It must make an angle of 45° to the vertical

44. Two identical small, charged spheres are a certain distance apart, and each initially experiences an electrostatic force of magnitude F due to the other. With time, the charge gradually diminishes on both spheres. What is the magnitude of the electrostatic force when each sphere has lost half its initial charge?

A. 1/16 F
B. 1/8 F
C. 1/4 F
D. 2 F

45. A proton, moving in a uniform magnetic field, moves in a circle perpendicular to the field lines and takes time T for each circle. If the proton's speed tripled, what would now be its time to go around each circle?

A. T/3
B. T
C. 6T
D. 3T

46. As measurements of the electrostatic field strength are taken at points that progressively approach a negatively-charged particle, the field vectors will point:

A. away from the particle and have a constant magnitude
B. away from the particle and have progressively decreasing the magnitude
C. towards the particle and have progressively increasing the magnitude
D. towards the particle and have progressively decreasing the magnitude

47. Every proton in the universe is surrounded by its own:

I. electric field II. gravitational field III. magnetic field

A. I only
B. II only
C. III only
D. I, II and III

48. A charge $Q = 3.1 \times 10^{-5}$ C is fixed in space while another charge $q = -10^{-6}$ C is 6 m away. Charge q is slowly moved 4 m in a straight line directly toward the charge Q. How much work is required to move charge q? (Use Coulomb's constant $k = 9 \times 10^9$ N·m²/C²)

A. –0.09 J
B. –0.03 J
C. 0.16 J
D. 0.08 J

49. A point charge $Q = -600$ nC. What is the number of excess electrons in charge Q? (Use the charge of an electron $e = -1.6 \times 10^{-19}$ C)

A. 5.6×10^{12} electrons
B. 2.1×10^{10} electrons
C. 2.8×10^{11} electrons
D. 3.8×10^{12} electrons

50. In electricity, what quantity is analogous to the acceleration of gravity, g (i.e., a force per unit mass)?

A. Electric charge
B. Electric current
C. Electric field
D. Electromagnetic force

51. Which type of electromagnetic (EM) wave travels through space the slowest?

A. Visible light
B. Ultraviolet light
C. Gamma rays
D. EM waves travel at the same speed

52. As a proton moves in the direction of the electric field lines, it is moving from:

A. high potential to low potential and losing electric potential energy
B. high potential to low potential and gaining electric potential energy
C. low potential to high potential and gaining electric potential energy
D. low potential to high potential and retaining electric potential energy

53. Which of the following requires a measure of time?

A. Joule
B. Watt
C. Volt
D. Coulomb

54. If an object is characterized as electrically polarized:

A. its internal electric field is zero
B. it is a strong insulator
C. it is electrically charged
D. its charges have been rearranged

55. Two positive charges Q_1 and $Q_2 = 3.4 \times 10^{-10}$ C, are located 10^{-3} m from each other, and point P is precisely between them. What is the magnitude of the electric field at point P?

A. 0 N/C
B. 10^{-10} N/C
C. 6.8×10^{-7} N/C
D. 1.7×10^{-5} N/C

56. Two equally charged spheres of mass 1 g are 2 cm apart. When released, they begin to accelerate at 440 m/s^2. What is the magnitude of the charge on each sphere? (Use Coulomb's constant $k = 9 \times 10^9$ N·m^2/C^2)

A. 80 nC
B. 65 nC
C. 140 nC
D. 100 nC

57. Which of the following is an accurate statement?

A. A conductor cannot carry a net charge
B. The electric field at the surface of a conductor is not necessarily parallel to the surface
C. If a solid metal sphere carries a net charge, the charge distributes uniformly throughout
D. If a solid metal sphere carries a net charge, the charge will move to the sphere surface

58. Two equal and opposite charges a certain distance apart is an electric 'dipole.' A positive test charge $+q$ is placed as shown, equidistant from the two charges.

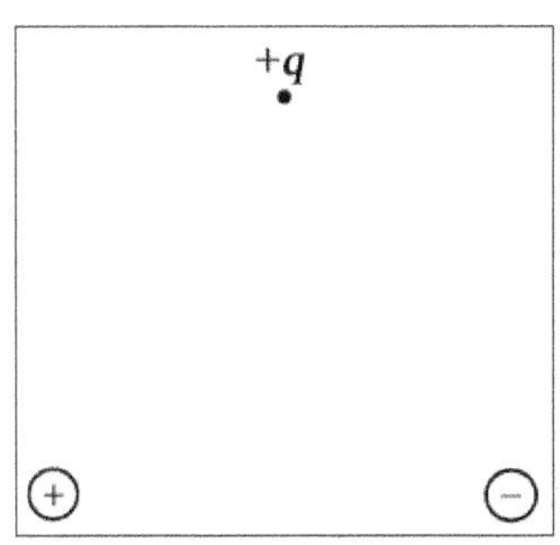

Which diagram gives the direction of the net force on the test charge?

A. ←
B. →
C. ↑
D. ↓

59. A charged particle moves and experiences no magnetic force. What can be concluded?

A. Either no magnetic field exists, or the particle is moving parallel to the field
B. No magnetic field exists in that region of space
C. The particle is moving at right angles to a magnetic field
D. The particle is moving parallel to a magnetic field

60. Find the magnitude of the electrostatic force between a +3 C point charge and a –12 C point charge if they are separated by 50 cm of space. (Use Coulomb's constant $k = 9 \times 10^9$ N·m^2/C^2)

A. 9.2×10^{12} N
B. 1.3×10^{12} N
C. 7.7×10^{12} N
D. 4.8×10^{12} N

Notes for active learning

Notes for active learning

Circuit Elements

1. What is the new resistance if the length of a specific wire is doubled and its radius is doubled?

A. It is √2 times as large
B. It is ½ as large
C. It stays the same
D. It is 2 times as large

2. A 6 Ω resistor is connected across the terminals of a 12 V battery. If 0.6 A of current flows, what is the internal resistance of the battery?

A. 2 Ω
B. 26 Ω
C. 20 Ω
D. 14 Ω

3. Three 8 V batteries are connected in series to power light bulbs A and B. The resistance of light bulb A is 60 Ω and the resistance of light bulb B is 30 Ω. How does the current through light bulb A compare with the current through light bulb B?

A. The current through light bulb A is less
B. The current through light bulb A is greater
C. The current through light bulb A is the same
D. The current through light bulb A is doubled that through light bulb B

4. A sphere with a radius 2 mm carries a 1 μC charge. What is the potential difference, $V_B - V_A$, between point B 3.5 m from the center of the sphere and point A 8 m from the center of the sphere? (Use Coulomb's constant $k = 9 \times 10^9\ \text{N·m}^2/\text{C}^2$)

A. –485 V
B. 1,140 V
C. –140 V
D. 1,446 V

5. Which of the following effect(s) capacitance of capacitors?

I. material between the conductors
II. distance between the conductors
III. the geometry of the conductors

A. I only
B. II only
C. III only
D. I, II and III

6. A proton with an initial speed of 1.5×10^5 m/s falls through a potential difference of 100 volts, gaining speed. What is the speed reached? (Use the mass of a proton = 1.67×10^{-27} kg and the charge of a proton = 1.6×10^{-19} C)

A. 2×10^5 m/s
B. 4×10^5 m/s
C. 8.6×10^5 m/s
D. 7.6×10^5 m/s

7. The current flowing through a circuit of constant resistance is doubled. What is the effect on the power dissipated by that circuit?

A. Decreases to one-half its original value
B. Decreases to one-fourth its original value
C. Quadruples its original value
D. Doubles its original value

8. A positively-charged particle is at rest in an unknown medium. What is the magnitude of the magnetic field generated by this particle?

A. Constant everywhere and dependent only on the mass of the medium
B. Less at points near to the particle compared to a distant point
C. Greater at points near to the particle compared to a distant point
D. Equal to zero

9. The heating element of a toaster is a long wire of some metal, often a metal alloy, which heats up when a 120 V potential difference is applied across it. Consider a 300 W toaster connected to a wall outlet. Which statement would result in an increase in the rate by which heat is produced?

A. Use a longer wire
B. Use a thicker wire
C. Use a thicker and longer wire
D. Use a thinner and longer wire

10. A 4 μC point charge and an 8 μC point charge are initially infinitely far apart. How much work is required to bring the 4 μC point charge to ($x = 2$ mm, $y = 0$ mm), and the 8 μC point charge to ($x = -2$ mm, $y = 0$ mm)? (Use Coulomb's constant $k = 9 \times 10^9\ \text{N}\cdot\text{m}^2/\text{C}^2$)

A. 32.6 J
B. 9.8 J
C. 72 J
D. 81 J

11. What current flows when a 400 Ω resistor is connected across a 220 V circuit?

A. 0.55 A
B. 1.8 A
C. 5.5 A
D. 0.18 A

12. Which statement is accurate for when different resistors are connected in parallel across an ideal battery?

A. Power dissipated in each is the same
B. Potential difference across each is the same
C. Current flowing in each is the same
D. Their equivalent resistance is equal to the average of the individual resistances

13. An electron was accelerated from rest through a potential difference of 990 V. What is its speed? (Use the mass of an electron = 9.11×10^{-31} kg, the mass of a proton = 1.67×10^{-27} kg, and the charge of a proton = 1.6×10^{-19} C)

A. 0.8×10^7 m/s
B. 3.7×10^7 m/s
C. 7.4×10^7 m/s
D. 1.9×10^7 m/s

14. A circular conducting loop with a radius of 0.5 m and a small gap filled with a 12 Ω resistor is oriented in the *xy*-plane. If a magnetic field of 1 T, making an angle of 30° with the *z*-axis, increases to 12 T in 5 s, what is the magnitude of the current flowing in the conductor?

A. 0.33 A
B. 0.13 A
C. 0.88 A
D. 1.5 A

15. For an electric motor with a resistance of 35 Ω that draws 10 A of current, what is the voltage drop?

A. 3.5 V
B. 25 V
C. 350 V
D. 3,500 V

16. A charged parallel-plate capacitor has an electric field E_0 between its plates. The bare nuclei of a stationary ^{1}H and ^{4}He are between the plates. Ignoring the force of gravity, how does the magnitude of the acceleration of the hydrogen nucleus a_H compare with the magnitude of the acceleration of the helium nucleus a_{He}? (Use mass of an electron = 9×10^{-31} kg, mass of a proton = 1.67×10^{-27} kg, mass of a neutron = 1.67×10^{-27} kg and charge of a proton = 1.6×10^{-19} C)

A. $a_H = 2a_{He}$
B. $a_H = 4a_{He}$
C. $a_H = \frac{1}{4}a_{He}$
D. $a_H = a_{He}$

17. Identical light bulbs are attached to identical batteries in three different ways (A, B, or C), as shown in the figure. What is the ranking (from lowest to highest) of the total power produced by the battery?

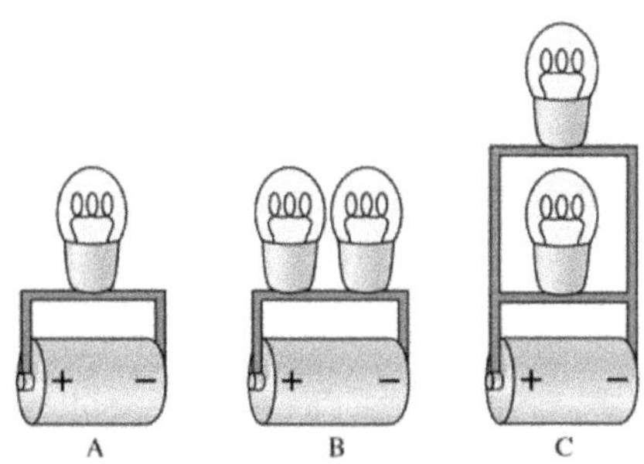

A. C, B, A
B. B, A, C
C. A, C, B
D. A, B, C

18. A parallel-plate capacitor consists of two parallel, square plates with dimensions 1 cm by 1 cm. What is the capacitance if the plates are separated by 1 mm, and the space between them is filled with Teflon? (Use the dielectric constant *k* for Teflon = 2.1 and the electric permittivity $\varepsilon_0 = 8.854 \times 10^{-12}$ F/m)

A. 0.83 pF
B. 2.2 pF
C. 0.46 pF
D. 1.9 pF

19. The resistor R has a variable resistance. Which statement is true when R is decreased? (Neglect the very small internal resistance r of the battery)

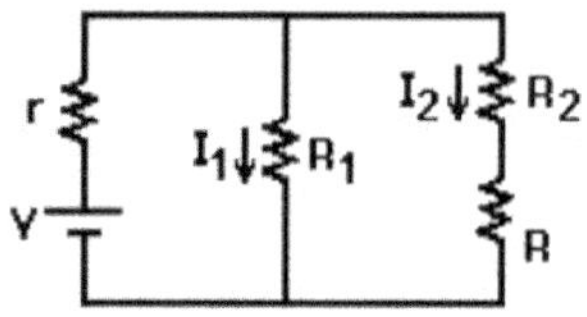

A. I_1 decreases, I_2 increases
B. I_1 increases, I_2 remains the same
C. I_1 remains the same, I_2 increases
D. I_1 remains the same, I_2 decreases

20. What physical quantity does the slope of the graph represent?

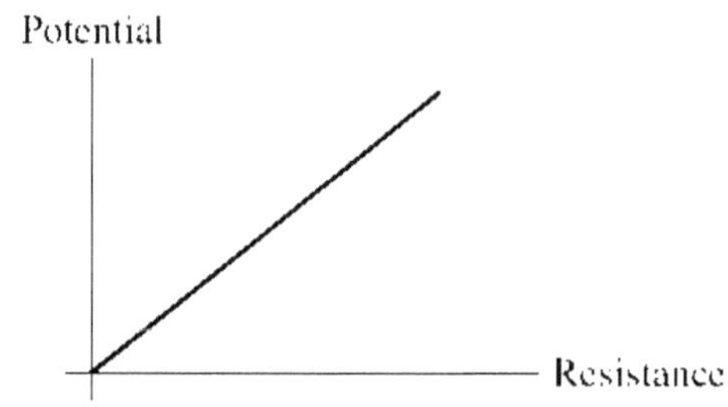

A. 1 / Current
B. Voltage
C. Current
D. Resistivity

21. An alternating current is supplied to an electronic component with a rating used only for voltages below 12 V. What is the highest V_{rms} that can be supplied to this component while staying below the voltage limit?

A. 6 V
B. $6\sqrt{2}$ V
C. $3\sqrt{2}$ V
D. $12\sqrt{2}$ V

22. A generator produces alternating current electricity with a frequency of 40 cycles per second. What is the maximum potential difference created by the generator if the RMS voltage is 150 V?

A. 54 V
B. 91 V
C. 212 V
D. 141 V

23. Kirchhoff's junction rule is a statement of:

A. Law of conservation of energy
B. Law of conservation of angular momentum
C. Law of conservation of momentum
D. Law of conservation of charge

24. Four identical capacitors are connected in parallel to a battery. If a total charge of Q flows from the battery, how much charge does each capacitor carry?

A. $Q/4$
B. Q
C. $4Q$
D. $16Q$

25. Which statement is correct for two conductors joined by a long copper wire?

A. The electric field at the surface of each conductor is the same
B. Each conductor must be at the same potential
C. Each conductor must have the same resistivity
D. A free charge must be present on either conductor

26. Electromagnetic induction occurs in a coil when there is a change in the:

A. coil's charge
B. current in the coil
C. magnetic field intensity in the coil
D. electric field intensity in the coil

27. When unequal resistors are connected in series across an ideal battery, the:

A. current flowing in each is the same
B. equivalent resistance of the circuit is less than that of the greatest resistor
C. power dissipated in each turn is the same
D. potential difference across each is the same

28. Electric current flows only from the point of:

A. equal potential
B. high pressure to the point of lower pressure
C. low pressure to the point of higher pressure
D. high potential to the point of lower potential

29. Consider the group of charges in this figure. All three charges have $Q = 3.8$ nC.

What is their electric potential energy?

(Use Coulomb's constant $k = 9.0 \times 10^9 \text{ N}\cdot\text{m}^2/\text{C}^2$)

A. 1.9×10^{-6} J
B. 1.0×10^{-5} J
C. 8.8×10^{-6} J
D. 9.7×10^{-6} J

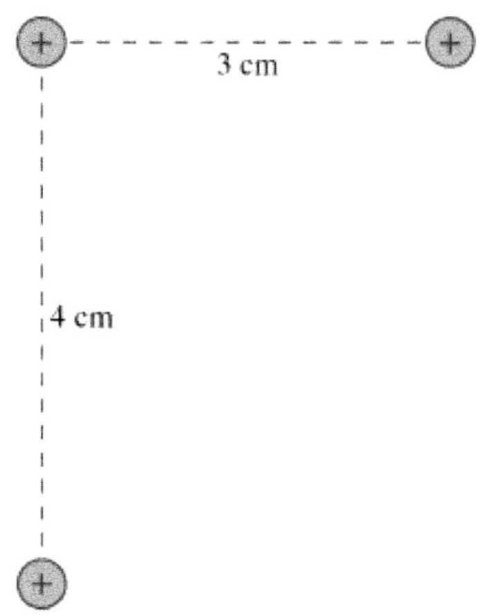

30. Positively-charged and negatively-charged particles are traveling on the same path perpendicular to a constant magnetic field. How do the forces experienced by the two particles differ if the magnitudes of the charges are equal?

A. Differ in direction, but not in magnitude
B. Differ in magnitude, but not in direction
C. No difference in magnitude or direction
D. Differ in both magnitude and direction

31. An electron moves in a direction opposite to an electric field. The potential energy of the system:

A. decreases, and the electron moves toward a region of lower potential
B. increases, and the electron moves toward a region of higher potential
C. decreases, and the electron moves toward a region of higher potential
D. remains constant, and the electron moves toward a region of higher potential

32. What is the name of a device that transforms electrical energy into mechanical energy?

A. Magnet
B. Transformer
C. Turbine
D. Motor

33. A hydrogen atom consists of a proton and an electron. If the orbital radius of the electron increases, the absolute magnitude of the potential energy of the electron:

A. remains the same
B. decreases
C. increases
D. depends on the potential of the electron

34. Copper wire A has a length L and a radius r. Copper wire B has a length of $2L$ and a radius of $2r$. Which of the following is true regarding the resistances across the ends of the wires?

A. The resistance of wire A is one-half that of wire B
B. The resistance of wire A is four times higher than that of wire B
C. The resistance of wire A is twice as high as that of wire B
D. The resistance of wire A is equal to that of wire B

35. When a negative charge is free, it tries to move:

A. toward infinity
B. away from infinity
C. from high potential to low potential
D. from low potential to high potential

36. Four 6 V batteries (in a linear sequence of A → B → C → D) are connected in series to power lights A and B. The resistance of light A is 50 Ω and the resistance of light B is 25 Ω. What is the potential difference at a point between battery C and battery D? (Assume that the potential at the start of the sequence is zero)

A. 4 volts
B. 12 volts
C. 18 volts
D. 26 volts

37. By what factor does the dielectric constant change when a material is introduced between the plates of a parallel-plate capacitor if the capacitance increases by a factor of 4?

A. ½
B. 4
C. 0.4
D. ¼

38. Two isolated copper plates, each area 0.4 m^2, carry opposite charges of magnitude 6.8×10^{-10} C. They are placed opposite each other in parallel alignment. What is the potential difference between the plates when their spacing is 4 cm? (Use the dielectric constant $k = 1$ in air and the electric permittivity $\varepsilon_0 = 8.854 \times 10^{-12}$ F/m)

A. 1.4 V
B. 4.1 V
C. 7.7 V
D. 3.2 V

39. The force on an electron moving in a magnetic field is largest when its direction is:

A. perpendicular to the magnetic field direction
B. at an angle greater than 90° to the magnetic field direction
C. at an angle less than 90° to the magnetic field direction
D. exactly opposite to the magnetic field direction

40. What is the quantity that is calculated in units of A·s?

A. Passivity
B. Capacitance
C. Potential
D. Charge

41. A proton with a speed of 1.7×10^5 m/s falls through a potential difference V and increases its speed to 3.2×10^5 m/s. Through what potential difference did the proton fall? (Use the mass of a proton = 1.67×10^{-27} kg and the charge of a proton = 1.6×10^{-19} C)

A. 880 V
B. 1,020 V
C. 384 V
D. 430 V

42. Three capacitors are connected to a battery, as shown. The capacitances are: $C_1 = 2C_2$ and $C_1 = 3C_3$. Which of the three capacitors stores the smallest amount of charge?

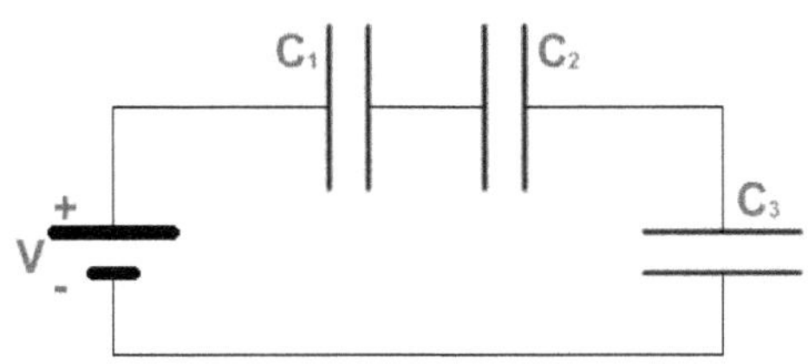

A. C_1
B. C_1 or C_3
C. C_2
D. The amount of charge is the same in the capacitors

43. Two isolated copper plates, each of area 0.6 m², carry opposite charges of magnitude 7.08×10^{-10} C. They are placed opposite each other in parallel alignment, with a spacing of 2 mm. What will be the potential difference between the plates when their spacing is increased to 6 cm? (Use the dielectric constant $k = 1$ in air and electric permittivity $\mathcal{E}_0 = 8.854 \times 10^{-12}$ F/m)

A. 8.0 V
B. 3.1 V
C. 4.3 V
D. 7.2 V

44. Electric current can only flow:

A. in a region of negligible resistance
B. through a potential difference
C. in a perfect conductor
D. in the absence of resistance

45. The metal detectors used to screen passengers at airports operate via:

A. Newton's Laws
B. Bragg's Law
C. Faraday's Law
D. Ohm's Law

46. A 7 μC negative charge is attracted to a large, well-anchored, positive charge. How much kinetic energy does the negatively-charged object gain if the potential difference through which it moves is 3.5 mV?

A. 24.5 nJ
B. 6.7 μJ
C. 36.7 μJ
D. 0.5 kJ

47. A wire of resistivity ρ is replaced in a circuit by a wire of the same material but four times as long. If the total resistance remains the same, the diameter of the new wire must be:

A. one-fourth the original diameter
B. two times the original diameter
C. the same as the original diameter
D. one-half the original diameter

48. The addition of resistors in series to a resistor in an existing circuit, while voltage remains constant, would result in [ ] in the original resistor.

A. an increase in current
B. a decrease in resistance
C. an increase in resistance
D. a decrease in current

49. In an experiment, a battery is connected to a variable resistor R, where resistance can be adjusted by turning a knob. The potential difference across the resistor and the current through it are recorded for different settings of the resistor knob. The battery is an ideal potential source in series with an internal resistor. The emf of the potential source is 9 V, and internal resistance is 0.1 Ω. What is the current if the variable resistor is at 0.5 Ω?

A. 15 A
B. 0.9 A
C. 4.5 A
D. 45 A

50. Two parallel plates initially uncharged are separated by 1.6 mm. What charge must be transferred from one plate to the other if 10 kJ of energy is to be stored in the plates? The area of each plate is 24 mm^2. (Use the dielectric constant $k = 1$ in air and the electric permittivity $\mathcal{E}_0 = 8.854 \times 10^{-12}$ F/m)

A. 78 μC
B. 15 mC
C. 52 μC
D. 29 μC

51. When a proton moves in the direction of the electric field, the potential energy of the system [ ] and moves toward [ ] electric potential. (Use the dielectric constant $k = 1$ in air and the electric permittivity $\mathcal{E}_0 = 8.854 \times 10^{-12}$ F/m)

A. increases ... increasing
B. decreases ... decreasing
C. increases ... decreasing
D. decreases ... increasing

52. Each plate of a parallel-plate air capacitor has an area of 0.004 m^2, and the separation of the plates is 0.02 mm. An electric field of 8.6×10^6 V/m is present between the plates. What is the energy density between the plates? (Use the electric permittivity $\mathcal{E}_0 = 8.854 \times 10^{-12}$ F/m)

A. 100 J/m^3
B. 400 J/m^3
C. 220 J/m^3
D. 330 J/m^3

53. What is the quantity calculated with units of $kg \cdot m^2/(s \cdot C^2)$?

A. Resistance
B. Capacitance
C. Potential
D. Resistivity

54. At a constant voltage, an increase in the resistance of a circuit results in:

A. no change in I or V
B. an increase in I
C. a decrease in I
D. constant power

55. A charge $+Q$ is located at one of the corners of a square. The absolute potential at the center of a square is 3 V. If a second charge $-Q$ is at one of the other three corners, what is the absolute potential at the square's center?

A. –6 V
B. 12 V
C. 6 V
D. 0 V

56. A uniform electric field has a strength of 6 N/C. What is the electric energy density of the field? (Use the electric permittivity $\mathcal{E}_0 = 8.854 \times 10^{-12}$ F/m)

A. 1.5×10^{12} J/m^3
B. 1.6×10^{-10} J/m^3
C. 2.3×10^{12} J/m^3
D. 2.7×10^{-11} J/m^3

57. Which change to a circuit element will always increase the current?

A. Increased voltage and decreased resistance
B. Decreased voltage and increased resistance
C. Increased voltage and increased resistance
D. Only a decrease in resistance, the voltage does not affect current

58. Three capacitors are arranged as shown. C_1 has a capacitance of 9 pF, C_2 has a capacitance of 18 pF, and C_3 has a capacitance of 24 pF. What is the voltage drop across the entire system if the voltage drop across C_2 is 240 V?

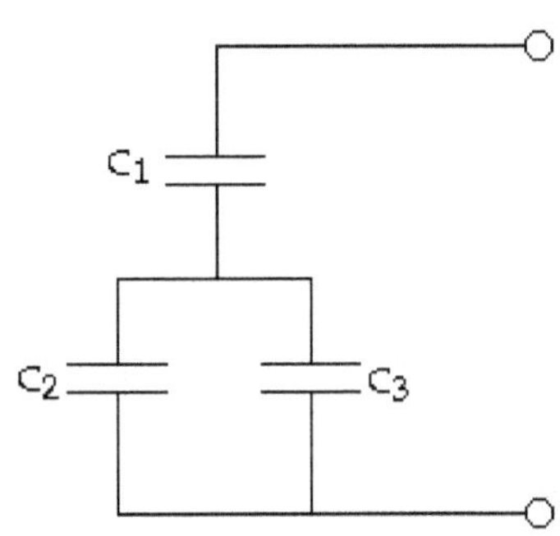

A. 430 V
B. 870 V
C. 1,200 V
D. 1,350 V

59. Doubling the capacitance of a capacitor that is holding a constant charge causes the energy stored in that capacitor to:

A. decrease to one-half
B. decrease to one-fourth
C. quadruple
D. double

60. When three resistors are added in series to a resistor in a circuit, the original resistor's voltage [ ] and its current [ ].

A. decreases ... increases
B. increases ... increases
C. decrease ... decreases
D. decreases ... remains the same

Notes for active learning

Notes for active learning

Sound

1. A 20 decibel (dB) noise is heard from a cricket at 30 m. How loud would it sound if the cricket were at 3 m?

A. 30 dB
B. 40 dB
C. $20 \times \sqrt{2}$ dB
D. 80 dB

2. A thunderclap occurs at 6 km from a stationary person. How soon does the person hear it? (Use the speed of sound in air $v = 340$ m/s)

A. 18 s
B. 30 s
C. 48 s
D. 56 s

3. Enrico Caruso, a famous opera singer, shattered a crystal chandelier with his voice. This demonstrates:

A. ideal frequency
B. resonance
C. a standing wave
D. sound refraction

4. A taut 2 m string is fixed at both ends and plucked. What is the wavelength corresponding to the third harmonic?

A. 2/3 m
B. 1 m
C. 4/3 m
D. 3 m

5. High-pitched sound has a high:

I. number of partial tones II. frequency III. speed

A. I only
B. II only
C. III only
D. I and II only

6. A light ray in air strikes a medium whose index of refraction is 1.5. If the angle of incidence is 60°, which of the following expressions gives the angle of refraction? (Use $n_{air} = 1$)

A. $\sin^{-1}(0.67 \sin 60°)$
B. $\sin^{-1}(1.5 \cos 60°)$
C. $\sin^{-1}(1.5 \sin 30°)$
D. $\sin^{-1}(0.67 \sin 30°)$

7. A string, 2 m in length, is fixed at both ends and tightened until the wave speed is 92 m/s. What is the frequency of the standing wave shown?

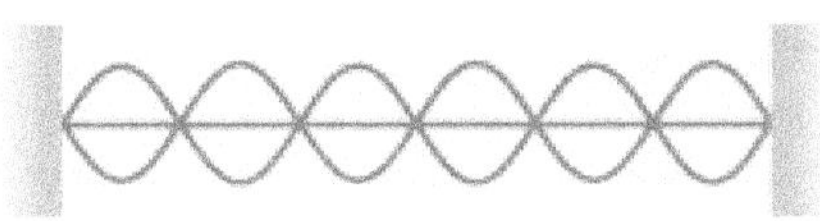

A. 46 Hz
B. 33 Hz
C. 240 Hz
D. 138 Hz

8. A 0.6 m uniform bar of metal with a diameter of 2 cm has a mass of 2.5 kg. A 1.5 MHz longitudinal wave is propagated along the length of the bar. A wave compression traverses the length of the bar in 0.14 ms. What is the wavelength of the longitudinal wave in the metal?

A. 2.9 mm
B. 1.8 mm
C. 3.2 mm
D. 4.6 mm

Questions **9-12** are based on the following:

The velocity of a wave on a wire or string is not dependent (to a close approximation) on frequency or amplitude and is given by $v^2 = T / \rho_L$. T is the tension in the wire. The linear mass density ρ_L (rho) is the mass per unit length of wire. Therefore, ρ_L is the product of the mass density and the cross-sectional area (A).

A sine wave is traveling to the right with a frequency of 250 Hz. Wire A is composed of steel and has a circular cross-section diameter of 0.6 mm, and tension of 2,000 N. Wire B is under the same tension and made of the same material as wire A but has a circular cross-section diameter of 0.3 mm. Wire C has the same tension as wire A and is made of a composite material. (Use the density of steel wire $\rho = 7$ g/cm^3 and the density of the composite material $\rho = 3$ g/cm^3)

9. How much does the tension need to increase the wave velocity on a wire by 30%?

A. 37%
B. 60%
C. 69%
D. 81%

10. What is the linear mass density of wire B compared to wire A?

A. √2 times
B. 2 times
C. 1/8
D. 1/4

11. What must the diameter of wire C be to have the same wave velocity as wire A?

A. 0.41 mm
B. 0.92 mm
C. 0.83 mm
D. 3.2 mm

12. How does the cross-sectional area change if the diameter increases by a factor of 4?

A. Increases by a factor of 16
B. Increases by a factor of 4
C. Increases by a factor of 2
D. Decreases by a factor of 4

13. A bird, emitting sounds with a frequency of 60 kHz, moves at a speed of 10 m/s toward a stationary observer. What is the frequency of the sound waves detected by the observer? (Use the speed of sound in air v = 340 m/s)

A. 55 kHz
B. 62 kHz
C. 68 kHz
D. 76 kHz

14. What is observed for a frequency heard by a stationary person when a sound source is approaching?

A. Equal to zero
B. The same as the source
C. Higher than the source
D. Lower than the source

15. Which of the following is a false statement?

A. The transverse waves on a vibrating string are different from sound waves
B. Sound travels much slower than light
C. Sound waves are longitudinal pressure waves
D. Sound can travel through a vacuum

16. Which of the following is a real-life example of the Doppler effect?

A. Changing pitch of the siren as an ambulance passes by the observer
B. Radio signal transmission
C. Sound becomes quieter as the observer moves away from the source
D. Human hearing is most acute at 2,500 Hz

17. Two sound waves have the same frequency and amplitudes of 0.4 Pa and 0.6 Pa, respectively. When they arrive at point X, what is the range of amplitudes for the sound at point X?

A. 0 – 0.4 Pa
B. 0.4 – 0.6 Pa
C. 0.2 – 1.0 Pa
D. 0.4 – 0.8 Pa

18. The intensity of the waves from a point source at a distance d from the source is I. What is the intensity at a distance $2d$ from the source?

A. I/2
B. I/4
C. 4I
D. 2I

19. Sound would be expected to travel most slowly in a medium that exhibited:

A. low resistance to compression and high density
B. high resistance to compression and low density
C. low resistance to compression and low density
D. high resistance to compression and high density

20. Which is valid for a resonating pipe that is open at both ends?

A. Displacement node at one end and a displacement antinode at the other end
B. Displacement antinodes at each end
C. Displacement nodes at each end
D. Displacement node at one end and a one-fourth antinode at the other end

21. In a pipe of length L that is open at both ends, the lowest tone to resonate is 200 Hz. Which of the following frequencies does not resonate in this pipe?

A. 400 Hz
B. 600 Hz
C. 500 Hz
D. 800 Hz

22. In general, a sound is conducted fastest through:

A. vacuum
B. gases
C. liquids
D. solids

23. If an electric charge is shaken up and down:

A. electron excitation occurs
B. a magnetic field is created
C. sound is emitted
D. its charge changes

24. What is the wavelength of a sound wave of frequency 620 Hz in steel, given that the speed of sound in steel is 5,000 m/s?

A. 1.8 m
B. 6.2 m
C. 8.1 m
D. 2.6 m

25. If the sound from a constant sound source radiates equally in all directions, as the distance doubles, what amount is the intensity of the sound reduced?

A. ¼
B. 1/16
C. $1/\sqrt{2}$
D. ½

26. Why does the intensity of waves from a sound source decrease with the square of the distance from the source?

A. The medium through which the waves travel absorbs the energy of the waves
B. The waves speed up as they travel away from the source
C. The waves lose energy as they travel
D. The waves spread out as they travel

Questions **27-30** are based on the following:

Steven is preparing a mailing tube that is 1.5 m long and 4 cm in diameter. The tube is open at one end and sealed at the other. Before he inserted his documents, the mailing tube fell to the floor and produced a note. (Use the speed of sound in air $v = 340$ m/s)

27. What is the wavelength of the fundamental?

A. 0.04 m
B. 6 m
C. 0.75 m
D. 1.5 m

28. If the tube were filled with helium, in which sound travels at 960 m/s, what is the fundamental frequency?

A. 160 Hz
B. 320 Hz
C. 80 Hz
D. 640 Hz

29. What is the wavelength of the fifth harmonic?

A. 3.2 m
B. 1.2 m
C. 2.4 m
D. 1.5 m

30. What is the frequency of the note that Steven heard?

A. 57 Hz
B. 85 Hz
C. 30 Hz
D. 120 Hz

31. A 4 g string, 0.34 m long, is under tension. The string vibrates in the third harmonic. What is the wavelength of the standing wave in the string? (Use the speed of sound in air $v = 344$ m/s)

A. 0.56 m
B. 0.33m
C. 0.23 m
D. 0.61 m

32. Two pure tones are sounded together, and a particular beat frequency is heard. What happens to the beat frequency if the frequency of one of the tones is increased?

A. Increases
B. Decreases
C. Remains the same
D. Increase or decrease

33. What are the wavelengths of the three lowest tones produced by a closed pipe of length L?

A. $4L$, $4/3L$, $4/5L$
B. $2L$, L, $2/3L$
C. $2L$, L, $½L$
D. $4L$, $2L$, L

34. Mary hears the barely perceptible buzz of a mosquito one meter from her ear in a quiet room. How much energy does a mosquito produce in 200 s? (Note: an almost inaudible sound has a threshold value of 9.8×10^{-12} W/m^2)

A. 6.1×10^{-8} J
B. 1.3×10^{-8} J
C. 6.4×10^{-10} J
D. 2.5×10^{-8} J

35. How long does it take for a light wave to travel 1 km through the water with a refractive index of 1.33? (Use the speed of light $c = 3 \times 10^8$ m/s)

A. 4.4×10^{-6} s
B. 4.4×10^{-9} s
C. 2.8×10^{-9} s
D. 2.8×10^{-12} s

36. In designing a music hall, an acoustical engineer deals mainly with:

A. wave interference
B. resonance
C. forced vibrations
D. modulation

37. Which curve in the figure represents wave speed variation (v) as a function of tension (T) for transverse waves on a stretched string?

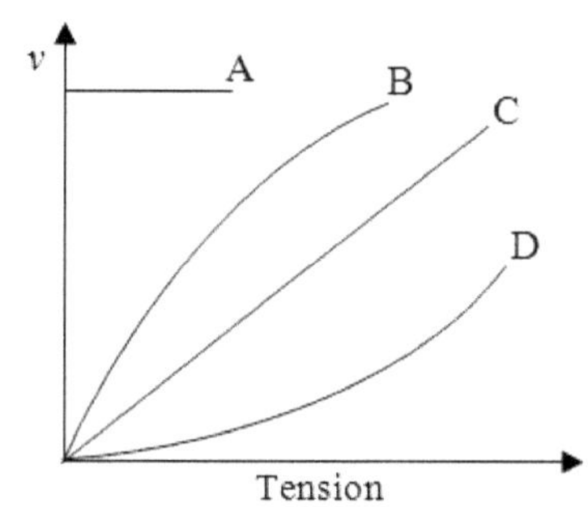

A. A
B. B
C. C
D. D

38. A string, 4 meters in length, is fixed at both ends and tightened until the wave speed is 20 m/s. What is the frequency of the standing wave shown?

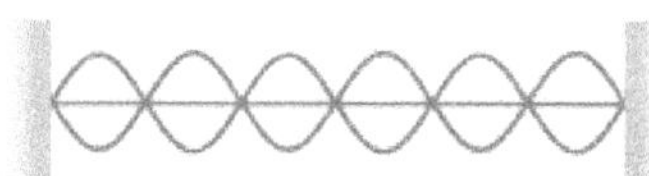

A. 13 Hz
B. 8.1 Hz
C. 5.4 Hz
D. 15.4 Hz

39. Compared to the velocity of a 600 Hz sound, the velocity of a 300 Hz sound through air is:

A. one-half as great
B. the same
C. twice as great
D. four times as great

40. Consider a string having a linear mass density of 0.40 g/m stretched to 0.50 m by the tension of 75 N, vibrating at the 6th harmonic. It excites an open pipe into the second overtone. What is the length of the pipe?

A. 0.25 m
B. 0.1 m
C. 0.20 m
D. 0.6 m

41. A string of length L is under tension, and the speed of a wave in the string is v. What is the speed of a wave in a string of the same mass under the same tension but twice as long?

A. $v\sqrt{2}$
B. $2v$
C. $v/2$
D. $v/\sqrt{2}$

42. If a guitar string has a fundamental frequency of 500 Hz, which of the following frequencies can set the string into resonant vibration?

A. 450 Hz
B. 760 Hz
C. 1,500 Hz
D. 2,250 Hz

43. When a light wave passes from a lower refractive index to a medium with a higher refractive index, some of the incident light is refracted, while some are reflected. What is the angle of refraction?

A. Greater than the angle of incidence and less than the angle of reflection
B. Less than the angle of incidence and greater than the angle of reflection
C. Greater than the angles of incidence and reflection
D. Less than the angles of incidence and reflection

44. The speed of a sound wave in air depends on:

I. the air temperature II. the wavelength III. the frequency

A. I only
B. II only
C. III only
D. I and II only

45. Which of the following statements is false?

A. The speed of a wave and the speed of the vibrating particles that constitute the wave are different entities
B. Waves transport energy and matter from one region to another
C. In a transverse wave, the particle motion is perpendicular to the velocity vector of the wave
D. Not all waves are mechanical

46. A 2.5 g string, 0.75 m long, is under tension. The string produces a 700 Hz tone when it vibrates in the third harmonic. What is the wavelength of the tone in the air? (Use the speed of sound in air $v = 344$ m/s)

A. 0.65 m
B. 0.57 m
C. 0.33 m
D. 0.5 m

47. Suppose that a source of sound is emitting waves uniformly in all directions. If an observer moves to a point twice as far from the source, what is the sound frequency?

A. $\sqrt{2}$ as large
B. Twice as large
C. Unchanged
D. Half as large

48. A 2.5 kg rope is stretched between supports 8 m apart. If one end of the rope is tweaked, how long will it take for the resulting disturbance to reach the other end? Assume that the tension in the rope is 40 N.

A. 0.71 s
B. 0.62 s
C. 0.58 s
D. 0.47 s

49. An office machine makes a rattling sound with an intensity of 10^{-5} W/m^2 when perceived by an office worker sitting 3 m away. What is the sound level in decibels for the sound of the machine? (Use the threshold of hearing $I_0 = 10^{-12}$ W/m^2)

A. 10 dB
B. 35 dB
C. 70 dB
D. 95 dB

50. A taut 1 m string is plucked. Point B is midway between both ends, and a finger is placed on point B such that a waveform exists with a node at B. What is the lowest frequency that can be heard? (Use the speed of waves on the string $v = 3.8 \times 10^4$ m/s)

A. 4.8×10^5 Hz
B. 3.8×10^4 Hz
C. 9.7×10^3 Hz
D. 7.4×10^3 Hz

51. For a light wave traveling in a vacuum, which of the following properties is true?

A. Increased *f* results in increased amplitude
B. Increased *f* results in decreased speed
C. Increased *f* results in an increased wavelength
D. Increased *f* results in a decreased wavelength

52. Which wave is different from the others (i.e., does not belong to the same grouping)?

A. Pressure wave
B. Radio wave
C. Ultrasonic wave
D. Infrasonic wave

53. Two speakers are 2 m apart, and both produce a sound wave (in-phase) with a wavelength of 0.8 m. A microphone is placed an equal distance from both speakers to determine the intensity of the sound at various points. What point is precisely halfway between the two speakers? (Use the speed of sound $v = 340$ m/s)

A. Both an antinode and a node
B. Neither an antinode nor a node
C. A node
D. An antinode

54. The siren of an ambulance blares at 1,200 Hz when the ambulance is stationary. What frequency does a stationary observer hear after this ambulance passes her while traveling at 30 m/s? (Use the speed of sound $v = 342$ m/s)

A. 1,240 Hz
B. 1,128 Hz
C. 1,103 Hz
D. 1,427 Hz

55. Compared to the wavelength of a 600 Hz sound, the wavelength of a 300 Hz sound in air is:

A. one-half as long
B. the same
C. one-fourth as long
D. twice as long

56. An organ pipe that is open at both ends is tuned to a given frequency. A second pipe with both ends open resonates with twice this frequency. What is the ratio of the length of the first pipe to the second pipe?

A. 0.5
B. 1
C. 2
D. 2.5

57. The frequency of the third harmonic of the C_4 string of a piano is 783.7 Hz. The fundamental frequency of the G_5 string is 782.4 Hz. When the key for C_4 is held down so that the string can vibrate, and the G_5 key is stricken loudly, the third harmonic of the C_4 string is excited. Then, when striking the G_5 key again more softly, the volume of the two strings is matched. What phenomenon is demonstrated when the G_5 string is used to excite the vibration of the C_4 string?

A. Resonance
B. Dispersion
C. Beats
D. Interference

58. Crests of an ocean wave pass a pier every 10 s. If the waves are moving at 4.5 m/s, what is the wavelength of the ocean waves?

A. 38 m
B. 16 m
C. 45 m
D. 25 m

59. Which statement explains why sound travels faster in water than in air?

A. Sound shifts to increased frequency
B. Sound shifts to decreased density
C. Density of water increases more quickly than its resistance to compression
D. Density of water increases more slowly than its resistance to compression

60. When visible light is incident upon transparent glass, the electrons in the atoms in the glass:

I. convert the light energy into internal energy
II. resonate
III. vibrate

A. I only
B. II only
C. III only
D. I and II only

Notes for active learning

Light and Geometrical Optics

1. What is the minimum thickness of a soap film that reflects a given wavelength of light?

A. ¼ the wavelength
B. ½ the wavelength
C. One wavelength
D. Two wavelengths

2. As the angle of an incident ray of light increases, the angle of the reflected ray:

A. increases
B. decreases
C. stays the same
D. increases or decreases

3. At what distance from a concave spherical mirror (with a focal length of 100 cm) must a woman stand to see an upright image of herself that is twice her actual height?

A. 100 cm
B. 50 cm
C. 300 cm
D. 25 cm

4. If a person's eyeball is too long from front to back, what is the name of the condition that the person likely suffers?

A. Hyperopia
B. Astigmatism
C. Presbyopia
D. Myopia

5. According to the relationship between frequency and energy of light ($E = hf$), which color of light has more energy?

A. Red
B. Yellow
C. Blue
D. Orange

6. A candle 18 cm tall sits 4 m from a diverging lens with a focal length of 3 m. What is the size of the image?

A. 6.3 cm
B. 7.7 cm
C. 2.9 cm
D. 13.5 cm

Questions **7-8** are based on the following:

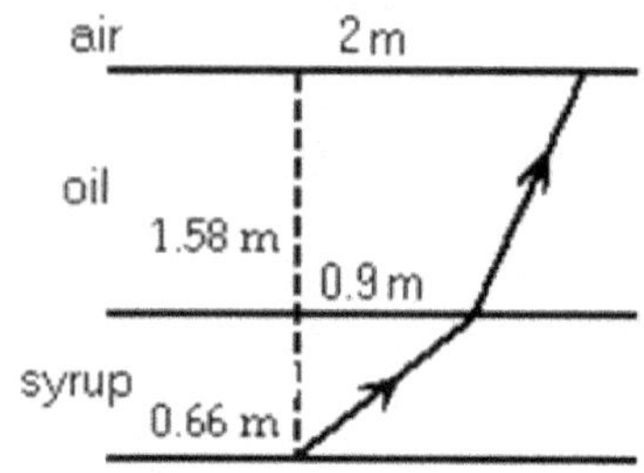

A tank holds a layer of oil 1.58 m thick that floats on a layer of syrup that is 0.66 m thick. Both liquids are clear and do not intermix. A ray, which originates at the bottom of the tank on a vertical axis (see figure), crosses the oil-syrup interface at a point 0.9 m to the right of the vertical axis. The ray continues and arrives at the oil-air interface, 2 m from the axis and at the critical angle. (Use the refractive index $n = 1$ for air)

7. The index of refraction of the oil is closest to:

A. 1.39
B. 1.56
C. 1.75
D. 1.82

8. What is the index of refraction of the syrup?

A. 1.53
B. 1.46
C. 1.17
D. 1.24

9. Which of the following cannot be explained with the wave theory of light?

A. Photoelectric effect
B. Interference
C. Polarization
D. Diffusion

10. The use of wavefronts and rays to describe optical phenomena is known as:

A. dispersive optics
B. reflector optics
C. wave optics
D. geometrical optics

11. In investigating a new type of optical fiber (index of refraction n = 1.26), a laser beam is incident on the flat end of a straight fiber in air, as shown in the figure. What is the maximum angle of incidence (θ_1) if the beam does not escape from the fiber?

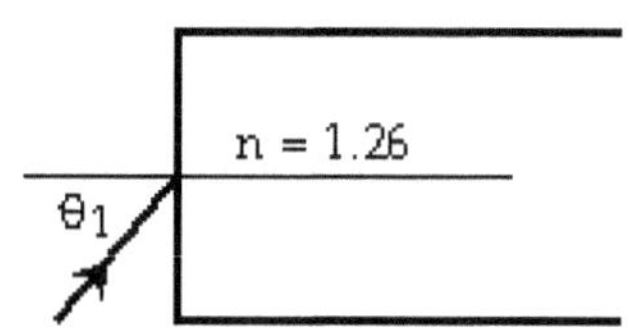

A. 36°
B. 43°
C. 58°
D. 50°

12. An object is 0.5 m from a converging lens with a power of 10 diopters. At what distance from the lens does the image appear?

A. 0.13 m
B. 0.47 m
C. 0.7 m
D. 1.5 m

13. A virtual image is:

I. produced by light rays
II. the brain's interpretations of light rays
III. found only on a concave mirror

A. I only
B. II only
C. III only
D. I and II only

14. If Karen stands in front of a convex mirror, at the same distance from it as its radius of curvature:

A. Karen does not see her image because it is focused at a different distance
B. Karen sees her image, and she appears the same size
C. Karen does not see her image, and she is not within its range
D. Karen sees her image, and she appears smaller

15. An object is viewed at various distances using a mirror with a focal length of 10 m. If the object is 20 m from the mirror, what best characterizes the image?

A. Inverted and real
B. Inverted and virtual
C. Upright and real
D. Upright and virtual

16. If an object is at a position beyond $2f$ of the focal point of a converging lens, the image is:

A. real, upright and enlarged
B. real, inverted and reduced
C. virtual, upright and reduced
D. real, inverted and enlarged

17. Which form of electromagnetic radiation has photons with the lowest energy?

A. X-rays
B. Ultraviolet radiation
C. Radio waves
D. Microwaves

18. If the index of refraction of a diamond is 2.43, a given wavelength of light travels:

A. 2.43 times faster in diamond than it does in air
B. 2.43 times faster in a vacuum than it does in diamond
C. 2.43 times faster in diamond than it does in a vacuum
D. 2.43 times faster in a vacuum than it does in air

19. An object is 15 cm to the left of a double-convex lens of focal length 20 cm. Where is the image of this object located?

A. 15 cm to the left of the lens
B. 30 cm to the left of the lens
C. 60 cm to the right of the lens
D. 60 cm to the left of the lens

20. A sheet of red paper appears black when it is illuminated with:

A. orange light
B. cyan light
C. red light
D. yellow light

21. Where is an object if the image produced by a lens appears close to its focal point?

A. Near the center of curvature of the lens
B. Far from the lens
C. Near the lens
D. Near the focal point

22. A light with the frequency 4.9×10^{14} Hz is produced by a source located 6 m from a converging lens with a focal length of 3 m. For a different frequency of light, the focal length of the lens is different than 3 m. This phenomenon is:

A. Dispersion
B. Incidence
C. Interference
D. Refraction

23. If an image appears at the same distance from a mirror as the object, the size of the image is:

A. quadrupled the size of the object
B. ¼ the size of the object
C. the same size as the object
D. twice the size of the object

24. When viewed straight down (90° to the surface), an incident light ray moving from water to air is refracted:

A. 37° away from the normal
B. 37° toward the normal
C. 28° toward the normal
D. 0°

25. Suppose that a beachgoer uses two lenses from a pair of disassembled polarized sunglasses and places one on top of the other. What would the beachgoer observe after rotating one lens 90° with respect to the normal position of the other lens and looking directly at the sun overhead?

A. Light with an intensity reduced to about 50% of what it would be with one lens
B. Light with an intensity that is the same as what it would be with one lens
C. Complete darkness, since no light would be transmitted
D. Light with an intensity reduced to about 25% of what it would be with one lens

26. A glass plate with an index of refraction of 1.45 is immersed in a liquid. The liquid is an oil with an index of refraction of 1.35. The surface of the glass is inclined at an angle of 54° with the vertical. A horizontal ray in the glass is incident on the interface of glass and liquid, and the incident horizontal ray refracts at the interface. The angle that the refracted ray in the oil makes with the horizontal is closest to:

A. 8.3°
B. 14°
C. 6°
D. 12°

27. Two plane mirrors make an angle of 30°. A light ray enters the system and is reflected once off each mirror. Through what angle is the ray turned?

A. 60°
B. 90°
C. 120°
D. 160°

28. Which of the following statements about light is TRUE?

A. A packet of light energy is known as a photon
B. Color can be used to determine the approximate energy of visible light
C. Light travels through space at a speed of 3.0×10^8 m/s
D. All the above

29. The angle of incidence:

A. may be greater than, less than, or equal to the angle of refraction
B. is always less than the angle of refraction
C. must equal the angle of refraction
D. is always greater than the angle of refraction

30. A 5-foot-tall woman stands next to a plane mirror on a wall. As she walks away from the mirror, her image:

A. is always a real image, no matter how far she is from the mirror
B. remains 5 feet tall
C. has a height of less than 5 feet
D. may or may not get smaller, depending on where she is positioned

31. If a spherical concave mirror has a radius of curvature R, its focal length is:

A. $2R$
B. R
C. $R/2$
D. $R/4$

32. Let n_1 be the index of refraction of the incident medium and let n_2 be the index of refraction of the refracting medium. Which of the following must be true if the angle that the refracted ray makes with the boundary (not with the normal) is less than the angle that the incident ray makes with the boundary?

A. $n_1 < n_2$
B. $n_1 > n_2$
C. $n_1 < 1$
D. $n_2 < 1$

33. If a person's eyeball is too short from front to back, the person is likely to suffer from:

A. nearsightedness
B. farsightedness
C. presbyopia
D. astigmatism

34. The shimmering that is observed over a hot surface is:

A. changing refraction from the mixing of warm and cool air
B. a mirage
C. heat rays
D. reflections from evaporating water vapor

35. When two parallel white rays pass through the outer edges of a converging glass lens, chromatic aberrations cause colors to appear on the screen in what order, from the top down?

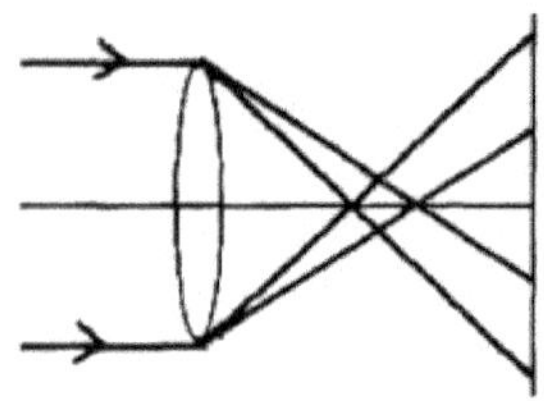

A. Blue, blue, red, red
B. Red, blue, blue, red
C. Blue, red, red, blue
D. Red, red, blue, blue

36. Two thin converging lenses are near each other so that the lens on the left has a focal length of 2 m and the one on the right has a focal length of 4 m. What is the focal length of the combination?

A. 1/4 m
B. 4/3 m
C. 3/4 m
D. 4 m

37. A cylindrical tank is 50 ft deep, 37.5 ft in diameter, and filled to the top with water. A flashlight shines into the tank from above. What is the minimum angle θ that its beam can make with the water surface if the beam is to illuminate part of the bottom? (Use the index of refraction $n = 1.33$ for water)

A. 25°
B. 31°
C. 37°
D. 53°

38. Which color of the visible spectrum has the shortest wavelength (400 nm)?

A. Violet
B. Green
C. Orange
D. Blue

39. An object is at a distance d in front of a plane mirror. The size of the image is:

A. dependent on where the observer is positioned when looking at the image
B. twice the size of the object
C. half the size of the object
D. the same as the object, independent of the position of the observer or distance d

40. If a single lens forms a virtual image of an object, then the:

I. image must be upright
II. the lens must be converging
III. the lens could be diverging or converging

A. I only
B. I and III only
C. III only
D. I and II only

41. When neon light passes through a prism, what is observed?

A. White light
B. Bright spots or lines
C. The same neon light
D. Continuous spectrum

42. The law of reflection holds for:

I. plane mirrors II. curved mirrors III. spherical mirrors

A. I only
B. II only
C. III only
D. I, II and III

43. The image formed by a single concave lens:

A. can be real or virtual but is always real when the object is at the focal point
B. can be real or virtual, depending on the object's distance compared to the focal length
C. is always virtual
D. is always real

44. A lens forms a virtual image of an object. Which of the following must be true of the image?

A. It is inverted
B. It is upright
C. It is larger than the object and upright
D. It is smaller than the object and inverted

45. Light with the lowest frequency (longest wavelength) detected by your eyes is perceived as:

A. red
B. green
C. yellow
D. orange

46. A 0.1 m tall candle is observed through a converging lens that is 3 m away and has a focal length of 6 m. The resulting image is:

A. 3 m from the lens on the opposite side of the object
B. 6 m from the lens on the opposite side of the object
C. 3 m from the lens on the same side as the object
D. 6 m from the lens on the same side as the object

47. Which statement about thin lenses is correct when considering only a single lens?

A. A diverging lens always produces a virtual, erect image
B. A diverging lens always produces a real, erect image
C. A diverging lens always produces a virtual, inverted image
D. A diverging lens always produces a real, inverted image

48. A double-concave lens has equal radii of curvature of 15 cm. An object placed 14 cm from the lens forms a virtual image 5 cm from the lens. What is the index of refraction of the lens material?

A. 0.8
B. 1.4
C. 2
D. 2.6

49. The magnification *m* for an object reflected from a mirror is the ratio of what characteristic of the image to the object?

A. Center of curvature
B. Distance
C. Orientation
D. Angular size

50. Suppose Mike places his face in front of a concave mirror. Which of the following statements is correct?

A. Mike's image is diminished in size
B. Mike's image is always inverted
C. No matter where Mike places himself, a virtual image is formed
D. If Mike positions himself between the center of curvature and the focal point of the mirror, he will not be able to see his image

51. Single-concave spherical mirrors produce images that:

A. are always smaller than the actual object
B. are always the same size as the actual object
C. are always larger than the actual object
D. could be smaller than, larger than, or the same size as the actual object, depending on the object's placement

52. When two converging lenses of equal focal length are used together, the effective combined focal length is less than the focal length of either of the lenses. The combined power of the two lenses used together is:

A. greater than the power of either individual lens
B. the same as the power of either individual lens
C. less than the power of either individual lens
D. greater than the sum of the powers of both individual lens

53. The index of refraction is based on the ratio of the speed of light in:

A. water to the speed of light in the transparent material
B. a vacuum to the speed of light in the transparent material
C. two different transparent materials
D. air to the speed of light in the transparent material

54. An object is located 2.2 m in front of a plane mirror. The image formed by the mirror appears:

A. 4.4 m behind the mirror's surface
B. 2.2 m in front of the mirror's surface
C. 4.4 m in front of the mirror's surface
D. 2.2 m behind the mirror's surface

55. An upright object is 40 cm from a concave mirror with a radius of 50 cm. The image is:

A. virtual and inverted
B. virtual and upright
C. real and inverted
D. real and upright

56. In the figure, a ray in glass arrives at the glass-water interface at an angle of 48° with the normal. The refracted ray makes an angle of 68° with the normal. If another ray in the glass makes an angle of 29° with the normal, what is the angle of refraction in the water? (Use the index of refraction of water $n = 1.33$)

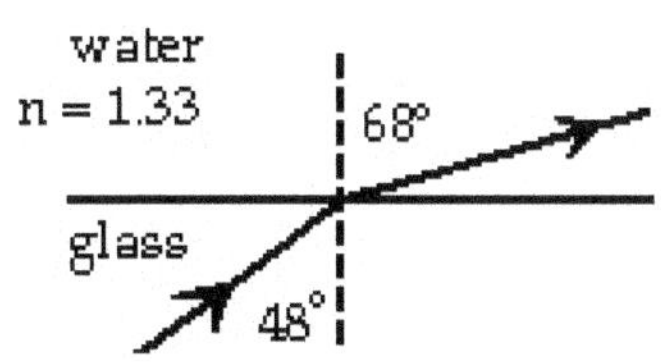

A. 29°
B. 37°
C. 31°
D. 46°

57. In a compound microscope:

A. the image of the objective serves as the object for the eyepiece
B. magnification is provided by the objective and not by the eyepiece; the eyepiece merely increases the brightness of the image viewed
C. magnification is provided by the objective and not by the eyepiece; the eyepiece merely increases the resolution of the image viewed
D. both the objective and the eyepiece form real images

58. Except for air, the refractive index of transparent materials is:

A. equal to 1
B. less than or equal to 1
C. less than 1
D. greater than 1

59. The radius of curvature of the curved side of a plano-convex lens made of glass is 33 cm. What is the focal length of the lens? (Use the index of refraction for glass $n = 1.64$)

A. –28 cm
B. 28 cm
C. 38 cm
D. 52 cm

60. The part of the electromagnetic spectrum most absorbed by water is:

A. lower frequencies in the visible
B. higher frequencies in the visible
C. infrared
D. ultraviolet

Notes for active learning

Thermodynamics

1. Compared to the initial value, what is the resulting pressure for an ideal gas compressed isothermally to one-third of its initial volume?

A. Equal
B. Three times larger
C. Larger, but less than three times larger
D. More than three times larger

2. A uniform hole in a brass plate has a diameter of 1.2 cm at 25 °C. What is the diameter of the hole when the plate is heated to 225 °C? (Use the coefficient of linear thermal expansion for brass = 19×10^{-6} K^{-1})

A. 2.2 cm
B. 2.8 cm
C. 1.2 cm
D. 1.6 cm

3. A student heats 90 g of water using 50 W of power, with 100% efficiency. How long does it take to raise the temperature of the water from 10 °C to 30 °C? (Use the specific heat of water $c = 4.186$ J/g·°C)

A. 232 s
B. 81 s
C. 59 s
D. 151 s

4. A runner generates 1,260 W of thermal energy. If heat is to be dissipated only by evaporation, how much water does she shed in 15 minutes of running? (Use latent heat of vaporization of water $L_v = 22.6 \times 10^5$ J/kg)

A. 500 g
B. 35 g
C. 350 g
D. 50 g

5. Phase changes occur as temperature:

I. decreases   II. increases   III. remains the same

A. I only
B. II only
C. III only
D. I and II only

6. How much heat is needed to melt a 55 kg sample of ice at 0 °C? (Use the latent heat of fusion for water L_f = 334 kJ/kg and the latent heat of vaporization L_v = 2,257 kJ/kg)

A. 0 kJ
B. 1.8×10^4 kJ
C. 3×10^5 kJ
D. 4.6×10^6 kJ

7. Metals are both good heat conductors and good electrical conductors because of the:

A. relatively high densities of metals
B. high elasticity of metals
C. ductility of metals
D. looseness of outer electrons in metal atoms

8. Solar houses are designed to retain the heat absorbed during the day to release the stored heat during the night. A botanist produces steam at 100 °C during the day and then allows the steam to cool to 0 °C and freeze during the night. How many kilograms of water are needed to store 200 kJ of energy for this process? (Use the latent heat of vaporization of water $L_v = 22.6 \times 10^5$ J/kg, the latent heat of fusion of water $L_f = 33.5 \times 10^4$ J/kg, and the specific heat capacity of water $c = 4{,}186$ J/kg·K)

A. 0.066 kg
B. 0.103 kg
C. 0.482 kg
D. 1.18 kg

9. The heat required to change a substance from the solid to the liquid state is referred to as the heat of:

A. condensation
B. freezing
C. fusion
D. vaporization

10. A rigid container holds 0.2 kg of hydrogen gas. How much heat is needed to change the temperature of the gas from 250 K to 280 K? (Use specific heat of hydrogen gas = 14.3 J/g·K)

A. 46 kJ
B. 72 kJ
C. 56 kJ
D. 86 kJ

11. An aluminum electric tea kettle with a mass of 500 g is heated with a 500 W heating coil. How many minutes are required to heat 1 kg of water from 18 °C to 98 °C in the tea kettle? (Use the specific heat of aluminum = 900 J/kg·K and specific heat of water = 4,186 J/kg·K)

A. 16 min
B. 12 min
C. 8 min
D. 4 min

12. Heat is added at a constant rate to a pure substance in a closed container. The temperature of the substance as a function of time is shown in the graph. If L_f = latent heat of fusion and L_v = latent heat of vaporization, what is the value of the ratio L_v / L_f for this substance?

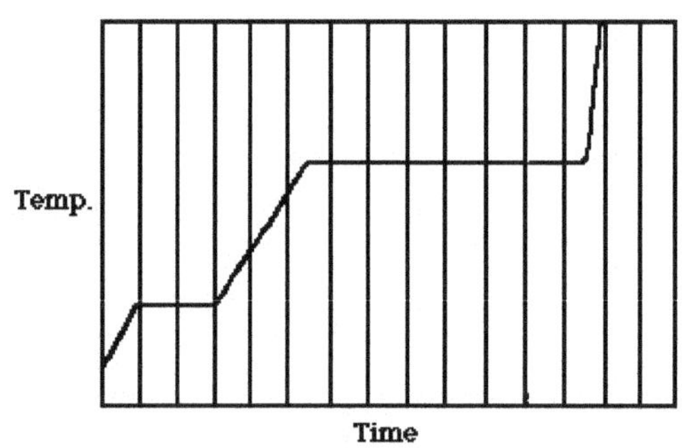

A. 3.5
B. 7.2
C. 4.5
D. 5.0

13. The moderate temperatures of islands throughout the world have much to do with the water's:

A. high evaporation rate
B. high specific heat capacity
C. vast supply of thermal energy
D. poor conductivity

14. A 4.5 g lead BB moving at 46 m/s penetrates a woodblock and comes to rest inside the block. If the BB absorbs half of the kinetic energy, what is the change in the temperature of the BB? (Use the specific heat of lead = 128 J/kg·K)

A. 2.8 K
B. 3.6 K
C. 1.6 K
D. 4.1 K

15. The heat required to change a substance from the liquid to the vapor state is referred to as the heat of:

A. melting
B. condensation
C. vaporization
D. fusion

16. A Carnot engine operating between a reservoir of liquid mercury at its melting point and a colder reservoir extracts 18 J of heat from the mercury. It does 5 J of work during each cycle. What is the temperature of the colder reservoir? (Use the melting temperature of mercury = 233 K)

A. 168 K
B. 66 K
C. 57 K
D. 82 K

17. A 920 g empty iron pan is put on a stove. How much heat must the iron pan absorb in joules to raise its temperature from 18 °C to 96 °C? (Use the specific heat for iron = 113 cal/kg·°C and the conversion of1 cal = 4.186 J)

A. 50,180 J
B. 81,010 J
C. 63,420 J
D. 33,940 J

18. When a solid melts, what change occurs in the substance?

A. Heat energy dissipates
B. Heat energy enters
C. Temperature increases
D. Temperature decreases

19. Which of the following is an accurate statement about the work done for a cyclic process conducted in a gas? (Use P for pressure and V for volume on the graph)

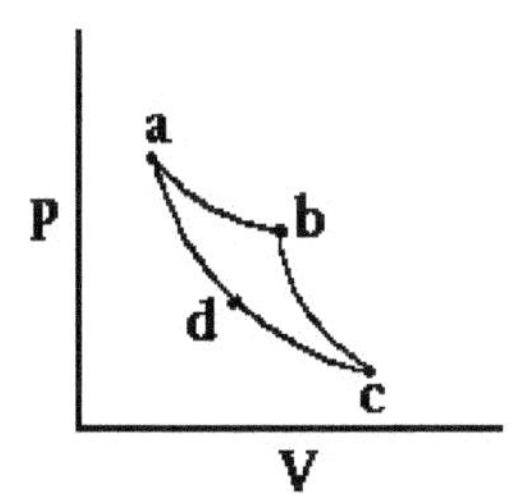

A. It is equal to the area under *ab* minus the area under *dc*
B. It is equal to the area under the curve *adc*
C. It is equal to the area enclosed by the cyclic process
D. It equals zero

20. Substance A has higher specific heat than substance B. With other factors equal, which substance requires more energy to be heated to the same temperature?

A. Substance A
B. Substance B
C. Both require the same amount of heat
D. Depends on the density of each substance

21. A 6.5 g meteor hits the Earth at a speed of 300 m/s. If the meteor's kinetic energy is entirely converted to heat, how much does its temperature rise? (Use the specific heat of the meteor = 120 cal/kg·°C and the conversion of 1 cal = 4.186 J)

A. 134 °C
B. 68 °C
C. 120 °C
D. 90 °C

22. When a liquid freezes, what change occurs in the substance?

A. Heat energy dissipates
B. Heat energy enters
C. Temperature increases
D. Temperature decreases

23. A monatomic ideal gas ($C_v = 3/2$ R) undergoes an isothermal expansion at 300 K as the volume increases from 0.05 m^3 to 0.2 m^3. The final pressure is 130 kPa. What is the heat transfer of the gas? (Use the ideal gas constant R = 8.314 J/mol·K)

A. –14 kJ
B. 36 kJ
C. 14 kJ
D. –21 kJ

24. What is the maximum temperature rise expected for 1 kg of water falling from a waterfall with a vertical drop of 30 m? (Use the acceleration due to gravity $g = 9.8$ m/s^2 and the specific heat of water = 4,186 J/kg·K)

A. 0.1 °C
B. 0.06 °C
C. 0.15 °C
D. 0.07 °C

25. When 0.75 kg of water at 0 °C freezes, what is the change in entropy of the water? (Use the latent heat of fusion of water $L_f = 33{,}400$ J/kg)

A. –92 J/K
B. –18 J/K
C. 44 J/K
D. 80 J/K

26. When a bimetallic bar made of a copper and iron strip is heated, the copper part of the bar bends toward the iron strip. The reason for this is:

A. copper expands more than iron
B. iron expands more than copper
C. iron gets hotter before copper
D. copper gets hotter before iron

27. In a flask, 110 g of water is heated using 60 W of power, with perfect efficiency. How long does it take to raise the temperature of the water from 20 °C to 30 °C? (Use the specific heat of water c = 4,186 J/kg·K)

A. 132 s
B. 57 s
C. 9.6 s
D. 77 s

28. When a liquid evaporates, what change occurs in the substance?

A. Heat energy dissipates
B. Heat energy enters
C. Temperature increases
D. Temperature decreases

29. A flask of liquid nitrogen is at –243 °C. If the nitrogen is heated until the average energy of the particles is doubled, what is the new temperature?

A. 356 °C
B. –356 °C
C. –213 °C
D. 134 °C

30. If a researcher is attempting to determine how much the temperature of a particular piece of material would rise when a known amount of heat is added to it, knowing which of the following quantities would be most helpful?

A. Density
B. Coefficient of linear expansion
C. Initial temperature
D. Specific heat

31. A substance has a density of 1,800 kg/m^3 in the liquid state. At atmospheric pressure, the substance has a boiling point of 170 °C. The vapor has a density of 6 kg/m^3 at the boiling point at atmospheric pressure. What is the change in the internal energy of 1 kg of the substance as it vaporizes at atmospheric pressure? (Use the heat of vaporization L_v= 1.7 × 10^5 J/kg)

A. 180 kJ
B. 170 kJ
C. 6 kJ
D. 12 kJ

32. If an aluminum rod that is at 5 °C is heated until it has twice the thermal energy, its temperature is:

A. 10 °C
B. 56 °C
C. 278 °C
D. 283 °C

33. A thermally isolated system comprises a hot piece of aluminum and a cold piece of copper, with the aluminum and the copper in thermal contact. The specific heat capacity of aluminum is more than double that of copper. Which object experiences the greater temperature change during the time the system takes to reach thermal equilibrium?

A. Both experience the same magnitude of temperature change
B. The mass of each is required
C. The copper
D. The aluminum

34. In liquid water of a given temperature, the water molecules move randomly at different speeds. Electrostatic forces of cohesion tend to hold them. Occasionally one molecule gains enough energy through multiple collisions to pull away and escape from the liquid. Which of the following is an illustration of this phenomenon?

A. When a large steel suspension bridge is built, gaps are left between the girders
B. When a body gets too warm, it produces sweat to cool itself down
C. Increasing the atmospheric pressure over a liquid causes the boiling temperature to decrease
D. If the snow begins to fall when Mary is skiing, she feels colder than before it started to snow

35. A 2,200 kg sample of water at 0 °C is cooled to –30 °C and freezes in the process. How much heat is liberated during this process? (Use heat of fusion for water L_f = 334 kJ/kg, heat of vaporization L_v = 2,257 kJ/kg and specific heat for ice = 2,050 J/kg·K)

A. 328,600 kJ
B. 190,040 kJ
C. 637,200 kJ
D. 870,100 kJ

36. Object 1 has three times the specific heat capacity and four times the mass of Object 2. The same amount of heat is transferred to the two objects. If Object 1 changes by an amount of ΔT, what is the change in Object 2?

A. 12ΔT
B. 3ΔT
C. ΔT
D. (3/4)ΔT

37. The process whereby heat flows using molecular collisions is:

A. radiation
B. inversion
C. conduction
D. convection

38. The graph shows a PV diagram for 5.1 g of oxygen gas in a sealed container. T_1 is 20 °C. What are the values for temperatures of T_3 and T_4, respectively? (Use the gas constant R= 8.314 J/mol·K)

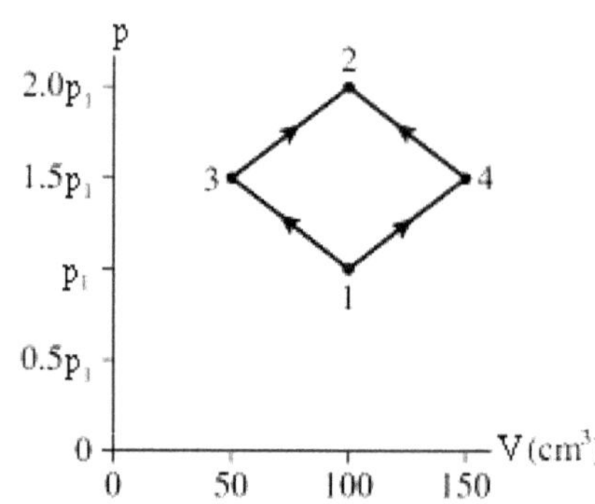

A. –53 °C and 387 °C
B. –14 °C and 34 °C
C. 210 °C and 640 °C
D. 12 °C and 58 °C

39. On a chilly day, a piece of steel feels much colder to the touch than a piece of plastic. This is due to the difference in which one of the following physical properties of these materials?

A. Emissivity
B. Thermal conductivity
C. Density
D. Specific heat

40. What is the term for a process when a gas is allowed to expand as heat is added to it at constant pressure?

A. Isochoric
B. Isobaric
C. Adiabatic
D. Isothermal

41. A Carnot engine is used as an air conditioner to cool a house in the summer. The air conditioner removes 20 kJ of heat per second from the house and maintains the inside at 293 K, while the outside temperature is 307 K. What is the power required for the air conditioner?

A. 2.3 kW
B. 3.22 kW
C. 1.6 kW
D. 0.96 kW

42. Heat energy is measured in units of:

I. Joules II. calories III. work

A. I only
B. II only
C. I and II only
D. III only

43. The process in which heat flows by the mass movement of molecules from one place to another is:

I. conduction II. convection III. radiation

A. I only
B. II only
C. III only
D. I and II only

44. The process whereby heat flows in the absence of a medium is:

A. radiation
B. inversion
C. conduction
D. convection

45. The figure shows 0.008 mol of gas that undergoes the process $1 \rightarrow 2 \rightarrow 3$. What is the volume of V_3?

(Use the ideal gas constant R = 8.314 J/mol·K and the conversion of 1 atm = 101,325 Pa)

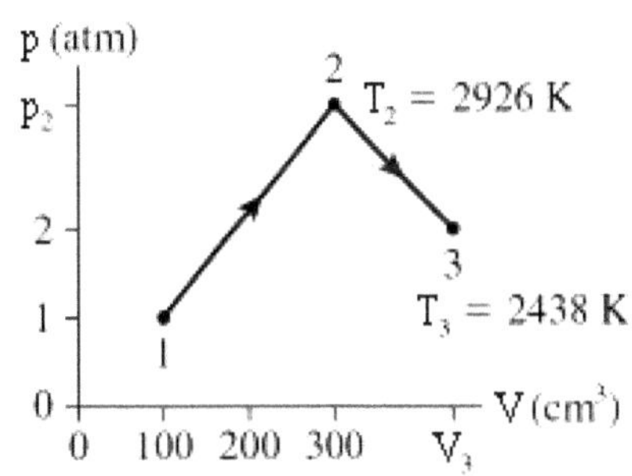

A. 435 cm^3
B. 568 cm^3
C. 656 cm^3
D. 800 cm^3

46. When a gas expands adiabatically:

A. it does no work
B. work is done on the gas
C. the internal (thermal) energy of the gas decreases
D. the internal (thermal) energy of the gas increases

47. Why is it that when a swimmer gets out of a swimming pool and stands in a breeze dripping wet, he feels much colder than when he dries off?

A. This is a physiological effect resulting from the skin's sensory nerves
B. To evaporate a gram of water from skin requires heat, and most of this heat flows from his body
C. The moisture on his skin has a good thermal conductivity
D. Water has a relatively small specific heat

48. Which heat flow method requires the movement of energy through solid matter to a new location?

I. Conduction   II. Convection   III. Radiation

A. I only
B. II only
C. III only
D. I and II only

49. An ideal gas is compressed via an isobaric process to one-third of its initial volume. Compared to the initial pressure, the resulting pressure is:

A. more than three times greater
B. nine times greater
C. three times greater
D. the same

50. Which of the following would be the best radiator of thermal energy?

A. A metallic surface
B. A black surface
C. A white surface
D. A shiny surface

51. A brass rod is 59.1 cm long, and an aluminum rod is 39.3 cm long when both rods are at an initial 0 °C. The rods are placed at 1.1 cm between them. The distance between the far ends of the rods is maintained at 99.5 cm. The temperature is raised until the two rods are barely in contact.

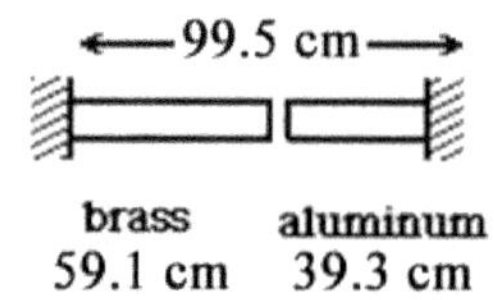

In the figure, what is the temperature at which contact of the rods barely occurs? (Use a coefficient of linear expansion of brass = 2×10^{-5} K^{-1} and the coefficient of linear expansion of aluminum = 2.4×10^{-5} K^{-1})

A. 424 °C
B. 588 °C
C. 518 °C
D. 363 °C

52. At room temperature, a person loses energy to the surroundings at the rate of 60 W. If an equivalent food intake compensates for this energy loss, how many kilocalories does he need to consume every 24 hours? (Use conversion of 1 cal = 4.186 J)

A. 1,240 kcal
B. 1,660 kcal
C. 600 kcal
D. 880 kcal

53. By what primary heat transfer mechanism does one end of an iron bar become hot when the other end is placed in a flame?

A. Convection
B. Forced convection
C. Radiation
D. Conduction

Questions **54-55** are based on the following:

Two experiments were performed to determine the calorimetric properties of alcohol with a melting point of –10 °C. In the first trial, a 220 g cube of frozen alcohol, at the melting point, is added to 350 g of water at 26 °C in a Styrofoam container. When thermal equilibrium is reached, the alcohol-water solution is at 5 °C. In the second trial, an identical cube of alcohol is added to 400 g of water at 30 °C, and the temperature at thermal equilibrium is 10 °C. (Use the specific heat of water = 4,190 J/kg·K and assume no heat exchange between the Styrofoam container and the surroundings).

54. What is the specific heat capacity of the alcohol?

A. 2,150 J/kg·K
B. 2,475 J/kg·K
C. 1,175 J/kg·K
D. 1,820 J/kg·K

55. What is the heat of fusion of the alcohol?

A. 7.2×10^3 J/kg
B. 1.9×10^5 J/kg
C. 5.2×10^4 J/kg
D. 10.3×10^4 J/kg

56. The silver coating on the glass surface of a Thermos bottle reduces the energy that is transferred by:

I. conduction II. convection III. radiation

A. I only
B. II only
C. III only
D. I and II only

57. A person consumes a snack containing 16 kcal. What is the power this food produces if it is to be expended during exercise in 5 hours? (Use the conversion 1 cal = 4.186 J)

A. 0.6 W
B. 3.7 W
C. 9.7 W
D. 96.3 W

58. If 50 kcal of heat is added to 5 kg of water, what is the resulting temperature change? (Use the specific heat of water = 1 kcal/kg·°C)

A. 10 °C
B. 20 °C
C. 5 °C
D. 40 °C

59. How much heat is needed to melt a 30 kg sample of ice at 0 °C? (Use latent heat of fusion for water L_f = 334 kJ/kg and latent heat of vaporization for water L_v = 2,257 kJ/kg)

A. 0 kJ
B. 5.6×10^4 kJ
C. 1×10^4 kJ
D. 2.4×10^6 kJ

60. A 6 kg aluminum rod is originally at 12 °C. If 160 kJ of heat is added to the rod, what is its final temperature? (Use the specific heat capacity of aluminum = 910 J/kg·K)

A. 32 °C
B. 41 °C
C. 54 °C
D. 23 °C

Notes for active learning

Notes for active learning

Atomic Nucleus and Electronic Structure

1. Which statement(s) about alpha particles is/are FALSE?

I. They are a harmless form of radiation
II. They have low penetrating power
III. They have high ionization power

A. I only
B. II only
C. III only
D. I and II only

2. What is the term for nuclear radiation that is identical to an electron?

A. Positron
B. Gamma ray
C. Beta minus particle
D. Alpha particle

3. Protons are accelerated in a particle accelerator. By what factor does their de Broglie wavelength change when the speed of the protons is doubled? (Note: consider this situation non-relativistic)

A. Increases by √2
B. Decreases by √2
C. Increases by 2
D. Decreases by 2

4. The Bohr model of the atom was able to explain the Balmer series because:

A. electrons were allowed to exist only in specific orbits and nowhere else
B. differences between the energy levels of orbits matched that between the energy levels the line spectra
C. smaller orbits require electrons to have more negative energy to match the angular momentum
D. differences between the energy levels of the orbits were half the differences between the energy levels of the line spectra

5. Radioactivity is the tendency for an element to:

A. become ionized easily
B. be dangerous to living things
C. emit radiation
D. emit protons

6. Which is the missing species in the nuclear equation: ${}^{100}_{44}Ru + {}^{0}_{-1}e^- \rightarrow$ ___?

A. ${}^{100}_{45}Ru$
B. ${}^{100}_{43}Ru$
C. ${}^{101}_{44}Ru$
D. ${}^{100}_{43}Tc$

7. The term nucleon refers to:

A. the nucleus of a specific isotope
B. both protons and neutrons
C. positrons emitted from an atom that undergoes nuclear decay
D. electrons emitted from a nucleus in a nuclear reaction

8. An isolated ^{9}Be atom spontaneously decays into two alpha particles. What can be concluded about the mass of the ^{9}Be atom?

A. The mass is less than twice the mass of the ^{4}He atom, but not equal to the mass of ^{4}He
B. No conclusions can be made about the mass
C. The mass is exactly twice the mass of the ^{4}He atom
D. The mass is greater than twice the mass of the ^{4}He atom

9. Which of the following isotopes contains the most neutrons?

A. $^{178}_{84}\text{Po}$
B. $^{178}_{87}\text{Fr}$
C. $^{181}_{86}\text{Rn}$
D. $^{170}_{83}\text{Bi}$

10. Which of the following correctly balances this nuclear fission reaction?

$$^{1}_{0}\text{n} + ^{235}_{92}\text{U} \rightarrow ^{131}_{53}\text{I} + \_\_ + 3\,^{1}_{0}\text{n}$$

A. $^{102}_{39}\text{Y}$
B. $^{102}_{36}\text{Kr}$
C. $^{104}_{39}\text{Y}$
D. $^{105}_{36}\text{Kr}$

11. What is the frequency of the light emitted by atomic hydrogen according to the Balmer formula where n = 12? (Use the Balmer series constant B = 3.6×10^{-7} m and the speed of light $c = 3 \times 10^8$ m/s)

A. 5.3×10^6 Hz
B. 8.1×10^{14} Hz
C. 5.9×10^{13} Hz
D. 1.2×10^{11} Hz

12. Gamma rays require the heaviest shielding of the common types of nuclear radiation because they have the:

A. heaviest particles
B. lowest energy
C. most intense color
D. highest energy

13. Transitioning from state n = 1 to state n = 2, the hydrogen atom must [ ] a photon of [ ]. (Use Planck's constant $h = 4.14 \times 10^{-15}$ eV·s, speed of light $c = 3 \times 10^8$ m/s and Rydberg constant $R = 1.097 \times 10^7 \text{ m}^{-1}$)

A. absorb … 10.2 eV
B. absorb … 8.6 eV
C. emit … 8.6 eV
D. emit … 10.2 eV

14. Rubidium $^{87}_{37}Rb$ is a naturally-occurring nuclide that undergoes β^- decay. What is the resultant nuclide from this decay?

A. $^{86}_{36}Rb$
B. $^{87}_{38}Kr$
C. $^{87}_{38}Sr$
D. $^{87}_{36}Kr$

15. Which of the following statements best describes the role of neutrons in the nucleus?

A. The neutrons stabilize the nucleus by attracting protons
B. The neutrons stabilize the nucleus by balancing charge
C. The neutrons stabilize the nucleus by attracting other nucleons
D. The neutrons stabilize the nucleus by repelling other nucleons

16. A Geiger–Muller counter detects radioactivity by:

A. ionizing argon gas in a chamber which produces an electrical signal
B. analyzing the mass and velocity of each particle
C. developing film which is exposed by radioactive particles
D. slowing the neutrons using a moderator and then counting the secondary charges produced

17. What percentage of the radionuclides in a given sample remains after three half-lives?

A. 25%
B. 12.5%
C. 6.25%
D. 33.3%

18. The Lyman series is formed by electron transitions in hydrogen that:

A. begin on the n = 2 shell
B. end on the n = 2 shell
C. begin on the n = 1 shell
D. end on the n = 1 shell

19. Most of the volume of an atom is occupied by:

A. neutrons
B. space
C. electrons
D. protons

20. Alpha and beta minus particles are deflected in opposite directions in a magnetic field because:

I. they have opposite charges
II. alpha particles contain nucleons, and beta minus particles do not
III. they spin in opposite directions

A. I only
B. II only
C. III only
D. I and II only

21. The conversion of mass to energy is measurable only in:

A. chemiluminescent transformations
B. spontaneous chemical reactions
C. endothermic reactions
D. nuclear reactions

22. What is the term given to a radioactive substance that undergoes 3.7×10^{10} disintegrations per second?

A. Rem
B. Rad
C. Curie
D. Roentgen

23. An isolated ^{235}U atom spontaneously undergoes fission into two approximately equal-sized fragments. What is missing from the product side of the reaction?

$^{235}U \rightarrow {}^{141}Ba + {}^{92}Kr + \_\_\_?$

A. A neutron
B. Two neutrons
C. Two protons and two neutrons
D. Two protons and a neutron

24. How many protons and neutrons are in $^{34}_{16}S$?

A. 18 neutrons and 34 protons
B. 16 neutrons and 18 protons
C. 16 protons and 34 neutrons
D. 16 protons and 18 neutrons

25. The radioactive isotope Z has a half-life of 12 hours. What is the fraction of the original amount remaining after two days?

A. 1/16
B. 1/8
C. 1/4
D. 1/2

26. Which of the following nuclear equations correctly describes alpha emission?

A. $^{238}_{92}U \rightarrow {}^{242}_{94}Pu + {}^{4}_{2}He$
B. $^{238}_{92}U \rightarrow {}^{4}_{2}He$
C. $^{238}_{92}U \rightarrow {}^{234}_{90}Th + {}^{4}_{2}He$
D. $^{238}_{92}U \rightarrow {}^{235}_{90}Th + {}^{4}_{2}He$

27. A hydrogen atom transitions downward from the n = 20 state to the n = 5 state. Find the wavelength of the emitted photon. (Use Planck's constant $h = 4.14 \times 10^{-15}$ eV·s, the speed of light $c = 3 \times 10^8$ m/s and the Rydberg constant $R = 1.097 \times 10^7 \text{ m}^{-1}$)

A. 1.93 μm
B. 2.82 μm
C. 1.54 μm
D. 2.43 μm

28. A nuclear equation is balanced when the:

A. same elements are on both sides of the equation
B. sums of the atomic numbers of the particles and atoms are the same on each side of the equation
C. sum of the mass numbers of the particles and the sum of atoms are the same on both sides of the equation
D. sum of the mass and atomic numbers of the particles and atoms are the same on each side of the equation

29. A blackbody is an ideal system that:

A. absorbs 50% of the light incident upon it and emits 50% of the radiation it generates
B. absorbs 0% of the light incident upon it and emits 100% of the radiation it generates
C. absorbs 100% of the light incident upon it and emits 100% of the radiation it generates
D. emits 100% of the light it generates and absorbs 50% of the radiation incident upon it

30. Recent nuclear bomb tests have created an extra-high level of atmospheric ^{14}C. When future archaeologists date samples without knowing these nuclear tests, will the dates they calculate be correct?

A. Correct, because biological materials do not gather ^{14}C from bomb tests
B. Correct, since the ^{14}C decays within the atmosphere at the natural rate
C. Incorrect, they would appear too old
D. Incorrect, they would appear too young

31. When an isotope releases gamma radiation, the atomic number:

A. and the mass number remain the same
B. and the mass number decrease by one
C. and the mass number increase by one
D. remains the same, and the mass number increases by one

32. If ^{14}Carbon is a beta emitter, what is the likely product of radioactive decay?

A. ^{22}Silicon
B. ^{13}Boron
C. ^{14}Nitrogen
D. ^{12}Carbon

33. In a balanced nuclear equation, the:

I. sum of the mass numbers on both sides must be equal
II. sum of the atomic numbers on both sides must be equal
III. daughter nuclide appears on the right side of the arrow

A. I only
B. II only
C. III only
D. I, II and III

34. What is the term for the number that characterizes an element and indicates the number of protons found in the nucleus of the atom?

A. Mass number
B. Atomic number
C. Atomic mass
D. Neutron number

35. Hydrogen atoms can emit four spectral lines with visible colors from red to violet. These four visible lines emitted by hydrogen atoms are produced by electrons that:

A. end in the ground state
B. end in the n = 3 level
C. end in the n = 2 level
D. start in the ground state

36. The electron was discovered through experiments with:

A. quarks
B. foil
C. light
D. electricity

Questions **37-39** are based on the following:

The image shows a beam of radiation passing between two electrically-charged plates.

I. a
II. b
III. c

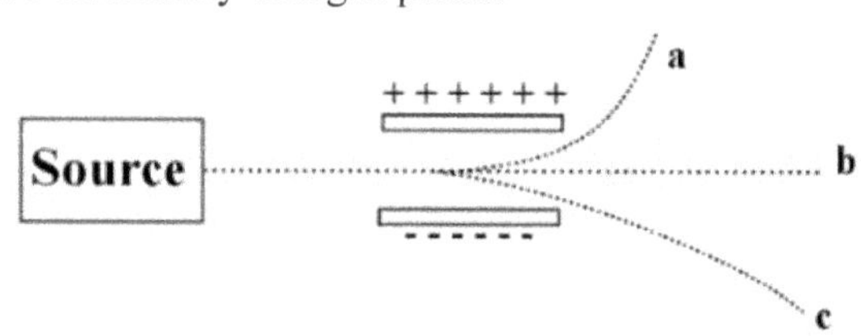

37. Which of the beams is due to an energetic light wave?

A. I only
B. II only
C. III only
D. I and II only

38. Which of the beams is/are composed of particles?

A. I only
B. II only
C. I, II and III
D. I and III only

39. Which of the beams is due to a positively-charged helium nucleus?

A. I only
B. II only
C. III only
D. I, II and III

40. Lithium atoms can absorb photons transitioning from the ground state (at –5.37 eV) to an excited state with one electron removed from the atom, corresponding to the zero-energy state. What is the wavelength of light from this transition? (Use Planck's constant $h = 4.14 \times 10^{-15}$ eV·s and the speed of light $c = 3 \times 10^8$ m/s)

A. 6.6×10^{-6} m
B. 2.3×10^{-7} m
C. 3.6×10^{6} m
D. 4.2×10^{5} m

41. All of the elements with atomic numbers of 84 and higher are radioactive because:

A. strong attractions between their nucleons make them unstable
B. their atomic numbers are larger than their mass numbers
C. strong repulsions between their electrons make them unstable
D. strong repulsions between their protons make their nuclei unstable

42. Which of the following statements about β particles is FALSE?

A. They have a smaller mass than α particles
B. They have high energy and a charge
C. They are created when neutrons become protons and vice versa
D. They are a harmless form of radioactivity

43. The material used in nuclear bombs is ^{239}Pu, with a half-life of about 20,000 years. What approximate amount of time must elapse for a buried stockpile of this substance to decay to 3% of its original ^{239}Pu mass?

A. 0.8 thousand years
B. 65 thousand years
C. 90 thousand years
D. 101 thousand years

44. Which statement regarding Planck's constant is true?

A. It relates mass to the amount of energy that can be emitted
B. It sets a lower limit to the amount of energy that can be absorbed or emitted
C. It sets an upper limit to the amount of energy that can be absorbed
D. It sets an upper limit to the amount of energy that can be absorbed or emitted

45. The decay rate of a radioactive isotope will NOT be increased by increasing the:

I. surface area II. pressure III. temperature

A. I only
B. II only
C. III only
D. I, II and III

46. Why are some smaller nuclei such as ^{14}Carbon often radioactive?

I. The attractive force of the nucleons has a limited range
II. The neutron to proton ratio is too large or too small
III. Most smaller nuclei are not stable

A. I only
B. II only
C. III only
D. II and III only

47. Scandium ^{44}Sc decays by emitting a positron. What is the resulting nuclide produced by this decay?

A. $^{43}_{21}Sc$
B. $^{45}_{21}Sc$
C. $^{44}_{20}Ca$
D. $^{43}_{20}Ca$

48. A scintillation counter detects radioactivity by:

A. analyzing the mass and velocity of each electron
B. ionizing argon gas in a chamber which produces an electrical signal
C. emitting light from a NaI crystal when radioactivity passes through the crystal
D. developing film which is exposed by radioactive particles

49. Which of the following is indicated by each detection sound by a Geiger counter?

A. One half-life
B. One nucleus decaying
C. One neutron being emitted
D. One positron being emitted

50. According to the Pauli Exclusion Principle, how many electrons in an atom may have a particular set of quantum numbers?

A. 1
B. 2
C. 3
D. 4

51. The atomic number of an atom identifies the number of:

A. excited states
B. electron orbits
C. neutrons
D. protons

52. Which of the following correctly characterizes gamma radiation?

A. High penetrating power; charge = –1; mass = 0 amu
B. Low penetrating power; charge = –1; mass = 0 amu
C. High penetrating power; charge = 0; mass = 0 amu
D. High penetrating power; charge = 0; mass = 4 amu

53. The rest mass of a proton is 1.0072764669 amu, and that of a neutron is 1.0086649156 amu. The ^{4}He nucleus weighs 4.002602 amu. What is the total binding energy of the nucleus? (Use the speed of light $c = 3 \times 10^8$ m/s and 1 amu = 1.6606×10^{-27} kg)

A. 2.7×10^{-11} J
B. 4.4×10^{-12} J
C. 1.6×10^{-7} J
D. 2.6×10^{-12} J

54. Which is the correct electron configuration for ground state boron ($Z = 5$)?

A. $1s^21p^22s$
B. $1s^22p^23s$
C. $1s^22p^3$
D. $1s^22s^22p$

55. The Sun produces 3.85×10^{26} J each second. How much mass does it lose per second from nuclear processes alone? (Use the speed of light $c = 3 \times 10^8$ m/s)

A. 9.8×10^1 kg
B. 2.4×10^9 kg
C. 4.3×10^9 kg
D. 1.1×10^8 kg

56. The damaging effects of radiation on the body are a result of:

A. extensive damage to nerve cells
B. transmutation reactions in the body
C. the formation of radioactive particles in the body
D. the formation of unstable ions or radicals in the body

57. How does the emission of a gamma-ray affect the radioactive atom?

I. The atomic mass increases
II. The atom has a smaller amount of energy
III. The atom gains energy for further radioactive particle emission

A. I only
B. II only
C. III only
D. I and II only

58. The nuclear particle, which is described by the symbol $^1_0$n is a(n):

A. neutron
B. gamma ray
C. beta particle
D. electron

59. Heisenberg's uncertainty principle states that:

A. at times, a photon appears to be a particle, and at other times it appears to be a wave
B. whether a photon is a wave, or a particle cannot be determined with certainty
C. the position and the momentum of a particle cannot be simultaneously known with absolute certainty
D. the properties of an electron cannot be known with absolute certainty

Notes for active learning

Particle Physics

1. Which of the following statements about hadrons are correct?

I. Hadrons are composed of leptons
II. Hadrons are composed of quarks
III. Protons and neutrons are hadrons, but electrons are not

A. I only
B. III only
C. II and III only
D. II and III only

2. Which of the following statements about hadrons are correct?

I. All hadrons interact by the strong nuclear force
II. Electrons, protons, and neutrons are commonly-occurring hadrons
III. Hadrons are composed of quarks

A. I only
B. II only
C. I and II only
D. I and III only

3. Leptons can interact by which of the following forces?

A. strong nuclear force and weak nuclear force
B. weak nuclear force, electromagnetic force, and gravitation
C. strong nuclear force, weak nuclear force, and electromagnetic force
D. strong nuclear force, weak nuclear force, electromagnetic force, and gravitation

4. Which of the following particles do NOT contain any quarks?

I. alpha particle II. positron III. neutron

A. I only
B. II only
C. I and III only
D. II and III only

5. What are the possible charges of a quark (not an antiquark)?

A. $-2/3\ e$, $+1/3\ e$
B. $-1/3\ e$, $+2/3\ e$
C. $-e$, 0, e
D. $-2/3\ e$, $-1/3\ e$, $+1/3\ e$, $+2/3\ e$

6. How many quarks are in a deuteron, ${}^{2}_{1}\text{H}$?

A. 1
B. 3
C. 4
D. 6

7. How many quarks are in a tritium isotope, ^3_1H?

A. 1
B. 4
C. 5
D. 9

8. How does the range of an exchange force depend on the mass of the exchange particle?

I. The range does not depend on the mass of the exchange particle
II. The range is shorter for a massive exchange particle than for a light exchange particle
III. The range is longer for a massive exchange particle than for a light exchange particle

A. I only
B. II only
C. III only
D. I and II only

9. An electron and a positron, both essentially at rest, annihilate each other, emitting two identical photons in the process. What is the wavelength of these photons? The mass of an electron or positron is 9.11×10^{-31} kg. (Use $h = 6.626 \times 10^{-34}$ J·s, $c = 3.00 \times 10^8$ m/s and $hc = 1240$ eV· nm)

A. 2.42 pm
B. 3.72 pm
C. 1.21×10^{-12} pm
D. 1.57 pm

10. A proton is a specific combination of three quarks. What are the electric charges on these quarks, expressed in terms of e?

A. $+2/3\ e$, $+2/3\ e$, and $-1/3\ e$
B. $-2/3\ e$, $+2/3\ e$, and $-1/3\ e$
C. $+2/3\ e$, $+2/3\ e$, and $+1/3\ e$
D. $-2/3\ e$, $-2/3\ e$, and $-1/3\ e$

11. A positron (or antielectron) is made to stop in a sample of matter. Soon after, how many gamma rays are observed, and with what energy each?

A. none
B. one, 0.511 MeV each
C. two, 0.511 MeV each
D. one, 13.6 eV each

12. What is the minimum energy required to produce a proton-antiproton pair? (Use the rest mass of a proton = 938 MeV/c^2)

A. 0 MeV
B. 938 MeV
C. 1876 MeV
D. 2814 MeV

13. A η meson with rest energy of 548 MeV decays into two gamma rays. What is the wavelength of the gamma rays? (Use $h = 6.63 \times 10^{-34}$ J·s, $c = 3.0 \times 10^8$ m/s and $hc = 1{,}240$ eV nm).

A. 2.3×10^{-15} m
B. 2.3×10^{-16} m
C. 4.5×10^{-15} m
D. 4.5×10^{-16} m

14. What is the speed of a proton with kinetic energy equal to 2 GeV?

A. 0.78 *c*
B. 0.84 c
C. 0.88 *c*
D. 0.95 c

15. What is the speed of a 2 MeV electron?

A. 0.75 c
B. 0.84 c
C. 0.90 *c*
D. 0.97 *c*

16. What is the temperature corresponding to the thermal energy of 50 GeV, the particle energy in the accelerator at SLAC?

A. 3.9×10^{11} K
B. 5.8×10^{14} K
C. 7.8×10^{11} K
D. 7.8×10^{14} K

17. How far is the galaxy from Earth if it recedes from Earth at a speed $v = 0.5\ c$? (Use $H_0 = 0.022$ m/s/lightyear for Hubble's constant).

A. 22.7 lightyears
B. 1.8×10^4 lightyears
C. 2.3×10^5 lightyears
D. 6.8×10^9 lightyears

18. A certain galaxy is 2.7×10^9 lightyears from Earth. How fast is it receding from Earth? (Use $H_0 = 0.022$ m/(s·lightyears) for Hubble's constant).

A. 0.2 c
B. 0.3 c
C. 0.4 c
D. 0.5 c

19. What is the minimum energy required to produce an electron-positron pair? (Use the rest mass of an electron = 0.511 MeV/c^2).

A. 0.511 MeV
B. 1.022 MeV
C. 1.533 MeV
D. 2.044 MeV

20. The rest energy of π meson is about 140 MeV. Using the uncertainty principle, estimate the range of strong force. ($h = 6.63 \times 10^{-34}$ J·s, $hc = 1.24$ eV μm).

A. 0.35×10^{-15} m
B. 0.7×10^{-15} m
C. 1.4×10^{-15} m
D. 2.8×10^{-15} m

21. The rest energy of a Z^0 boson is 91 GeV. Using the uncertainty principle, what is the approximate interaction time when it mediates a weak interaction? (Use $h = 6.63 \times 10^{-34}$ J·s, and $hc = 1.24$ eV μm).

A. 3.6×10^{-26} s
B. 2.6×10^{-27} s
C. 7.2×10^{-27} s
D. 5.2×10^{-28} s

22. A μ^+ and μ^-at rest annihilate to produce two gamma rays that fly in opposite directions. What is the wavelength of each gamma-ray? (Use the rest energy of a muon = 106 MeV, $h = 6.63 \times 10^{-34}$ J·s and $hc = 1.24$ eV μm).

A. 1.17×10^{-14} m
B. 5.63×10^{-15} m
C. 2.48×10^{-15} m
D. 1.24×10^{-15} m

23. The rest energy of W^+ boson is about 80 GeV. Using the uncertainty principle, estimate the range of weak interaction. (Use $h = 6.63 \times 10^{-34}$ J·s and $hc = 1.24$ eV μm).

A. 1.3×10^{-15} m
B. 2.5×10^{-16} m
C. 2.5×10^{-17} m
D. 2.5×10^{-18} m

Notes for active learning

Notes for active learning

Quantum Mechanics

1. The work function of a certain metal is 1.90 eV. What is the longest wavelength of light that can cause photoelectron emission from this metal?

A. 64 nm
B. 98 nm
C. 247 nm
D. 653 nm

2. Which of the following statements is correct if the frequency of the light in a laser beam is doubled while the number of photons per second in the beam is fixed?

I. The power in the beam does not change
II. The intensity of the beam doubles
III. The energy of individual photons does not change

A. I only
B. II only
C. III only
D. I and II only

3. Upon being struck by 240 nm photons, a material ejects electrons with a maximum kinetic energy of 2.58 eV. What is the work function of this material?

A. 1.17 eV
B. 2.04 eV
C. 2.60 eV
D. 3.26 eV

4. A high-energy photon collides with matter and creates an electron-positron pair. What is the minimum frequency of the photon? (Use the $m_{electron} = 9.11 \times 10^{-31}$ kg, $c = 3.00 \times 10^8$ m/s, and $h = 6.626 \times 10^{-34}$ J·s)

A. greater than 1.24×10^{12} Hz
B. greater than 2.47×10^{16} Hz
C. greater than 2.47×10^{20} Hz
D. greater than 2.47×10^{22} Hz

5. What is the longest wavelength of light that can cause photoelectron emission from metal with a work function of 2.20 eV?

A. 216 nm
B. 372 nm
C. 484 nm
D. 564 nm

6. In 1932, C. D. Anderson:

A. set the limits on the probability of measurement accuracy
B. predicted the positron from relativistic quantum mechanics
C. discovered the positron using a cloud chamber
D. was the first to produce diffraction patterns of electrons in crystals

7. What is the energy of an optical photon of frequency 6.43×10^{14} Hz (Use $h = 6.626 \times 10^{-34}$ J·s and 1 eV = 1.60×10^{-19} J)

A. 1.04 eV
B. 1.86 eV
C. 2.66 eV
D. 3.43 eV

8. A photocathode has a work function of 2.4 eV. The photocathode is illuminated with monochromatic radiation, whose photon energy is 3.4 eV. What is the maximum kinetic energy of the photoelectrons produced? (Use 1 eV = 1.60×10^{-19} J)

A. 3.4×10^{-20} J
B. 1.6×10^{-19} J
C. 4.6×10^{-19} J
D. 5.8×10^{-19} J

9. A photocathode whose work function is 2.9 eV is illuminated with white light with a continuous wavelength band from 400 nm to 700 nm. What is the range of the wavelength band in this white light illumination for which photoelectrons are NOT produced?

A. 360 to 440 nm
B. 400 to 480 nm
C. 430 to 500 nm
D. 430 to 700 nm

10. Photon A has twice the momentum of photon B as both travels in a vacuum. Which of the following statements about these photons is correct?

A. Both photons have the same speed
B. Both photons have the same wavelength
C. Photon A is traveling twice as fast as photon B
D. The energy of photon A is half as great as the energy of photon B

11. What is the energy of the photon emitted when an electron drops from the n = 20 state to the n = 7 state in a hydrogen atom?

A. 0.244 eV
B. 0.288 eV
C. 0.336 eV
D. 0.404 eV

12. Protons are being accelerated in a particle accelerator. What is the de Broglie wavelength when the energy of the protons is doubled if the protons are non-relativistic (i.e., their kinetic energy is much less than mc^2)?

A. increases by a factor of 2
B. increases by a factor of 3
C. decreases by a factor of 2
D. decreases by a factor of $\sqrt{2}$

13. A certain photon, after being scattered from a free electron at rest, moves at an angle of 120° with respect to the incident direction. If the wavelength of the incident photon is 0.591 nm, what is the wavelength of the scattered photon? (Use $m_{electron} = 9.11 \times 10^{-31}$ kg, $c = 3.00 \times 10^8$ m/s and $h = 6.626 \times 10^{-34}$ J·s)

A. 0.0 nm
B. 0.180 nm
C. 0.252 nm
D. 0.595 nm

14. Increasing the *brightness* of a beam of light without changing its color increases the:

A. speed of the photons
B. frequency of the light
C. number of photons per second traveling in the beam
D. energy of each photon

15. What is the frequency of the light emitted by atomic Hydrogen according to the Balmer formula with $m = 4$ and $n = 9$?

A. 820 Hz
B. 1.65×10^{14} Hz
C. 1,820 Hz
D. 1,640 Hz

16. One of the emission lines described by the original version of the Balmer formula has a wavelength of 377 nm. What is the value of n in the Balmer formula that gives this emission line?

A. 5
B. 7
C. 9
D. 11

17. What is the wavelength of the most intense light emitted by a giant star of surface temperature 5000 K? (Use the constant in Wien's law = 0.00290 m·K)

A. 366 nm
B. 448 nm
C. 580 nm
D. 490 nm

18. In 1928, Paul Dirac:

A. set the limits on the probability of measurement accuracy
B. developed a wave equation for matter waves
C. suggested the existence of matter waves
D. predicted the positron from relativistic quantum mechanics

19. A photocathode has a work function of 2.4 eV. The photocathode is illuminated with monochromatic radiation, whose photon energy is 3.4 eV. What is the maximum kinetic energy of the photoelectrons produced?

A. 0.9×10^{-19} J
B. 1.6×10^{-19} J
C. 2.8×10^{-19} J
D. 4.2×10^{-19} J

20. In a Compton scattering experiment, which scattering angle produces the greatest change in wavelength?

A. 0° **C.** 90°
B. 45° **D.** 180°

21. A photocathode has a work function of 2.4 eV. The photocathode is illuminated with monochromatic radiation, whose photon energy is 3.5 eV. What is the wavelength of the illuminating radiation?

A. 280 nm **C.** 350 nm
B. 325 nm **D.** 395 nm

22. If the wavelength of a photon is doubled, what happens to its energy?

A. It is reduced to one-half of its original value **C.** It is doubled
B. It stays the same **D.** It is increased to four times its original value

23. A photocathode has a work function of 2.8 eV. The photocathode is illuminated with monochromatic radiation. What is the threshold frequency for the monochromatic radiation required to produce photoelectrons?

A. 1.2×10^{14} Hz **C.** 4.6×10^{14} Hz
B. 2.8×10^{14} Hz **D.** 6.8×10^{14} Hz

24. If the de Broglie wavelength of an electron is 380 nm, what is the speed of this electron? (Use $m_{electron}$ = 9.11×10^{-31} kg and $h = 6.626 \times 10^{-34}$ J·s)

A. 0.6 km/s **C.** 3.4 km/s
B. 1.9 km/s **D.** 4.6 km/s

25. What is the wavelength of the scattered photon if a photon of wavelength 1.50×10^{-10} m is scattered at an angle of 90° in the Compton effect?

A. 1.29×10^{-10} m **C.** 1.84×10^{-10} m
B. 1.52×10^{-10} m **D.** 2.42×10^{-10} m

26. When the surface of a metal is exposed to blue light, electrons are emitted. Which of the following increases if the intensity of the blue light increases?

I. the maximum kinetic energy of the ejected electrons
II. the number of electrons ejected per second
III. the time lag between the onset of the absorption of light and the ejection of electrons

A. I only **C.** III only
B. II only **D.** I and II only

27. A proton has a speed of 7.2 x 10^4 m/s. What is the energy of a photon with the same wavelength as the de Broglie wavelength of this proton? (Use $m_{proton} = 1.67 \times 10^{-27}$ kg and $c = 3.00 \times 10^8$ m/s)

A. 80 keV
B. 120 keV
C. 160 keV
D. 230 keV

28. In the Compton effect, as the scattering angle increases monotonically from 0° to 180°, the frequency of the X-rays scattered at that angle:

A. decreases by the √2
B. decreases monotonically
C. increases by the √2
D. increases monotonically

29. In the spectrum of Hydrogen, the lines obtained by setting m = 1 in the Rydberg formula are the Lyman series. What is the wavelength of the spectral line of the 15th member of the Lyman series?

A. 91.6 nm
B. 126.2 nm
C. 244.6 nm
D. 368.2 nm

30. If the frequency of the light in a laser beam is doubled while the number of photons per second in the beam is fixed, which of the following statements is correct?

I. The energy of individual photons doubles
II. The wavelength of the individual photons doubles
III. The intensity of the beam doubles

A. I only
B. II only
C. III only
D. I and III only

31. The spacing of the surface planes of a crystal is 159.0 pm. A beam directed at the surface of the crystal undergoes first-order diffraction at an angle of 58° from the normal. What is the energy of the neutrons if the diffraction is done with a beam of monenergistic neutrons? (Use 1.67×10^{-27} kg for the mass of a neutron)

A. 0.0155 eV
B. 0.0106 eV
C. 0.0909 eV
D. 0.1450 eV

32. A photocathode whose work function is 2.5 eV is illuminated with white light with a continuous wavelength band from 360 nm to 700 nm. What is the stopping potential for this white light illumination?

A. 0.95 V
B. 1.45 V
C. 1.90 V
D. 2.6 V

33. If the momentum of an electron is 1.95×10^{-27} kg·m/s, what is its de Broglie wavelength? (Use $h = 6.626 \times 10^{-34}$ J·s)

A. 86.2 nm
B. 130.6 nm
C. 240.8 nm
D. 340.0 nm

34. A Hydrogen atom is excited to the n = 9 level, and its decay to the n = 6 level is detected on a photographic plate. What is the frequency of the light photographed?

A. 3,810 Hz
B. 7,240 Hz
C. 5.08×10^{13} Hz
D. 3.28×10^{-9} Hz

35. How much energy is carried by a photon of light having a frequency of 110 GHz? (Use $h - 6.626 \times 10^{-34}$ J·s)

A. 7.3×10^{-23} J
B. 2.9×10^{-25} J
C. 1.7×10^{-26} J
D. 1.1×10^{-21} J

36. How many of the infinite number of Balmer spectrum lines are in the visible spectrum range?

A. 0
B. 2
C. 4
D. 6

37. Each photon in a beam of light has an energy of 4.20 eV. What is the wavelength of this light? ($c = 3.00 \times 10^8$ m/s, $h = 6.626 \times 10^{-34}$ J·s and 1 eV = 1.60×10^{-19} J)

A. 118.0 nm
B. 296.0 nm
C. 365.0 nm
D. 462.0 nm

38. Electrons are emitted from a surface when the light of wavelength 500.0 nm is shone on the surface, but electrons are not emitted for longer wavelengths of light. What is the work function of the surface?

A. 0.5 eV
B. 1.6 eV
C. 2.5 eV
D. 3.8 eV

39. In the Bohr theory, the orbital radius depends upon the principal quantum number in what way?

A. n
B. 1/n
C. n^2
D. n^3

40. The uncertainty in the position of a proton is 0.053 nm. What is the minimum uncertainty in its speed? (Use 1.67×10^{-27} kg as the proton mass)

A. 0.6×10^3 m/s
B. 1.2×10^3 m/s
C. 2.4×10^3 m/s
D. 3.6×10^3 m/s

41. What is the wavelength of the light emitted by atomic Hydrogen according to the Balmer formula with m = 9 and n = 11?

A. 8,500 nm
B. 14,700 nm
C. 22,300 nm
D. 31,900 nm

42. A certain particle's energy is known within 10^{-18} J. What is the minimum uncertainty in its arrival time at a detector?

A. 5.08×10^{-12} s
B. 4.25×10^{-13} s
C. 3.88×10^{-14} s
D. 1.05×10^{-16} s

43. Upon being struck by 240.0 nm photons, a material ejects electrons with a maximum kinetic energy of 2.58 eV. What is the work function of this material?

A. 1.20 eV
B. 2.82 eV
C. 2.60 eV
D. 3.46 eV

44. What is the de Broglie wavelength of a 1.30 kg missile moving at 28.10 m/s. (Use $h = 6.626 \times 10^{-34}$ J·s)

A. 1.85×10^{-37} m
B. 2.40×10^{-36} m
C. 1.81×10^{-35} m
D. 3.37×10^{-35} m

45. A beam of light falling on a metal surface causes electrons to be ejected from the surface. If the frequency of the light now doubles, which of the following statements is always true?

A. The number of electrons ejected per second doubles
B. Twice as many photons hit the metal surface as before
C. The kinetic energy of the ejected electrons doubles
D. None of the above statements is always true

46. What is the longest wavelength of a photon emitted by a hydrogen atom, for which the initial state is n = 3?

A. 486 nm
B. 510 nm
C. 540 nm
D. 656 nm

47. Which of the following always increases if the brightness of a beam of light is increased without changing its color?

I. the speed of the photons
II. the average energy of each photon
III. the number of photons

A. I only
B. II only
C. III only
D. I and II only

48. In a particular case of Compton scattering, a photon collides with a free electron and scatters backward. The wavelength after the collision is exactly double the wavelength before the collision. What is the wavelength of the incident photon? (Use $m_{electron} = 9.11 \times 10^{-31}$ kg, $c = 3.00 \times 10^8$ m/s and $h = 6.626 \times 10^{-34}$ J·s)

A. 3.4×10^{-12} m
B. 4.8×10^{-12} m
C. 5.6×10^{-12} m
D. 6.8×10^{-12} m

49. Two sources emit beams of microwaves. The microwaves from source A have a frequency of 15 GHz, and the microwaves from source B have a frequency of 30 GHz. This is all the information available for the two beams. Which of the following statements about these microwave beams must be correct?

A. The intensity of beam B is twice as great as the intensity of beam A
B. A photon in beam B has the same energy as a photon in beam A
C. Beam B carries twice as many photons per second as beam A
D. A photon in beam B has twice the energy of a photon in beam A

50. What is the shortest wavelength of a photon that can be emitted by a hydrogen atom, for which the initial state is n = 3?

A. 102.6 nm
B. 97.3 nm
C. 820.0 nm
D. 121.6 nm

51. The radius of a typical nucleus is about 5.0×10^{-15} m. Assuming this to be the uncertainty in the position of a proton in the nucleus, what is the uncertainty in the proton's energy? (Use 1.67×10^{-27} kg as the proton mass)

A. 0.06 MeV
B. 0.25 MeV
C. 0.4 MeV
D. 0.8 MeV

52. A photocathode has a work function of 2.4 eV. The photocathode is illuminated with monochromatic radiation, and electrons are emitted with a stopping potential of 1.1 volts. What is the wavelength of the illuminating radiation? (Use $c = 3.00 \times 10^8$ m/s, $h = 6.626 \times 10^{-34}$ J·s and 1 eV = 1.60×10^{-19} J)

A. 300 nm
B. 350 nm
C. 390 nm
D. 420 nm

53. The Compton effect directly demonstrated which property of electromagnetic radiation?

I. energy content
II. particle nature
III. momenta

A. I only
B. II only
C. I and II only
D. II and III only

54. What is the wavelength of the matter-wave associated with an electron moving with a speed of 2.5×10^7 m/s? (Use $m_{electron} = 9.11 \times 10^{-31}$ kg and $h = 6.626 \times 10^{-34}$ J·s)

A. 17 pm
B. 29 pm
C. 39 pm
D. 51 pm

55. A laser produces a beam of 4000 nm light. A shutter allows a pulse of light, for 30.0 ps to pass. What is the uncertainty in the energy of a photon in the pulse? ($h = 6.626 \times 10^{-34}$ J·s and 1 eV = 1.60×10^{-19} J)

A. 2.6×10^{-2} eV
B. 4.2×10^{-3} eV
C. 6.8×10^{-4} eV
D. 2.2×10^{-5} eV

Notes for active learning

Special Relativity

1. Consider three galaxies, Alpha, Beta, and Gamma. An astronomer in Beta sees each of the other two galaxies moving away from him in opposite directions at $0.70c$. At what speed would an observer in Alpha see the galaxy Gamma moving?

A. $0.94c$
B. $0.84c$
C. $1.2c$
D. $0.65c$

2. A super high-speed train moves in a direction parallel to its length with a speed that approaches the speed of light. The height of the train, as measured by a stationary observer on the ground:

A. decreases slightly
B. increases slightly
C. approaches zero
D. does not change due to the motion

3. A spaceship visits Alpha Centauri and returns to Earth. Alpha Centauri is 4.367 light-years from Earth. If the spaceship travels at one-half the speed of light for substantially all its expedition, how long was the ship gone according to an observer on Earth?

A. 2.184 years
B. 2.090 years
C. 4.367 years
D. 17.468 years

4. A muon at rest decays in 2.2 μs. Moving at 99% the speed of light, it would be seen to "live" for how long?

A. 0.31 μs
B. 0.218 μs
C. 2.178 μs
D. 16 μs

5. A spaceship approaching an asteroid at a speed of $0.6c$ launches a scout rocket with a speed of $0.4c$. At what speed is the scout rocket approaching the asteroid from the point of view of the asteroid?

A. $0.52c$
B. $1.15c$
C. $0.66c$
D. $0.81c$

6. An astronaut is resting on a bed inclined at an angle above the floor of a spaceship, with his head elevated near the front of the capsule while his feet are lower and facing the rear of the capsule. From the point of view of an observer who sees the spaceship moving near the speed of light parallel to the floor, the angle the bed makes with the floor:

A. could be greater or smaller than the angle observed by the astronaut depending on whether the rocket is moving to the right or the left
B. is smaller than the angle observed by the astronaut
C. is the same as the angle observed by the astronaut
D. is greater than the angle observed by the astronaut

7. As measured in Earth's rest frame, a spaceship traveling at $0.9640c$ takes 10.5 years to travel between two planets, not moving relative to each other. How many years does the trip take as measured by someone on the spaceship?

A. 2.79
B. 5.43
C. 8.42
D. 19.25

8. A hydrogen atom and a neutron fuse to form deuterium. If the energy released is 3.6×10^{-13} J, what is the mass of the deuterium? (Use the H atom nucleus as a single proton.).

A. 0.85×10^{-20} kg
B. 1.6×10^{-27} kg
C. 3.3×10^{-27} kg
D. 4.1×10^{-27} kg

9. Steve is moving past Earth at $0.99c$ and notices his heart beating 88 times/minute. What does his doctor on Earth observe his heart rate to be?

A. 8 beats/min
B. 12 beats/min
C. 22 beats/min
D. 28 beats/min

10. Two spaceships approach one another, each at a speed of $0.31c$ relative to a stationary observer on Earth. In terms of the speed of light, what speed does an observer on one spaceship record for the other approaching spaceship?

A. $0.71c$
B. $0.38c$
C. $0.57c$
D. $0.64c$

11. The special theory of relativity predicts an upper limit to the speed of a particle. It thus follows that there is an upper limit on which of the following property of a particle:

A. the kinetic energy
B. the total energy
C. the linear momentum
D. none of the above

12. Two fixed navigation beacons mark the approach lane to a star. The beacons are in line with the star and are 49 million meters apart. A spaceship approaches the star with a relative velocity of $0.50c$ and passes the beacons. The passage of the ship between the beacons is timed by an observer on the ship. What is the time interval of the passage?

A. 240 ms
B. 280 ms
C. 360 ms
D. 140 ms

13. A pion is an unstable particle with a mean lifetime of 2.55×10^{-8} s. This is the time interval between its creation in a nuclear process and its extinction into decay products, as measured in a frame of reference at rest with respect to the pion. How far does it travel in its lifetime relative to Earth? (Use the value that an average pion travels at $0.230c$ relative to Earth).

A. 1.81 m
B. 1.09 m
C. 2.46 m
D. 3.83 m

14. Astronaut Yuri is space-traveling from planet Y to planet Z at a speed of $0.60c$. When he is halfway between the planets, 1 light-hour from each as measured in the frame of the planets, nuclear devices are detonated. The explosions are simultaneous in the frame of the planets. What is the difference in the arrival time of the flashes from the explosions as observed by Yuri?

A. 45 min
B. 90 min
C. 125 min
D. 300 min

15. From Earth, humans see the Enterprise approaching from the west at $0.800c$ and the Klingons approaching from the east at $0.900c$. What speed, in terms of the speed of light, does the crew of the Enterprise measure for the Klingon ship?

A. $0.721c$
B. $0.378c$
C. $0.547c$
D. $0.988c$

16. Sofia is a passenger on a spaceship. After the speed of the spaceship has increased, she observes:

A. her watch moving slower
B. her watch moving faster
C. her watch got bigger
D. nothing unusual about the behavior of her watch

17. Two fixed navigation beacons mark the approach lane to a star. The beacons are in line with the star and are 40 million meters apart. A spaceship approaches the star with a relative velocity of $0.30c$ and passes the beacons. The passage of the ship between the beacons is timed by observers on the beacons. What is the time interval of the passage?

A. 610 ms
B. 560 ms
C. 444 ms
D. 240 ms

18. Calculate the kinetic energy of an electron moving at $0.737c$. (Use the rest energy of an electron = 511 keV)

A. 184 keV
B. 245 keV
C. 292 keV
D. 338 keV

19. In a common rest frame, event A happens on planet A, and event B occurs on planet B 100 years later. If the space-time interval between these two events is 90 light-years, how far apart are the 2 planets in their common rest frame?

A. 18 light-years
B. 22 light-years
C. 32 light-years
D. 44 light-years

20. At what speed are lengths contracted to half and times dilated by a factor of 2?

A. $0.866c$
B. $0.986c$
C. $1.007c$
D. $1.134c$

21. Suppose one twin takes a ride in a spaceship, traveling at a very high speed to a distant star and then back again while the other twin remains on Earth. Compared to the twin who remained on Earth, the twin who rode in the spaceship is:

A. younger than the Earth-twin
B. the same age as the Earth-twin
C. older than the Earth-twin
D. The ages cannot be determined from the information provided

22. Two satellites with equal rest masses of 100 kg travel toward each other in deep space. One is traveling at $0.650c$ and the other at $0.850c$. The satellites collide and somehow manage to stick. What is the rest mass of the combined object after the collision?

A. 100 kg
B. 208 kg
C. 246 kg
D. 312 kg

23. What is the momentum of a proton when it is moving with a speed of $0.60c$? (Use the mass of a proton $m_{\text{proton}} = 1.67 \times 10^{-27}$ kg)

A. 1.4×10^{-19} kg·m/s
B. 2.6×10^{-19} kg·m/s
C. 5.3×10^{-19} kg·m/s
D. 3.8×10^{-19} kg·m/s

24. How many joules of energy are required to accelerate a 1.0 kg rock from rest to a speed of $0.866c$? (Use speed of light $c = 3.0 \times 10^8$ m/s)

A. 2.6×10^{16} J
B. 9.0×10^{16} J
C. 6.8×10^{17} J
D. 4.3×10^{16} J

25. A spaceship traveling at $0.50c$ away from Earth launches a secondary rocket in the forward direction at $0.50c$ relative to the spaceship. As measured from the frame of Earth, how fast is the secondary rocket moving away from Earth?

A. $0.92c$
B. $0.56c$
C. $0.80c$
D. $0.72c$

26. Observer A sees a pendulum oscillating back and forth in a relativistic rocket and measures its period to be T_A. Observer B moves along with the rocket and measures the period of the pendulum to be T_B. What is true about these two-time measurements?

A. $T_A > T_B$
B. $T_A = T_B$
C. $T_A < T_B$
D. T_A could be greater or smaller than T_A depending on the direction of the motion

27. A spaceship travels to Alpha Centauri. The nearby star Alpha Centauri is 4.367 light-years from Earth. If the spaceship travels at one-half the speed of light for essentially all its expedition, how long was the ship gone, according to an observer on the spaceship?

A. 4.5 years
B. 7.6 years
C. 8.7 years
D. 11.3 years

28. A particle physicist observes cosmic rays creating a new particle high in the atmosphere, and the speed of this particle is measured at $0.997c$. It is unstable and is observed to decay in an average of 37.0 μs. If this particle were at rest in the laboratory, what would be its average lifetime?

A. 8.4 μs
B. 7.3 μs
C. 16.4 μs
D. 2.9 μs

29. If, instead of the correct formula for kinetic energy, someone carelessly uses the classical expression for kinetic energy for a particle moving at half the speed of light, by what magnitude percent will their calculation be in error? (Use the speed of light $c = 3.0 \times 10^8$ m/s)

A. 8%
B. 12%
C. 19%
D. 27%

30. In their typical rest frame, two stars are 90.0 light-years apart. If they appear to be 68.7 light-years apart from a spaceship, how fast is the spaceship moving?

A. $0.323c$
B. $0.646c$
C. $0.246c$
D. $0.810c$

31. Robert is riding in a spaceship with no windows, radios, or other means for him to observe or measure what is outside. How can he determine if the ship is stopped or moving at constant velocity?

A. He determines if the ship is moving by measuring the apparent velocity of light in the spaceship
B. He can determine if the ship is moving by checking his precision timepiece. If it is running slow, the ship is moving
C. He can determine if the ship is moving by lying down and measuring his height. If he is shorter than usual, the ship is moving
D. He cannot determine if the ship is stopped or moving at a constant velocity

32. Two satellites with equal rest masses are traveling toward each other in deep space. One is traveling at $0.550c$ and the other at $0.750c$. What is the speed of the combined object after the collision if the satellites collide and stick?

A. $0.108c$
B. $0.124c$
C. $0.245c$
D. $0.175c$

33. A spaceship approaches Earth with a speed of $0.50c$. A passenger in the spaceship measures his heartbeat as 70 beats per minute. What is his heartbeat rate according to an observer that is at rest relative to Earth?

A. 61 beats per minute
B. 55 beats per minute
C. 68 beats per minute
D. 76 beats per minute

34. A particle in a 453-m long particle accelerator is moving at $0.875c$. How long does the particle accelerator appear to the particle?

A. 148 m
B. 219 m
C. 264 m
D. 462 m

35. Someone in Earth's rest frame says that a spaceship's trip between two planets took 10.0 years, while an astronaut on the spaceship says that the trip took 5.78 years. What is the speed of the spaceship, assuming the planets are essentially at rest?

A. $0.244c$
B. $0.515c$
C. $0.816c$
D. $1.17c$

36. A super-speed train moves in a direction parallel to its length with a speed that approaches the speed of light. The length of the train, as measured by a stationary observer on the ground:

A. approaches infinity
B. increases due to the motion
C. decreases due to the motion
D. is not affected by the motion

37. A spaceship enters the solar system moving toward the sun at a constant speed relative to the sun. By its clock, the time elapsed between the time it crosses the orbit of Jupiter and the time it crosses the orbit of Mars is 50.0 minutes. As measured in the coordinate system of the sun, how many minutes did it take for the spaceship to travel that distance? (Use the radius of the Jupiter orbit = 778×10^9 m and the orbit of Mars = 228×10^9 m)

A. 28.4 minutes
B. 44.8 minutes
C. 58.6 minutes
D. 66.2 minutes

38. Two fixed navigation beacons mark the approach lane to a star. The beacons are in line with the star and are 65×10^6 meters apart. A spaceship approaches the star with a relative velocity of $0.90c$ and passes the beacons. As the ship passes the first beacon, the ship emits a short radar pulse toward the second beacon, and the radar echo is received at the ship. What is the time interval between the emission of the radar pulse and the reception of the radar echo as measured on the spaceship?

A. 210 ms
B. 420 ms
C. 130 ms
D. 100 ms

39. An electron that was initially at rest reaches a speed of $0.648c$. Through what potential difference would the electron need to be accelerated to reach this speed? (Use the rest energy of an electron = 511 keV)

A. 120 keV
B. 160 keV
C. 190 keV
D. 230 keV

40. A certain unstable particle has a mean lifetime of 1.52×10^{-6} s. This is the time interval between its creation in a nuclear process and its extinction into decay products, as measured in a frame of reference at rest with respect to the particle. A scientist on Earth observes an average such particle to travel 342 m in its lifetime. What is the speed of the particle relative to Earth? (Use the speed of light $c = 3.00 \times 10^8$ m/s)

A. $0.844c$
B. $0.486c$
C. $0.600c$
D. $0.754c$

41. The relativistic kinetic energy formula is valid:

I. for speeds near the speed of light
II. for subatomic particles, such as electrons and protons
III. at all speeds

A. I only
B. II only
C. III only
D. I and II only

42. A spaceship with a constant velocity of $0.800c$ relative to Earth travels to the star 4.30 light-years from Earth. Measured by a passenger on the ship, what distance does the spaceship travel?

A. 1.32 light-years
B. 2.58 light-years
C. 3.82 light-years
D. 4.60 light-years

43. A spaceship moves away from an asteroid with a speed of $0.80c$ relative to the asteroid. The spaceship then fires a missile with a speed of $0.50c$ relative to the spaceship. What is the speed of the missile measured by astronauts on the asteroid if the missile is fired away from the asteroid?

A. $0.43c$
B. $0.93c$
C. $0.18c$
D. $0.37c$

44. What is the total energy of an electron moving with a speed of $0.950c$? (Use the speed of light $c = 3.0 \times 10^8$ m/s and $m_{el} = 9.11 \times 10^{-31}$ kg)

A. 2.6×10^{-13} J
B. 7.4×10^{-14} J
C. 1.8×10^{-13} J
D. 3.6×10^{-14} J

45. A spaceship carrying a light clock moves at a speed of $0.960c$ relative to an observer on Earth. If the clock on the ship advances by 1.00 s as measured by the space travelers aboard the ship, how long did that advance take as measured by the observer on Earth?

A. 0.84 s
B. 1.9 s
C. 2.4 s
D. 3.6 s

46. In an "atom smasher," two particles collide head-on at relativistic speeds. The velocity of the first particle is $0.741c$ to the left, and the velocity of the second particle is $0.543c$ to the right (both speeds are measured in Earth's rest frame). How fast are the particles moving with respect to each other?

A. $0.916c$
B. $1.164c$
C. $0.744c$
D. $0.836c$

47. A particle is moving at approximately $0.86c$. When expressed as a percentage error for the difference between the erroneous and correct expressions relative to the correct one, what percentage is the Newtonian expression for momentum in error?

A. 49%
B. 38%
C. 57%
D. 64%

48. A spaceship, traveling at $0.100c$ away from a stationary space platform, launches a secondary rocket towards the station, with a speed of $0.560c$ relative to the spaceship. What is the speed of the secondary rocket relative to the space platform?

A. $0.364c$
B. $0.487c$
C. $0.524c$
D. $0.588c$

49. An electron slowing from $0.998c$ to $0.500c$ loses how much momentum?

A. 2.17 MeV/c
B. 4.36 MeV/c
C. 5.28 MeV/c
D. 7.77 MeV/c

50. A star is moving toward Earth at $0.90c$. The star emits light, which moves away from the star at the speed of light. What do Earth-based astronomers measure for the speed of this light?

A. $0.88c$
B. $1.1c$
C. $1.8c$
D. c

Notes for active learning

Diagnostic Tests

Diagnostic Test 1

This Diagnostic Test is designed to assess your proficiency on each topic. Use your test results and identify areas of strength and weakness to adjust your study plan and enhance your fundamental knowledge.

#	Answer:					Review	#	Answer:					Review
1:	A	B	C	D	E	___	31:	A	B	C	D	E	___
2:	A	B	C	D	E	___	32:	A	B	C	D	E	___
3:	A	B	C	D	E	___	33:	A	B	C	D	E	___
4:	A	B	C	D	E	___	34:	A	B	C	D	E	___
5:	A	B	C	D	E	___	35:	A	B	C	D	E	___
6:	A	B	C	D	E	___	36:	A	B	C	D	E	___
7:	A	B	C	D	E	___	37:	A	B	C	D	E	___
8:	A	B	C	D	E	___	38:	A	B	C	D	E	___
9:	A	B	C	D	E	___	39:	A	B	C	D	E	___
10:	A	B	C	D	E	___	40:	A	B	C	D	E	___
11:	A	B	C	D	E	___	41:	A	B	C	D	E	___
12:	A	B	C	D	E	___	42:	A	B	C	D	E	___
13:	A	B	C	D	E	___	43:	A	B	C	D	E	___
14:	A	B	C	D	E	___	44:	A	B	C	D	E	___
15:	A	B	C	D	E	___	45:	A	B	C	D	E	___
16:	A	B	C	D	E	___	46:	A	B	C	D	E	___
17:	A	B	C	D	E	___	47:	A	B	C	D	E	___
18:	A	B	C	D	E	___	48:	A	B	C	D	E	___
19:	A	B	C	D	E	___	49:	A	B	C	D	E	___
20:	A	B	C	D	E	___	50:	A	B	C	D	E	___
21:	A	B	C	D	E	___	51:	A	B	C	D	E	___
22:	A	B	C	D	E	___	52:	A	B	C	D	E	___
23:	A	B	C	D	E	___	53:	A	B	C	D	E	___
24:	A	B	C	D	E	___	54:	A	B	C	D	E	___
25:	A	B	C	D	E	___	55:	A	B	C	D	E	___
26:	A	B	C	D	E	___	56:	A	B	C	D	E	___
27:	A	B	C	D	E	___	57:	A	B	C	D	E	___
28:	A	B	C	D	E	___	58:	A	B	C	D	E	___
29:	A	B	C	D	E	___	59:	A	B	C	D	E	___
30:	A	B	C	D	E	___	60:	A	B	C	D	E	___

Page intentionally left blank

1. A chipmunk is running along a wire with constant acceleration. If it has an initial velocity of 0.4 m/s and a final velocity of 1.8 m/s after 4 s, how far does the chipmunk run in that time?

A. 2.3 m
B. 4.4 m
C. 6.7 m
D. 9.9 m
E. 7.2 m

2. An object is moving to the right in a straight line. What happens to the object if the net force acting on the object is directed to the right, but the magnitude of the force decreases with time?

A. It continues moving to the right with a constant speed
B. It stops and then begins moving to the left with increasing speed
C. It continues moving to the right with increasing speed
D. It continues moving to the right with decreasing speed
E. It moves to the left with increasing speed

3. What is the specific heat capacity if 150 kcal of heat raises the temperature of 3.0 kg of material by 200 °C?

A. 0.35 kcal/kg·°C
B. 0.5 kcal/kg·°C
C. 0.75 kcal/kg·°C
D. 1.15 kcal/kg·°C
E. 0.25 kcal/kg·°C

4. It takes 40 J of work to push a large suitcase 4 m across a floor. Assuming the suitcase is pushed in the same direction as it moves, what is the magnitude of the force on the suitcase?

A. 10 N
B. 4 N
C. 16 N
D. 26 N
E. 34 N

5. A simple pendulum with a length of 58 cm has a period of 2.5 s on Mars. What is the acceleration g due to gravity on Mars?

A. 0.7 m/s^2
B. 3.7 m/s^2
C. 6.8 m/s^2
D. 9.8 m/s^2
E. 17.2 m/s^2

6. What is the source of electromagnetic waves?

A. Crystalline fluctuations
B. Vibrating nuclei
C. Accelerating electric charges
D. Changes in atomic energy levels
E. All of the above

7. How far does a wavefront of an acoustical wave travel in 1 s if it has an f of 800 Hz and a λ of 0.5 m?

A. 400 m
B. 150 m
C. 75 m
D. 300 m
E. 600 m

8. What is the value of a resistor if, when a 12 V battery is connected to a resistor, 2 A of current flows through the resistor?

A. 1.5 ohms
B. 6 ohms
C. 7.5 ohms
D. 3 ohms
E. 24 ohms

9. What is the result of connecting two identical storage batteries in parallel in a circuit?

A. Half the voltage and half the total charge compared to a circuit from a single battery
B. Half the voltage, but the same total charge compared to a circuit from a single battery
C. The same voltage and the same total charge compared to a circuit from a single battery
D. The same voltage, but half the total charge compared to a circuit from a single battery
E. The same voltage, but twice the total charge compared to a circuit from a single battery

10. What is the power of the combination for 5 D and 3 D thin lenses that are positioned near each other?

A. 1/8 D
B. 5/3 D
C. 3/5 D
D. 8 D
E. 8/3 D

11. How long does it take for a rotating object to speed up from 15.0 rad/s to 33.3 rad/s if it has a uniform angular acceleration of 3.45 rad/s^2?

A. 3.45 s
B. 5.30 s
C. 8.35 s
D. 14.60 s
E. 20.80 s

12. A 1,450 kg cannon fires a 90 kg cannonball at 30 m/s. Assuming that frictional forces are negligible, and the cannonball is fired horizontally, what is the recoil velocity of the cannon?

A. 5.1 m/s
B. 1.2 m/s
C. 3.4 m/s
D. 1.9 m/s
E. 2.8 m/s

13. An engineer expends 900 J to lift a block to a height h. He then repeats the task using a simple non-motorized pulley system that reduces the input force by half. How much work must the engineer perform using the pulley system to lift the block to height h?

A. 100 J
B. 225 J
C. 900 J
D. 450 J
E. 1,400 J

14. A massless, ideal spring projects horizontally from a wall and connects to a 0.1 kg mass. The mass is oscillating in one dimension, such that it moves 0.4 m from one end of its oscillation to the other. What is the amplitude of the oscillation if it experiences 20 complete oscillations in 60 s?

A. 0.2 m
B. 0.4 m
C. 0.8 m
D. $\sqrt{0.4}$ m
E. 4 m

15. What is the wave speed of the standing wave shown if it is oscillating at 900 Hz on a string?

A. 280 m/s
B. 170 m/s
C. 360 m/s
D. 570 m/s
E. 420 m/s

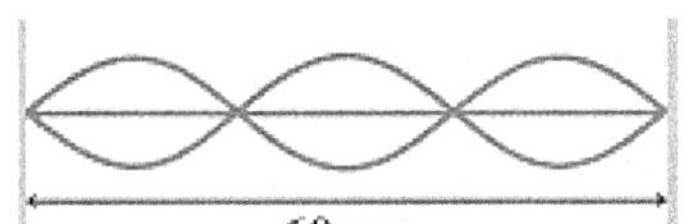

16. What is the density of water?

A. 1×10^3 kg/cm^3
B. 1×10^1 kg/cm^3
C. 1×10^2 kg/m^3
D. 1×10^{-3} g/m^3
E. 1×10^3 kg/m^3

17. Two stationary positive charges of $Q = 10^{-10}$ C are fixed and separated by a distance of $d = 4$ cm. What is the net electrostatic force on a charge $q = -10^{-9}$ C that is placed at a distance of $d / 2$ from each of the charges?

A. 5×10^{-10} N
B. 1.5×10^{-9} N
C. 0 N
D. 2.5×10^{-10} N
E. 1×10^{-9} N

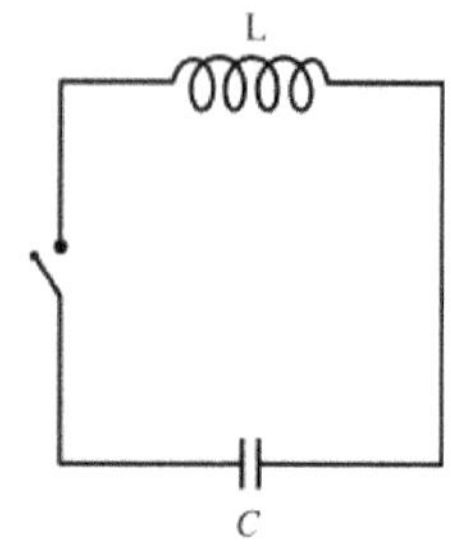

18. A capacitor, initially having a charge of Q on the left plate and a charge of $-Q$ on the right plate, is connected to a switch and an inductor.

Assuming that the resistance of the circuit is zero, which statement is true when the switch is closed?

A. Current flows back and forth through the inductor, with the magnitude of the current steadily decreasing and eventually becoming zero
B. Current flows until the left plate of the capacitor has a charge of $-Q$, and the current flows in the opposite direction, reversing again when the left plate has a charge of $+Q$. The cycle repeats
C. Charge flows out of the capacitor until the left plate is no longer charged, and then current ceases
D. Because the switch is closed, the current flows clockwise with a constant magnitude.
E. Current flows back and forth through the inductor, with the magnitude of current increased at a constant rate

19. What is observed when sunlight passes through a prism?

I. Bright spots and lines II. White light III. Continuous spectrum

A. I only
B. II only
C. III only
D. I and II only
E. I and III only

20. A solid uniform sphere of mass 120.0 kg and radius 1.7 m starts from rest and rolls without slipping down an inclined plane of vertical height 5.3 m; the sphere starts at the top of the ramp. What is the angular speed of the sphere at the bottom of the inclined plane? The moment of inertia of a solid sphere is $(2/5)mR^2$.

A. 0.81 rad/s
B. 1.7 rad/s
C. 2.9 rad/s
D. 4.3 rad/s
E. 5.1 rad/s

21. A heat engine performs the reversible cycle *abca* with 9 moles of an ideal gas. Path *ca* is an adiabatic process. The temperatures at points *a* and *b* are 300 K and 500 K, respectively. The volume at point *c* is 0.2 m^3. The adiabatic constant of the gas is 1.6. What is the heat absorbed by the gas in path *ca*?

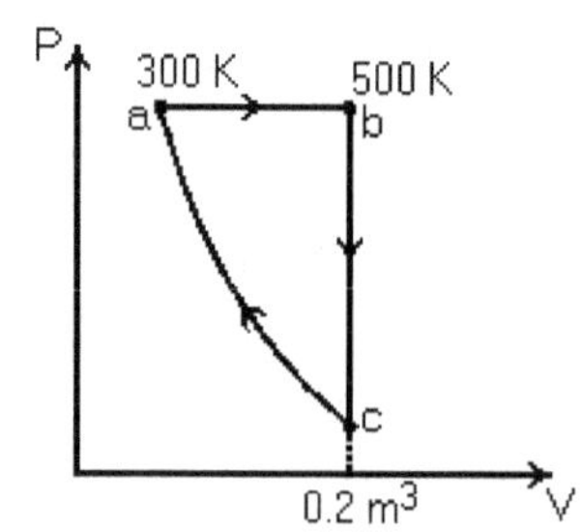

A. –10 kJ
B. 16 kJ
C. 10 kJ
D. 0 kJ
E. –16 kJ

22. Forces A, B and C act on a body. A fourth force, F, keeps the body in equilibrium. What is the x component of the force F?

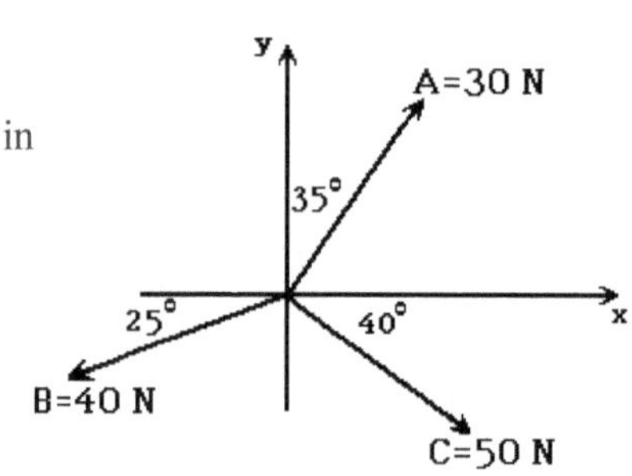

A. –20 N
B. –28 N
C. 20 N
D. 28 N
E. –32 N

23. In an organ pipe (cylindrical 0.2 m long tube 0.02 m in diameter, open at both ends), what is the λ of the 4th harmonic?

A. 0.08 m
B. 0.1 m
C. 0.05 m
D. 0.2 m
E. 0.8 m

24. What is the velocity of a fluid in a tube with a diameter of 3 mm for a pump that has a piston of 15 cm in diameter that moves 2 cm/s?

A. 50 m/s
B. 21 cm/s
C. 25 m/s
D. 50 cm/s
E. 5 cm/s

25. What is the essential difference between microwaves and blue light?

A. One is a form of radiation while the other is not
B. Blue light is a beam of photons, and microwaves are positrons
C. One has a positive charge while the other does not
D. One undergoes refraction while the other does not
E. There is no essential difference other than frequency and wavelength

26. The resistance of a variable resistor can be adjusted. A battery is connected to a variable resistor. The potential difference across the resistor and the current through it are recorded at variable resistance. Assume that the battery is an ideal potential source in series with an internal resistor. Under which circumstances is it valid to ignore the internal resistance of the battery?

I. small external resistance   II. large external resistance   III. low total current

A. I only
B. II only
C. III only
D. I and III only
E. II and III only

27. Which type of mirror produces an image that is always virtual, erect the same height as the object?

A. concave
B. convex
C. plane
D. spherical
E. none of the above

28. Total constructive interference is observed when two waves of the same frequency and wavelength are at a:

A. 45° phase difference
B. 90° phase difference
C. 180° phase difference
D. –90° phase difference
E. 0° phase difference

29. What is being described when a manufacturer advertises that their car goes *from zero to sixty in 7 s*?

A. Average acceleration
B. Instantaneous acceleration
C. Average speed
D. Change in speed
E. Displacement

30. Two forces are acting on an object as shown, with magnitudes $F_1 = 80$ N and $F_2 = 24$ N. What third force causes the object to be in equilibrium?

A. 3.3 N up
B. 3.3 N down
C. 56 N up
D. 56 N down
E. 124 N up

31. Compared to falling on a wooden floor, a wine glass may not break when it falls on a carpeted floor because of the:

I. lesser impulse in stopping
II. longer time to stop
III. decreased acceleration due to gravity

A. I only
B. II only
C. III only
D. I and II only
E. I and III only

32. A 6.5 kg bag of groceries is carried 1.1 m above the ground at a constant velocity across a 3.8 m room. How much work was done on the bag in this process?

A. 183 J
B. 155 J
C. 0 J
D. 84 J
E. 22.7 J

33. A wave that has a lower frequency has a:

A. higher amplitude
B. shorter period
C. slower velocity
D. shorter wavelength
E. longer wavelength

34. A rod has a length of 2.0000 m at 20 °C. The rod length increases to 2.0005 m when the temperature increases to 40 °C. What is the coefficient of thermal expansion α of the material for the rod?

A. 1.25×10^{-5}/K
B. 5×10^{-3}/K
C. 0.75×10^{-5}/K
D. 2.5×10^{-5}/K
E. 0.75×10^{-3}/K

35. An object is viewed at various distances using a concave mirror with a focal length of 12 m. Relative to the mirror, where is the image when the object is 6 m in front of the mirror?

A. 12 m behind
B. 12 m in front
C. 6 m behind
D. 6 m in front
E. 24 m behind

36. What effect does the addition of resistors in parallel have on a circuit?

A. Decreases the current
B. Increases the current
C. Decreases the voltage
D. Increases the voltage
E. No change in the current

37. The electric field at point P due to a point charge of Q a distance R away has a magnitude of E. Which statement is true to double the magnitude of the field at P?

A. Reduce the charge to $Q/2$
B. Reduce the distance to R/2
C. Double the distance to 2R while reducing the charge to $Q/2$
D. Double the charge to $2Q$
E. Double the charge to $2Q$ while reducing the distance to R/2

38. If 20% of the volume of a floating buoy is above the surface of a liquid, then the density of the buoy is what percent of the density of the surrounding liquid?

A. 40%
B. 90%
C. 60%
D. 70%
E. 80%

39. A pipe of length L that is open at both ends resonates at its fundamental frequency. Which statement about the sound is correct?

A. The λ is $L/2$, and there is a displacement node at the midpoint of the pipe
B. The λ is $L/2$, and there is a displacement antinode at the midpoint of the pipe
C. The λ is $2L$, and there is a displacement node at the midpoint of the pipe
D. The λ is $2L$, and there is a displacement antinode at the midpoint of the pipe
E. The λ is $4L$, and there is a displacement node at the midpoint of the pipe

40. Wave interference occurs for:

I. sound waves II. water waves III. light waves

A. I only
B. II only
C. III only
D. I and III only
E. I, II and III

41. What is the spring constant value if a force of 30 N stretches a spring 0.75 m from equilibrium?

A. 85 N/m
B. 25 N/m
C. 35 N/m
D. 40 N/m
E. 30 N/m

42. In a perfectly inelastic collision, two moving objects of unequal mass (A and B) collide and immediately come to rest. Before colliding, object A was traveling at speed 5 times that of object B. Which is the ratio of the mass of object A to that of object B?

A. 50 : 1
B. 5 : 1
C. 1 : 5
D. 1 : 50
E. $\sqrt{5}$: 1

43. If the mass of an object does not change, a constant net force on the object produces constant:

I. velocity II. acceleration III. speed

A. I only
B. II only
C. III only
D. I and II only
E. I and III only

44. A chestnut falls straight down from a chestnut tree growing on a 20° slope. What is the chestnut's impact velocity component parallel to the ground if it hits the ground with a speed of 16 m/s?

A. 5.5 m/s
B. 8.6 m/s
C. 4.1 m/s
D. 11.4 m/s
E. 10.6 m/s

45. In a given reversible process, the temperature of an ideal gas is constant as the gas is compressed to a smaller volume. Which statement is correct?

A. The process is adiabatic
B. The pressure remains constant
C. The gas releases heat to its surroundings
D. The gas absorbs heat from its surroundings
E. None of the above

46. A heavy pile driver starting from rest smashes into a pile with a force that depends on the following:

I. distance the pile driver falls
II. initial potential energy of the driver
III. initial height of the driver

A. I only
B. II only
C. III only
D. I and III only
E. I, II and III

47. Which of the following is an accurate statement?

A. The fundamental frequency is the highest frequency at which a system naturally vibrates
B. The air in an organ pipe can vibrate at an infinite number of frequencies
C. A system, like a vibrating string, has only one possible frequency
D. For a singer to break a wine glass, the amplitude of sound must be greater than the amplitude of vibration for a wine glass
E. None of the above

48. The unit used to measure electrical resistance is:

A. watts
B. ohms
C. AC
D. volts
E. amps

49. A myopic girl wears eyeglasses that allow her to have clear distant vision. The power of the lenses of her eyeglasses is –3 diopters. Without eyeglasses, the far point of the girl is:

A. 0.32 m
B. 0.17 m
C. 0.75 m
D. 0.33 m
E. 0.50 m

50. A woman starts her car from rest and accelerates at a constant 2.5 m/s^2 for 8 s to get to her cruising speed. What is her velocity 8 s after beginning her trip?

A. 1.8 m/s
B. 5.5 m/s
C. 12.5 m/s
D. 7.5 m/s
E. 20 m/s

51. What is the velocity of the cars immediately after impact for a perfectly inelastic collision between a 2,500 kg car moving North at 7 m/s and a 2,000 kg car moving South at 14 m/s?

A. 2.3 m/s S
B. 9.4 m/s S
C. 5.4 m/s N
D. 9.4 m/s N
E. 11 m/s N

52. Grandfather clocks are adjusted by moving the weight at the bottom of the pendulum up or down. Which of the following weight adjustments would make a slow-running clock more accurate?

A. Move weight down
B. Move weight up
C. Remove half the mass from the weight
D. Add double the mass to the weight
E. Increase the amplitude of swing by a significant amount

53. Why does the paper rise when the air is blown above a paper strip?

A. Air above the paper moves slower, while the pressure remains constant
B. Air above the paper moves slower, and the pressure is higher
C. Air above the paper moves faster, while the pressure remains constant
D. Air above the paper moves slower, and the pressure is lower
E. Air above the paper moves faster, and the pressure is lower

54. If the internal resistance of the battery is 20 Ω, what current flows in a 4 V battery if the circuit is shorted?

A. 5 A
B. 3.2 A
C. 80 A
D. 0.2 A
E. 7.3 A

55. Three solid, uniform, cylindrically shaped flywheels, each of mass 65.0 kg and radius 1.47 m, rotate independently around a common axis. Two flywheels rotate in one direction at 3.83 rad/s; the other rotates in the opposite direction at 3.42 rad/s. What is the magnitude of the net angular momentum of the system?

A. 168.0 kg·m²/s
B. 298.0 kg·m²/s
C. 456.0 kg·m²/s
D. 622.0 kg·m²/s
E. 882.0 kg·m²/s

56. A glass flask has a volume of 450 ml at a temperature of 22 °C. The flask contains 442 ml of mercury at an equilibrium temperature of 22 °C. The temperature is raised until the mercury reaches the 450 ml reference mark. What is the temperature at which this occurs? (Use coefficients of volume expansion α of mercury = 18×10^{-5} K^{-1} and α of glass = 2×10^{-5} K^{-1})

A. 103 °C
B. 152 °C
C. 92 °C
D. 135 °C
E. 116 °C

57. Which graph represents a constant positive acceleration?

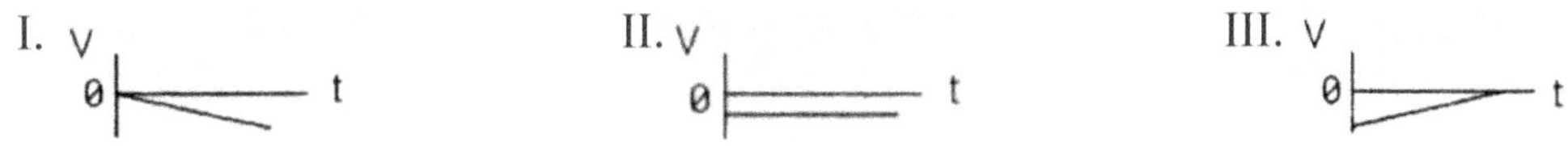

A. I only
B. II only
C. III only
D. II and III only
E. I, II and III

58. A Ferrari accelerates from 0 to 100 km/h in 4.8 s. What force does a 68 kg passenger experience during the acceleration?

A. 342 N
B. 311 N
C. 394 N
D. 82.0 N
E. 7 N

59. What happens to an object if a constant net torque is applied?

A. Rotates with increasing angular acceleration
B. Rotates with constant angular acceleration
C. Rotates with constant angular velocity
D. Rotates with increasing linear velocity
E. Increases its moment of inertia

60. What is the focal length of a lens if a candle is placed at 4 m from the lens and the image is 2 m from the other side of the lens?

A. –2 m
B. –4/3 m
C. 3/4 m
D. 2 m
E. 4/3 m

Use the answer key to check your answers. Review the explanations in detail, focusing on questions you didn't answer correctly. Note the topic of those questions. Complete this BEFORE taking the next Diagnostic Test.

Notes for active learning

Diagnostic Test 2

This Diagnostic Test is designed to assess your proficiency on each topic. Use your test results and identify areas of strength and weakness to adjust your study plan and enhance your fundamental knowledge.

#	Answer:					Review	#	Answer:					Review
1:	A	B	C	D	E	___	31:	A	B	C	D	E	___
2:	A	B	C	D	E	___	32:	A	B	C	D	E	___
3:	A	B	C	D	E	___	33:	A	B	C	D	E	___
4:	A	B	C	D	E	___	34:	A	B	C	D	E	___
5:	A	B	C	D	E	___	35:	A	B	C	D	E	___
6:	A	B	C	D	E	___	36:	A	B	C	D	E	___
7:	A	B	C	D	E	___	37:	A	B	C	D	E	___
8:	A	B	C	D	E	___	38:	A	B	C	D	E	___
9:	A	B	C	D	E	___	39:	A	B	C	D	E	___
10:	A	B	C	D	E	___	40:	A	B	C	D	E	___
11:	A	B	C	D	E	___	41:	A	B	C	D	E	___
12:	A	B	C	D	E	___	42:	A	B	C	D	E	___
13:	A	B	C	D	E	___	43:	A	B	C	D	E	___
14:	A	B	C	D	E	___	44:	A	B	C	D	E	___
15:	A	B	C	D	E	___	45:	A	B	C	D	E	___
16:	A	B	C	D	E	___	46:	A	B	C	D	E	___
17:	A	B	C	D	E	___	47:	A	B	C	D	E	___
18:	A	B	C	D	E	___	48:	A	B	C	D	E	___
19:	A	B	C	D	E	___	49:	A	B	C	D	E	___
20:	A	B	C	D	E	___	50:	A	B	C	D	E	___
21:	A	B	C	D	E	___	51:	A	B	C	D	E	___
22:	A	B	C	D	E	___	52:	A	B	C	D	E	___
23:	A	B	C	D	E	___	53:	A	B	C	D	E	___
24:	A	B	C	D	E	___	54:	A	B	C	D	E	___
25:	A	B	C	D	E	___	55:	A	B	C	D	E	___
26:	A	B	C	D	E	___	56:	A	B	C	D	E	___
27:	A	B	C	D	E	___	57:	A	B	C	D	E	___
28:	A	B	C	D	E	___	58:	A	B	C	D	E	___
29:	A	B	C	D	E	___	59:	A	B	C	D	E	___
30:	A	B	C	D	E	___	60:	A	B	C	D	E	___

Page intentionally left blank

1. Ignoring air resistance, if it takes 16 s for a package to strike the ground, how high above the ground was the package when it was thrown upward from a stationary helicopter at 15 m/s? (Use the acceleration due to gravity $g = 10\ m/s^2$)

A. 920 m
B. 740 m
C. 1,400 m
D. 1,040 m
E. 1,800 m

2. What is the mass of the car if a tow truck exerts a force of 3,000 N on a car and accelerates it at 2 m/s^2?

A. 500 kg
B. 1,000 kg
C. 1,500 kg
D. 2,000 kg
E. None of the above

3. A bimetallic strip, consisting of metal A on the top and metal B on the bottom, is rigidly attached to a wall at the left, as shown. The coefficient of linear thermal expansion of metal A is greater than that of metal B. Which statement is true if the strip is heated uniformly?

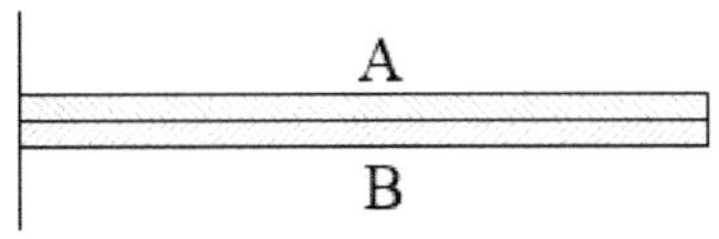

A. remains horizontal but increases in length
B. remains horizontal but decreases in length
C. curves upward
D. curves downward
E. bends in the middle

4. A massless, ideal spring has a spring constant of 4,600 N/m as it hangs from the ceiling. How much does its potential energy increase if a 30 kg mass is attached to it? (Use the acceleration due to gravity $g = 10\ m/s^2$)

A. 9.8 J
B. 6.3 J
C. 19.4 J
D. 12.4 J
E. 4.3 J

5. What is the speed of a water wave if a floating object oscillates up and down two complete cycles in 1 s as the water wave with a wavelength of 6 m passes?

A. 1 m/s
B. 0.17 m/s
C. 12 m/s
D. 18 m/s
E. 3 m/s

6. By what value does the sound level decrease if the intensity of the sound wave decreases by a factor of 10?

A. 0.1 dB
B. 10 dB
C. 1 dB
D. 100 dB
E. 0.01 dB

7. A piece of wood is floating in a bathtub. The second piece of wood sits on top of the first piece and does not touch the water. If the top piece is taken off and placed in the water, what happens to the water level in the tub?

A. It remains the same
B. It decreases
C. It increases
D. Requires the volumes of the two pieces of wood
E. Requires the densities of the two pieces of wood

8. A point charge Q_1 has a charge of –1 μC. What is the number of excess electrons in charge Q_1? (Use the charge of an electron $e = -1.60 \times 10^{-19}$ C)

A. 2×10^{11}
B. 2×10^{12}
C. 6.3×10^{12}
D. 6.3×10^{13}
E. 2×10^{13}

9. The capacitance of a capacitor depends on its:

I. energy stored II. charge III. potential difference across it

A. I only
B. II only
C. II and III only
D. I and III only
E. none of the above

10. What is the radius of curvature for a flat plane mirror?

A. Imaginary
B. Zero
C. Infinite
D. Negative
E. Undefined

11. A cylinder and a sphere are released simultaneously at the top of an inclined plane. Which reaches the bottom first when rolling down the inclined plane without slipping?

A. The one of smallest diameter
B. The one of greatest mass
C. The disk
D. The sphere
E. They reach the bottom at the same time

12. Consider two uniform solid spheres with the same diameter, but one has twice the mass of the other. What is the ratio of the larger moment of inertia to the smaller?

A. $\sqrt{2} : 1$
B. 4 : 1
C. 10 : 1
D. 2 : 1
E. 16 : 1

13. Using 1,000 J of work, an elevator is raised from the ground to the second floor in 40 s. How much power does the elevator use?

A. 100 W
B. 25 W
C. 50 W
D. 1,000 W
E. 200 W

14. A ball is attached to an ideal spring and oscillates with a period T. What is the new period produced by doubling the ball's mass?

A. $T\sqrt{2}$
B. T
C. T/2
D. 2T
E. $T/\sqrt{2}$

15. Larry drops a mailing tube to the floor, and the tube produces a musical note. The mailing tube is 1.5 m long with a cylindrical cross-sectional diameter of 3.5 cm. It is sealed at one end and open at the other. What is the wavelength of the third harmonic? (Use the speed of sound in air v = 340 m/s)

A. 1.5 m
B. 2 m
C. 3.5 m
D. 4 m
E. 7.5 m

16. A compressed gas with a total mass of 115 kg is stored in a spherical container with a radius of 0.6 m. What is the density of the compressed gas?

A. 147 kg/m^3
B. 183 kg/m^3
C. 151 kg/m^3
D. 127 kg/m^3
E. 267 kg/m^3

17. Electrons are made to flow in a wire when there is:

I. a potential difference across its ends
II. an imbalance between positive and negative charges in the wire
III. more final kinetic energy than initial kinetic energy in the wire

A. I only
B. II only
C. III only
D. I and II only
E. I and III only

18. A Coulomb per second is the same as a(n):

A. Newton
B. volt-second
C. ampere
D. watt
E. volt per second

19. What is observed when an object is placed precisely at its center of curvature in front of a concave mirror?

A. The image is not seen because it is focused at a different distance
B. The image is seen, but it appears smaller
C. The image is seen, but it appears larger
D. The image is seen as the same size, and it is upright
E. The image is seen as the same size, but it is inverted

20. What is the angular speed of a flywheel turning at 813.0 rpm?

A. 8.33 rad/s
B. 56.23 rad/s
C. 33.84 rad/s
D. 85.14 rad/s
E. 116.48 rad/s

21. The process shown on the pressure *vs.* volume diagram represents which type of expansion?

A. Isobaric
B. Isometric
C. Isothermal
D. Adiabatic
E. Isochoric

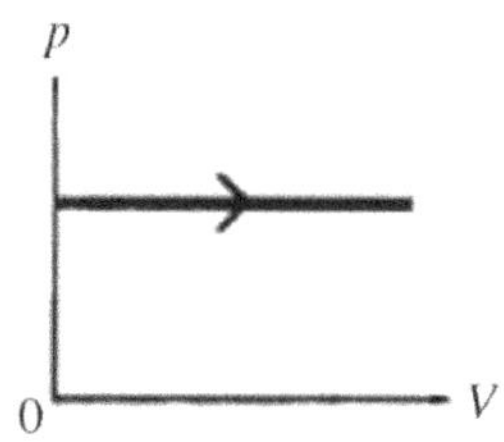

22. A block is on a frictionless table on Earth. The block accelerates at 4 m/s^2 when a 20 N horizontal force is applied. The block and table are then set on the Moon, where the acceleration due to gravity is 1.62 m/s^2. What is the weight of the block on the Moon?

A. 11 N
B. 9.7 N
C. 5.6 N
D. 6.2 N
E. 8.1 N

23. In the figure, the frequency is:

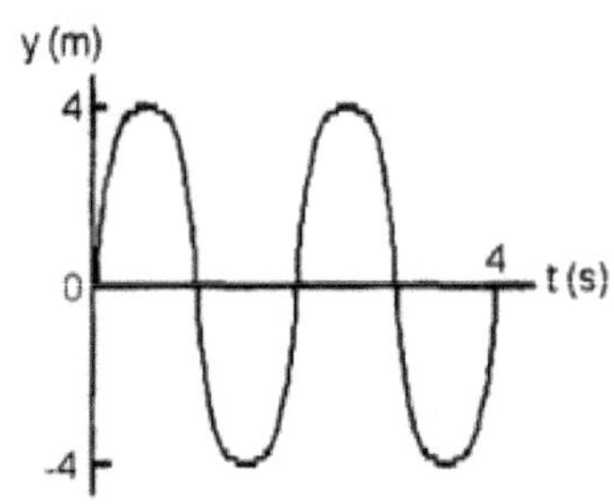

A. 2 Hz
B. 4 Hz
C. 0.5 Hz
D. 1 Hz
E. 1.5 Hz

24. Consider a tiny hole in the bottom of a tank 18 cm in diameter and filled with water to a height of 80 cm. What is the speed at which the water exits the tank through the hole? (Use the acceleration due to gravity $g = 9.8$ m/s^2)

A. 4 m/s
B. 17 m/s
C. 42 m/s
D. 36 m/s
E. 7 m/s

25. Two charges, $Q_1 = 3.4 \times 10^{-10}$ C and $Q_2 = 6.8 \times 10^{-9}$ C, are separated by 1 cm. Let F_1 be the magnitude of the electrostatic force felt by Q_1 due to Q_2 and let F_2 be the magnitude of the electrostatic force felt by Q_2 due to Q_1. What is the ratio of F_1 / F_2?

A. 2
B. 1
C. 16
D. 8
E. 4

26. A component with an 18 Ω resistor is rated for use at power levels not exceeding 16 W. How much current can safely flow through the component?

A. 0.12 A
B. 2.8 A
C. 1.31 A
D. 0.94 A
E. 12.1 A

27. How are the wavelength and frequency of light related?

A. Wavelength increases as the frequency increases
B. Wavelength increases as the frequency decreases
C. Wavelength is one-fourth of the frequency
D. Wavelength is four times the frequency
E. Wavelength is not related to the frequency

28. What is the approximate wavelength of a wave that has a speed of 360 m/s and a period of 4.2 s?

A. 85.7 m
B. 1.86 m
C. 1,512 m
D. 288.6 m
E. 422.1 m

29. A merry-go-round is 18 m in diameter, with a child sitting on the outer edge. If the merry-go-round makes 8.3 rev/min, what is the velocity of the child?

A. 1.9 m/s
B. 7.8 m/s
C. 3.9 m/s
D. 14.6 m/s
E. 18.3 m/s

30. What is the magnitude of the instantaneous force with which a ball hits a bat if a batter hits a ball pitched with a 1,700 N instantaneous force?

A. Greater than 1,700 N
B. Zero
C. Equal to √1,700 N
D. Less than 1,700 N
E. Equal to 1,700 N

31. A 2.8 kg block, moving on a frictionless surface with a speed (v_i) of 8.5 m/s, makes a perfectly elastic collision with a block of mass *m* at rest. After the collision, the 2.8 kg block recoils with a speed (v_f) of 1.1 m/s. What is the mass of block *m*?

A. 22.8 kg
B. 14.4 kg
C. 17.8 kg
D. 3.6 kg
E. 12.2 kg

32. If the height of a frictionless incline is *h*, determine the work done by the force due to gravity *F* as the crate of mass *m* slides down the incline.

A. $mgh \sin^2$
B. mgh
C. $mgh \cos^2$
D. $mgh \sin$
E. $mgh \tan$

33. What is the speed of a wave if it has a wavelength of 25 cm and a frequency of 1.6 kHz?

A. 400 m/s
B. 1,400 m/s
C. 0.4 m/s
D. 14 m/s
E. 28 m/s

34. A glass beaker of unknown mass contains 60 ml of water. The system absorbs 2,200 cals of heat, and the temperature rises 25 °C as a result. What is the mass of the beaker? (Use specific heat for glass = 0.18 cal/g·°C and specific heat for water = 1.0 cal/g·°C)

A. 675 g
B. 540 g
C. 460 g
D. 370 g
E. 156 g

35. What is the radius of curvature for a concave mirror with a focal length of 20 cm?

A. 20 cm
B. –40 cm
C. 40 cm
D. 10 cm
E. –20 cm

36. What is the power dissipated in the 2 Ω resistor in the circuit?

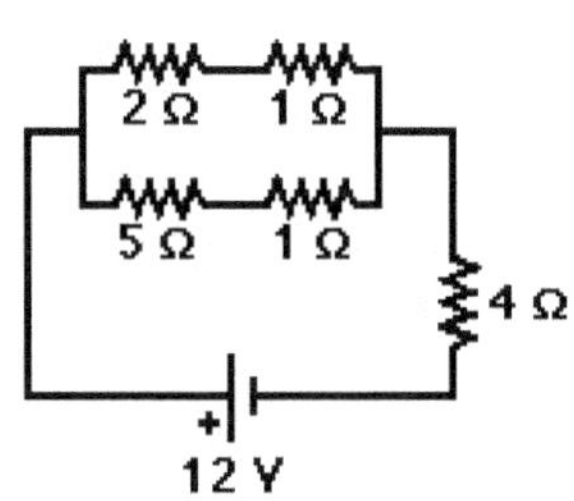

A. 6.3 W
B. 5.7 W
C. 3.5 W
D. 7.8 W
E. 8.5 W

37. What causes an object to become electrostatically charged?

A. Charge is created
B. Protons are transferred
C. Electrons are transferred
D. Protons, and electrons are transferred
E. Charge is destroyed

38. What is the gauge pressure 10 m below the surface of the ocean? (Use the acceleration due to gravity $g = 9.8$ m/s^2, the density of water $\rho = 10^3$ kg/m^3 and $P_{atm} = 1.01 \times 10^5$ Pa)

A. 1.4×10^5 Pa
B. 2.3×10^5 Pa
C. 9.8×10^4 Pa
D. 6.5×10^4 Pa
E. 5×10^6 Pa

39. If a 75 cm string, fixed at both ends between two adjacent nodes, supports a standing wave, what is the wavelength λ?

A. 15 cm
B. 300 cm
C. 7.5 cm
D. 200 cm
E. 150 cm

40. The speed of sound in steel is 4,900 m/s. What is the wavelength of a sound wave in the steel of frequency 640 Hz?

A. 1.3 m
B. 0.83 m
C. 7.7 m
D. 4.2 m
E. 2.6 m

41. Ignoring the force of friction, what is the work done by Stacey when she carries a 15 kg mass and walks along the x-axis for a distance of 100 m with a constant velocity of 2.5 m/s?

A. 0 J
B. 5 J
C. 30 J
D. 50 J
E. 450 J

42. Hannah is in karate and delivers a swift blow to sever a cement block in two with her bare hand. The magnitude is the same for the:

I. time of impact on both the block and Hannah's hand
II. force on both the block and Hannah's hand
III. impulse on both the block and Hannah's hand

A. I only
B. II only
C. III only
D. I and III only
E. I, II and III

43. Karen is driving a car along a road when she realizes, almost too late, that she needs to make a left-hand turn. Karen quickly turns the car's steering wheel, and the textbooks in the passenger seat go crashing against the passenger door. Which of the following are true?

I. The textbooks were pushed against the door by a centrifugal force
II. The textbooks were pushed against the door by a centripetal force
III. The forces acting on the textbooks when they crashed against the door were gravity, the normal force, and a force toward the right

A. I and III only
B. II and III only
C. III only
D. I, II and III
E. None of the above

44. How far does a car travel while accelerating if it accelerates from 5 m/s to 21 m/s at a rate of 4 m/s²?

A. 112 m
B. 35 m
C. 71 m
D. 52 m
E. 153 m

45. The coefficient of linear expansion of lead is 30×10^{-6} K^{-1}. What change in temperature causes a 10 m long lead bar to change in length by 3 mm?

A. 7.5 K
B. 20 K
C. 10 K
D. 5 K
E. 40 K

46. A constant force of 10 N is applied horizontally for 15 s on a 2 kg toy cart so that the cart begins to move along the level frictionless floor. What is the kinetic energy of the cart just after the 15 s?

A. 1,540 J
B. 2,300 J
C. 3,450 J
D. 2,985 J
E. 5,625 J

47. Mary has tuned the B string of a guitar to the fundamental frequency of an E (680 Hz). The E string is left untuned. When she plucks the B string and the untuned E string together, she hears a note that changes from loud to soft to loud twice per second. What is an allowable fundamental *f* on the untuned E string?

A. 2 Hz
B. 340 Hz
C. 680 Hz
D. 682 Hz
E. 1,360 Hz

48. A positive reaction charge Q is held fixed at the origin. A positive charge q is on the positive x-axis and released. Assuming no friction, which statement describes the theoretical acceleration of q after its release?

A. Decreases and then increases
B. Remains constant for a while and then decreases, eventually reaching zero
C. Decreases and eventually reaches zero
D. Decreases forever but never reaches zero
E. Acceleration remains unchanged

49. Which statement about a single thin lens is correct?

A. A diverging lens always produces a virtual upright image
B. A converging lens sometimes produces a real upright image
C. A diverging lens sometimes produces a virtual upright image
D. A converging lens always produces a real upright image
E. A diverging lens produces a virtual upright image only if the object is located between the lens and its focal point

50. What is the magnitude and direction of the combined vectors A and B if vector A has 5 units and is directed to the North (N) while vector B has 11 units and is directed to the South (S)?

A. 16 units, S
B. 16 units, N
C. 6 units, N
D. 6 units, S
E. 2.2 units, N

51. The conservation of momentum is most closely related to:

I. Newton's First Law II. Newton's Second Law III. Newton's Third Law

A. I only
B. II only
C. III only
D. I and II only
E. I and III only

52. At which point in a swinging pendulum's arc does it have the most energy?

I. At its lowest point, it has the most potential energy
II. The energy is the same at all points of the arc
III. At its highest point, it has the most kinetic energy

A. I only
B. II only
C. III only
D. I and II only
E. I and III only

53. If the amount of fluid flowing through a tube remains constant, how does the fluid speed change if the tube's diameter increases from 3 cm to 8 cm?

A. Increases by a factor of 5
B. Increases by a factor of 9
C. Decreases by a factor of 5
D. Decreases by a factor of 7
E. Remains the same

54. Voltage can be induced in a wire by:

I. changing the current in a nearby wire
II. moving a magnet near the wire
III. moving the wire near a magnet

A. I only
B. II only
C. III only
D. II and III only
E. I, II and III

55. A machinist turns the power on for a grinding wheel at time $t = 0$ s. The wheel accelerates uniformly from rest for 10.0 s and reaches the operating angular speed of 96.0 rad/s. The wheel is run at that angular velocity for 40.0 s, and then power is shut off. The wheel slows down uniformly at 1.5 rad/s^2 until the wheel stops. For how long after the power is shut off does it take the wheel to stop?

A. 56.0 s
B. 64.0 s
C. 72.0 s
D. 82.0 s
E. 90.0 s

56. A container of an ideal gas at standard temperature and pressure undergoes an isothermal expansion. How much work does it do if its entropy changes by 2.6 J/K? (Use standard temperature = 273 K and pressure = 1 atm)

A. 1.4×10^3 J
B. -1×10^3 J
C. 7.1×10^2 J
D. 0 J
E. -1.4×10^3 J

57. Jack stands on a bridge and throws a rock straight down. The rock leaves Jack's hand at $t = 0$. Which of the graphs represents the velocity of the stone as a function of time?

A.
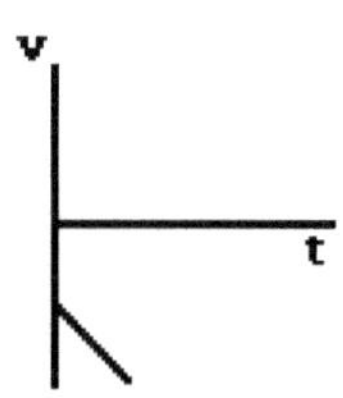

B.
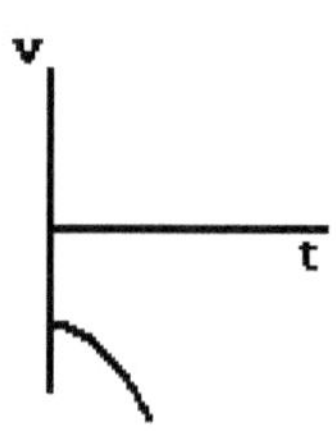

C.
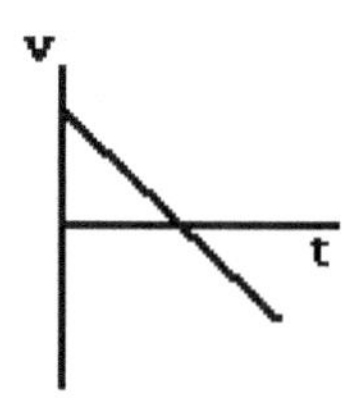

D.
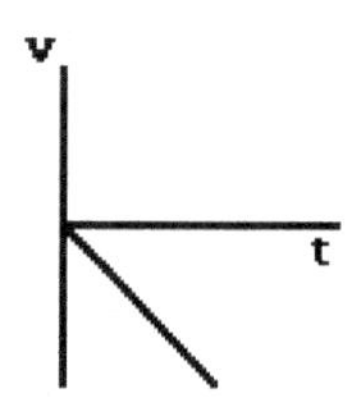

E.
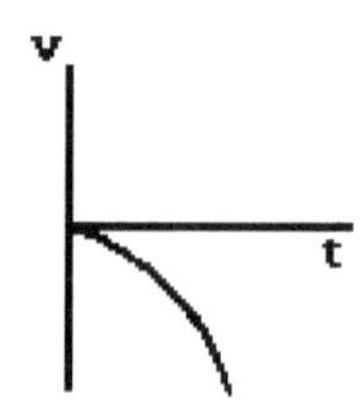

58. A 100 g ball of clay is thrown horizontally with a speed of 50 m/s toward a 900 g block resting on a frictionless surface. It hits the block and sticks. The clay exerts a constant force on the block during the 10 ms it takes for the clay to come to rest relative to the block. After 10 ms, the block and the clay slide along the surface as a single system. What is the force of the clay on the block during the collision?

A. 500 N
B. 2,250 N
C. 5,400 N
D. 4,500 N
E. 450 N

59. Cart A (2 kg) and Cart B (2.5 kg) run along a frictionless, level, one-dimensional track. Cart B is initially at rest, and Cart A travels 0.6 m/s toward the right when it encounters Cart B. After the collision, Cart A is at rest. What is the final velocity of Cart B?

A. 0.48 m/s
B. 0.36 m/s
C. 0.75 m/s
D. 0.64 m/s
E. 1.1 m/s

60. A girl of height h stands in front of a plane mirror. What must the minimum length of the mirror be so that she can view her entire body?

A. $\frac{1}{4}h$
B. $2h$
C. $\frac{1}{2}h$
D. h
E. Depends on her distance from the mirror

Use the answer key to check your answers. Review the explanations in detail, focusing on questions you didn't answer correctly. Note the topic of those questions.

Notes for active learning

Answer Keys and Detailed Explanations

Answer Keys

Translational Motion

1: D	11: C	21: D	31: B	41: C	51: A
2: B	12: B	22: A	32: C	42: A	52: A
3: B	13: A	23: B	33: D	43: D	53: B
4: A	14: C	24: D	34: A	44: B	54: D
5: C	15: B	25: B	35: C	45: A	55: A
6: C	16: D	26: A	36: D	46: D	56: A
7: D	17: A	27: C	37: D	47: B	57: C
8: B	18: B	28: D	38: B	48: C	58: C
9: A	19: D	29: D	39: C	49: D	59: B
10: B	20: C	30: C	40: A	50: C	60: D

Force and Motion

1: B	11: A	21: B	31: C	41: A	51: C
2: D	12: C	22: A	32: A	42: C	52: A
3: A	13: B	23: C	33: D	43: B	53: D
4: C	14: C	24: B	34: A	44: C	54: C
5: B	15: B	25: D	35: D	45: D	55: B
6: D	16: A	26: A	36: A	46: C	56: D
7: A	17: D	27: D	37: D	47: D	57: B
8: D	18: C	28: D	38: A	48: A	58: D
9: A	19: A	29: A	39: B	49: C	59: C
10: B	20: B	30: D	40: C	50: A	60: B
					61: A

Equilibrium and Momentum

1: A	11: D	21: D	31: C	41: B	51: D
2: D	12: C	22: B	32: D	42: A	52: D
3: D	13: D	23: A	33: C	43: C	53: A
4: C	14: B	24: D	34: B	44: B	54: D
5: D	15: D	25: D	35: D	45: D	55: B
6: B	16: D	26: A	36: C	46: C	56: C
7: D	17: B	27: D	37: B	47: B	57: D
8: A	18: D	28: C	38: C	48: A	58: A
9: B	19: B	29: A	39: D	49: D	59: D
10: C	20: C	30: B	40: A	50: C	60: B

Rotational Motion

1: D	11: A	21: B	31: D	41: C	51: D
2: B	12: D	22: D	32: D	42: D	52: B
3: A	13: C	23: B	33: B	43: C	53: D
4: D	14: B	24: A	34: C	44: C	54: D
5: C	15: D	25: A	35: D	45: B	55: A
6: C	16: D	26: A	36: B	46: C	56: B
7: D	17: D	27: A	37: C	47: B	57: B
8: A	18: A	28: C	38: B	48: D	58: C
9: C	19: D	29: C	39: B	49: B	
10: B	20: D	30: C	40: D	50: D	

Work and Energy

1: D	11: D	21: B	31: B	41: A	51: D
2: B	12: B	22: D	32: C	42: A	52: C
3: A	13: A	23: B	33: D	43: C	53: C
4: B	14: C	24: C	34: A	44: B	54: B
5: C	15: B	25: D	35: B	45: D	55: C
6: A	16: D	26: A	36: B	46: D	56: D
7: D	17: A	27: D	37: C	47: A	57: C
8: C	18: A	28: C	38: D	48: A	58: C
9: A	19: D	29: B	39: A	49: C	59: A
10: B	20: A	30: C	40: B	50: B	60: D

Periodic Motion

1: B	11: D	21: B	31: B	41: A	51: A
2: D	12: A	22: C	32: D	42: C	52: C
3: D	13: B	23: B	33: A	43: B	53: B
4: C	14: C	24: A	34: D	44: D	54: C
5: A	15: A	25: D	35: C	45: A	55: D
6: D	16: B	26: C	36: D	46: B	56: A
7: C	17: C	27: D	37: B	47: A	57: B
8: D	18: B	28: B	38: C	48: C	58: D
9: A	19: A	29: C	39: D	49: D	59: C
10: B	20: D	30: A	40: C	50: A	60: A

Fluids and Solids

1: C	11: A	21: C	31: C	41: A	51: A
2: D	12: A	22: C	32: A	42: D	52: B
3: A	13: A	23: D	33: B	43: B	53: D
4: A	14: D	24: B	34: D	44: C	54: D
5: D	15: C	25: D	35: C	45: A	55: C
6: C	16: D	26: B	36: D	46: C	
7: B	17: C	27: D	37: B	47: A	
8: D	18: B	28: D	38: D	48: B	
9: A	19: C	29: B	39: B	49: A	
10: C	20: D	30: B	40: D	50: D	

Electrostatics and Magnetism

1: C	11: D	21: A	31: A	41: B	51: D
2: D	12: A	22: C	32: B	42: C	52: A
3: A	13: C	23: A	33: C	43: A	53: B
4: C	14: D	24: B	34: D	44: C	54: D
5: B	15: D	25: A	35: C	45: B	55: A
6: A	16: A	26: D	36: A	46: C	56: C
7: D	17: B	27: C	37: B	47: D	57: D
8: B	18: D	28: C	38: A	48: A	58: B
9: D	19: B	29: D	39: C	49: D	59: A
10: A	20: C	30: A	40: A	50: C	60: B

Circuit Elements

1: B	11: A	21: B	31: C	41: C	51: B
2: D	12: B	22: C	32: D	42: D	52: D
3: C	13: D	23: D	33: B	43: A	53: A
4: D	14: B	24: A	34: C	44: B	54: C
5: D	15: C	25: B	35: D	45: C	55: D
6: A	16: A	26: C	36: C	46: A	56: B
7: C	17: B	27: A	37: B	47: B	57: A
8: D	18: D	28: D	38: C	48: D	58: D
9: B	19: C	29: B	39: A	49: A	59: A
10: C	20: C	30: A	40: D	50: C	60: C

Sound

1: B	11: B	21: C	31: C	41: A	51: D
2: A	12: A	22: D	32: D	42: C	52: B
3: B	13: B	23: B	33: A	43: D	53: D
4: C	14: C	24: C	34: D	44: A	54: C
5: B	15: D	25: A	35: A	45: B	55: D
6: A	16: A	26: D	36: A	46: D	56: C
7: D	17: C	27: B	37: B	47: C	57: A
8: A	18: B	28: A	38: D	48: A	58: C
9: C	19: A	29: B	39: B	49: C	59: D
10: D	20: B	30: A	40: C	50: B	60: C

Light and Geometrical Optics

1: A	11: D	21: B	31: C	41: B	51: D
2: A	12: A	22: A	32: B	42: D	52: A
3: B	13: D	23: C	33: B	43: C	53: B
4: D	14: D	24: D	34: A	44: B	54: D
5: C	15: A	25: C	35: C	45: A	55: C
6: B	16: B	26: C	36: B	46: D	56: B
7: C	17: C	27: A	37: C	47: A	57: A
8: D	18: B	28: D	38: A	48: C	58: D
9: A	19: D	29: A	39: D	49: B	59: D
10: D	20: B	30: B	40: B	50: D	60: C

Thermodynamics

1: B	11: B	21: D	31: B	41: D	51: C
2: C	12: A	22: A	32: D	42: C	52: A
3: D	13: B	23: B	33: B	43: B	53: D
4: A	14: D	24: D	34: B	44: A	54: B
5: C	15: C	25: A	35: D	45: D	55: D
6: B	16: A	26: A	36: A	46: C	56: C
7: D	17: D	27: D	37: C	47: B	57: B
8: A	18: B	28: B	38: A	48: A	58: A
9: C	19: C	29: C	39: B	49: D	59: C
10: D	20: A	30: D	40: B	50: B	60: B

Atomic Nucleus and Electronic Structure

1: A	11: B	21: D	31: A	41: D	51: D
2: C	12: D	22: C	32: C	42: D	52: C
3: D	13: A	23: B	33: D	43: D	53: B
4: B	14: C	24: D	34: B	44: B	54: D
5: C	15: C	25: A	35: C	45: D	55: C
6: D	16: A	26: C	36: D	46: B	56: D
7: B	17: B	27: D	37: B	47: C	57: B
8: D	18: D	28: D	38: C	48: C	58: A
9: C	19: B	29: C	39: C	49: B	59: C
10: A	20: A	30: D	40: B	50: A	

Particle Physics

1: C	6: D	11: C	16: B	21: C
2: D	7: D	12: C	17: D	22: A
3: B	8: B	13: C	18: A	23: B
4: B	9: A	14: D	19: B	
5: B	10: A	15: D	20: C	

Quantum Mechanics

1: D	11: A	21: C	31: B	41: C	51: D
2: B	12: D	22: A	32: A	42: D	52: B
3: C	13: D	23: D	33: D	43: C	53: B
4: C	14: C	24: B	34: C	44: C	54: B
5: D	15: B	25: B	35: A	45: D	55: D
6: C	16: D	26: B	36: C	46: D	
7: C	17: C	27: D	37: B	47: C	
8: B	18: D	28: B	38: C	48: B	
9: D	19: B	29: A	39: C	49: D	
10: A	20: D	30: D	40: B	50: A	

Special Relativity

1: A	11: D	21: A	31: D	41: C
2: D	12: B	22: D	32: D	42: B
3: D	13: A	23: D	33: A	43: B
4: D	14: B	24: B	34: B	44: A
5: D	15: D	25: C	35: C	45: D
6: D	16: D	26: A	36: C	46: A
7: A	17: C	27: B	37: C	47: A
8: C	18: B	28: D	38: D	48: B
9: B	19: D	29: C	39: B	49: D
10: C	20: A	30: B	40: C	50: D

Notes for active learning

Translational Motion – Detailed Explanations

1. D is correct.

$t = (v_f - v_i) / a$

$t = (60 \text{ mi/h} - 0 \text{ mi/h}) / (13.1 \text{ mi/h·s})$

$t = 4.6 \text{ s}$

Acceleration is in mi/h·s, so miles and hours cancel, and the answer is in units of seconds.

2. B is correct.

At the top of the parabolic trajectory, the vertical velocity $v_{yf} = 0$

The initial upward velocity is the vertical component of the initial velocity:

$v_{yi} = v \sin \theta$

$v_{yi} = (20 \text{ m/s}) \sin 30°$

$v_{yi} = (20 \text{ m/s})·(0.5)$

$v_{yi} = 10 \text{ m/s}$

$t = (v_{yf} - v_{yi}) / a$

$t = (0 - 10 \text{ m/s}) / (-10 \text{ m/s}^2)$

$t = (-10 \text{ m/s}) / (-10 \text{ m/s}^2)$

$t = 1 \text{ s}$

3. B is correct.

$\Delta d = 31.5 \text{ km} = 31{,}500 \text{ m}$

$1.25 \text{ hr} \times 60 \text{ min/hr} = 75 \text{ min}$

$\Delta t = 75 \text{ min} \times 60 \text{ s/min}$

$\Delta t = 4{,}500 \text{ s}$

$v_{avg} = \Delta d / \Delta t$

$v_{avg} = 31{,}500 \text{ m} / 4{,}500 \text{ s}$

$v_{avg} = 7 \text{ m/s}$

4. A is correct.

Instantaneous speed is the scalar magnitude of the velocity. It can only be positive or zero (because magnitudes cannot be negative).

5. C is correct.

$d = (v_f^2 - v_i^2) / 2a$

$d = [(21\text{ m/s})^2 - (5\text{ m/s})^2] / [2(3\text{ m/s}^2)]$

$d = (441\text{ m}^2/\text{s}^2 - 25\text{ m}^2/\text{s}^2) / 6\text{ m/s}^2$

$d = (416\text{ m}^2/\text{s}^2) / 6\text{ m/s}^2$

$d = 69\text{ m}$

6. C is correct.

$a = (v_f - v_i) / t$

$a = [0 - (-30\text{ m/s})] / 0.15\text{ s}$

$a = (30\text{ m/s}) / 0.15\text{ s}$

$a = 200\text{ m/s}^2$

To represent the acceleration in terms of g, divide a by 9.8 m/s^2:

$\#\text{ of } g = (200\text{ m/s}^2) / 9.8\text{ m/s}^2$

$\#\text{ of } g = 20\ g$

The initial velocity (v_i) is negative due to the acceleration of the car being a positive value.

Since the car is decelerating, its acceleration is opposite of its initial velocity.

7. D is correct.

When a bullet is fired, it is in projectile motion.

The only force in projectile motion (if air resistance is ignored) is the force of gravity.

8. B is correct.

When a car is slowing down, it is decelerating, equivalent to acceleration in the opposite direction.

9. A is correct

Uniform acceleration:

a = change in velocity / change in time

$a = \Delta v / \Delta t$

$\Delta v = a\Delta t$

$\Delta v = (20\text{ m/s}^2)\cdot(1\text{ s})$

$\Delta v = 20\text{ m/s}$

10. B is correct.

Uniform acceleration:

a = change in velocity / change in time

$a = \Delta v / \Delta t$

$a = (40\text{ m/s} - 15\text{ m/s}) / 10\text{ s}$

$a = (25\text{ m/s}) / 10\text{ s}$

$a = 2.5\text{ m/s}^2$

11. C is correct.

$t = d / v$

$t = (540\text{ mi}) / (65\text{ mi/h})$

$t = 8.3\text{ h}$

The time she can stop is the difference between the total allowed time and the time t to make the trip:

$t_{stop} = 9.8\text{ h} - 8.3\text{ h}$

$t_{stop} = 1.5\text{ h}$

12. B is correct.

Average velocity is the change in position with respect to time:

$v = \Delta x / \Delta t$

After one lap, the racecar's final position is the same as its initial position.

Thus, $x = 0$, which implies the average velocity of 0 m/s.

13. A is correct.

$d = v_i\Delta t + ½a\Delta t^2$

$d = (0.2\text{ m/s})\cdot(5\text{ s}) + ½(-0.05\text{ m/s}^2)\cdot(5\text{ s})^2$

$d = 1\text{ m} + ½(-0.05\text{ m/s}^2)\cdot(25\text{ s}^2)$

$d = 1\text{ m} + (-0.625\text{ m})$

$d = 0.375\text{ m} \approx 0.38\text{ m}$

Decelerating is set to negative.

The net displacement is the difference between the final and initial positions after 5 s.

14. C is correct.

a = change in velocity / change in time

$a = \Delta v / \Delta t$

15. B is correct.

Convert the final speed from km/h to m/s:

$$v_f = (210 \text{ km/h}) \times [(1{,}000 \text{ m}/1 \text{ km})] \times [(1 \text{ h}/3{,}600 \text{ s})]$$

$$v_f = 58.33 \text{ m/s}$$

Calculate the acceleration necessary to reach this speed:

$$a = (v_f^2 - v_i^2) / 2d$$

$$a = [(58.33 \text{ m/s})^2 - (0 \text{ m/s})^2] / 2(1{,}800 \text{ m})$$

$$a = (3{,}402.39 \text{ m}^2/\text{s}^2) / (3{,}600 \text{ m})$$

$$a = 0.95 \text{ m/s}^2$$

16. D is correct.

The distance the rocket travels during its acceleration upward is calculated by:

$$d_1 = ½at^2$$

$$d_1 = ½(22 \text{ m/s}^2)·(4 \text{ s})^2$$

$$d_1 = 176 \text{ m}$$

The distance from when the motor shuts off to when the rocket reaches maximum height can be calculated using the conservation of energy:

$$mgd_2 = ½mv^2$$

cancel m from each side of the expression

$$gd_2 = ½v^2$$

where,

$$v = at$$

$$gd_2 = ½(at)^2$$

$$d_2 = ½(at)^2 / g$$

$$d_2 = ½[(22 \text{ m/s}^2)·(4 \text{ s})]^2 / (10 \text{ m/s}^2)$$

Magnitudes are not vectors but scalars, so no direction is needed

$$d_2 = 387 \text{ m}$$

For the maximum elevation, add the two distances:

$$h = d_1 + d_2$$

$$h = 176 \text{ m} + 387 \text{ m}$$

$$h = 563 \text{ m}$$

17. A is correct.

Speed is a scalar (i.e., one-dimension physical property); velocity is a vector (i.e., magnitude and direction).

18. B is correct.

Acceleration due to gravity is constant and independent of mass.

19. D is correct.

As an object falls, its acceleration is constant due to gravity.

However, the magnitude of the velocity increases due to the acceleration of gravity, and the displacement increases because the object is going further *away* from its starting point.

20. C is correct.

The man is moving at constant velocity (no acceleration), so the net force is zero. The only objects interacting with the man directly are Earth and the floor of the elevator.

The cable does not touch the man; it pulls the elevator car up, and the elevator floor pushes on the man.

21. D is correct.

Horizontal velocity (v_x):

$$v_x = d_x / t$$

$$v_x = (44\text{ m}) / (2.9\text{ s})$$

$$v_x = 15.2\text{ m/s}$$

The x component of a vector is calculated by:

$$v_x = v \cos \theta$$

Rearrange the equation to determine the initial velocity of the ball:

$$v = v_x / \cos \theta$$

$$v = (15.2\text{ m/s}) / (\cos 45°)$$

$$v = (15.2\text{ m/s}) / 0.7$$

$$v = 21.4\text{ m/s}$$

22. A is correct.

Conservation of energy:

$$mgh = ½mv_f^2$$

cancel m from each side of the expression

$$gh = ½v_f^2$$

continued…

$(10\ \text{m/s}^2)h = ½(14\ \text{m/s})^2$

$(10\ \text{m/s}^2)h = ½(196\ \text{m}^2/\text{s}^2)$

$h = (98\ \text{m}^2/\text{s}^2) / (10\ \text{m/s}^2)$

$h = 9.8\ \text{m} \approx 10\ \text{m}$

23. B is correct.

$d = v_it + ½at^2$

$d = (20\ \text{m/s})\cdot(7\ \text{s}) + ½(1.4\ \text{m/s}^2)\cdot(7\ \text{s})^2$

$d = (140\ \text{m}) + ½(1.4\ \text{m/s}^2)\cdot(49\ \text{s}^2)$

$d = 174.3\ \text{m} \approx 174\ \text{m}$

24. D is correct.

Force is not a scalar because it has a magnitude and direction.

25. B is correct.

$d = ½at^2$

$d_A = ½at^2$

$d_B = ½a(2t)^2$

$d_B = ½a(4t^2)$

$d_B = 4 \times ½at^2$

$d_B = 4d_A$

26. A is correct.

$d = v_{average} \times \Delta t$

$d = ½(v_i + v_f)\Delta t$

$d = ½(5\ \text{m/s} + 30\ \text{m/s})\cdot(10\ \text{s})$

$d = 175\ \text{m}$

27. C is correct.

If there is no acceleration, then velocity is constant.

28. D is correct.

The gravitational force between two objects in space, each having masses of m_1 and m_2, is:

$$F_G = Gm_1m_2 / r^2$$

where G is the gravitational constant and r is the distance between the two objects

Doubling the distance between the two objects:

$$F_{G2} = Gm_1m_2 / (2r)^2$$

$$F_{G2} = Gm_1m_2 / (4r^2)$$

$$F_{G2} = \tfrac{1}{4}Gm_1m_2 / r^2$$

$$F_{G2} = \tfrac{1}{4}Gm_1m_2 / r^2$$

$$F_{G2} = \tfrac{1}{4}F_G$$

When the distance between the objects is doubled, the force (F_G) is one-fourth.

29. D is correct.

I: If the velocity is constant, the instantaneous velocity is always equal to the average velocity.

II and III: If the velocity increases, the average value of velocity over an interval must lie between the initial velocity and the final velocity. In going from its initial value to its final value, the instantaneous velocity must cross the average value at one point, regardless of whether the velocity is changing at a constant rate or changing irregularly.

30. C is correct.

velocity = acceleration × time

$$v = at$$

$$v = (10 \text{ m/s}^2)\cdot(10 \text{ s})$$

$$v = 100 \text{ m/s}$$

31. B is correct.

velocity = distance / time

$$v = d / t$$

d is constant, while t decreases by a factor of 3

32. C is correct.

The equation for distance, given a constant acceleration and both the initial and final velocity, is:

$$d = (v_i^2 + v_f^2) / 2a$$

Since the car is coming to rest, $v_f = 0$

$$d = v_i^2 / 2a$$

If the initial velocity is doubled while acceleration and final velocity remain unchanged, the distance traveled is:

$$d_2 = (2v_i)^2 / 2a$$

$$d_2 = 4(v_i^2 / 2a)$$

$$d_2 = 4d_1$$

Another method to solve this problem

$$d_1 = (29 \text{ mi/h})^2 / 2a$$

$$d_2 = (59 \text{ mi/h})^2 / 2a$$

$$d_2 / d_1 = [(59 \text{ mi/h})^2 / 2a] / [(29 \text{ mi/h })^2 / 2a]$$

$$d_2 / d_1 = (59 \text{ mi/h})^2 / (29 \text{ mi/h})^2$$

$$d_2 / d_1 = (3{,}481 \text{ mi/h}) / (841 \text{ mi/h})$$

$$d_2 / d_1 = 4$$

33. D is correct.

$$\text{speed}_{\text{average}} = \text{total distance / time}$$

$$\text{speed} = (400 \text{ m}) / (20 \text{ s})$$

$$\text{speed} = 20 \text{ m/s}$$

If this were velocity, it would be 0.

34. A is correct.

$$\Delta v = a\Delta t$$

$$(v_f - v_i) = a\Delta t$$

where, $v_f = 0$ m/s (when the car stops)

$a = -0.1$ m/s^2 (negative because deceleration), $\Delta t = 5$ s

$$v_i = v_f - a\Delta t$$

$$v_i = [(0 \text{ m/s}) - (-0.1 \text{ m/s}^2)]\cdot(5 \text{ s})$$

$$v_i = (0.1 \text{ m/s}^2)\cdot(5 \text{ s})$$

$$v_i = 0.5 \text{ m/s}$$

$\Delta v = a\Delta t$

35. C is correct.

If acceleration is constant, then the velocity *vs.* time graph is linear.

The average velocity is the average of the final and initial velocity.

$$v_{average} = v_f - v_i / \Delta t$$

If acceleration is not constant, then the velocity *vs.* time graph is nonlinear.

$$v_{average} \neq v_f - v_i / \Delta t$$

36. D is correct.

Find velocity of thrown rock:

$$v_{f1}^2 - v_i^2 = 2ad$$

$$v_{f1}^2 = v_i^2 + 2ad$$

$$v_{f1}^2 = (10 \text{ m/s})^2 + [2(9.8 \text{ m/s}^2)\cdot(300 \text{ m})]$$

$$v_{f1}^2 = 100 \text{ m}^2/\text{s}^2 + 5{,}880 \text{ m}^2/\text{s}^2$$

$$v_{f1}^2 = 5{,}980 \text{ m}^2/\text{s}^2$$

$$v_{f1} = 77.33 \text{ m/s}$$

$$t_1 = (v_f - v_i) / a$$

$$t_1 = (77.33 \text{ m/s} - 10 \text{ m/s}) / 9.8 \text{ m/s}^2$$

$$t_1 = (67.33 \text{ m/s}) / (9.8 \text{ m/s}^2)$$

$$t_1 = 6.87 \text{ s}$$

Find velocity of dropped rock:

$$v_{f2} = \sqrt{2ad}$$

$$v_{f2} = \sqrt{[(2)\cdot(9.8 \text{ m/s}^2)\cdot(300 \text{ m})]}$$

$$v_{f2} = 76.7 \text{ m/s}$$

$$t_2 = (76.7 \text{ m/s}) / (9.8 \text{ m/s}^2)$$

$$t_2 = 7.82 \text{ s}$$

$$\Delta t = (7.82 \text{ s} - 6.87 \text{ s})$$

$$\Delta t = 0.95 \text{ s}$$

37. D is correct.

From Newton's Second Law:

Force = mass × acceleration

$$F = ma$$

continued...

Rearrange to isolate acceleration:

$a = F / m$

Therefore, acceleration equals net force divided by mass.

For example, assuming a net force of 100 N for a 10-kg object:

$a = F / m$

$a = 100 \text{ N} / 10 \text{ kg}$

$a = 10 \text{ m/s}^2$

If the net force acting on that object doubles to 200 N, then:

$a = 200 \text{ N} / 10 \text{ kg}$

$a = 20 \text{ m/s}^2$

Force and acceleration are directly proportional, so doubling force doubles acceleration.**38. B is correct.**

Velocity is defined by speed and direction.

If either speed or direction, or both, change, the object experiences acceleration.

39. C is correct.

The acceleration is negative because it acts to slow the car down against the $+y$ direction.

It is unclear if the acceleration decreases in magnitude from the data provided.

40. A is correct.

Total distance is the area under the velocity-time curve with respect to the x-axis.

This graph can be broken up into sections; calculate the area under the curve.

$d_{total} = d_A + d_B + d_C + d_D$

$d_A = ½(4 \text{ m/s})·(2 \text{ s}) = 4 \text{ m}$

$d_B = ½(4 \text{ m/s} + 2 \text{ m/s})·(2 \text{ s}) = 6 \text{ m}$

$d_C = (2 \text{ m/s})·(4 \text{ s}) = 8 \text{ m}$

Since the total distance traveled needs to be calculated, the area under the curve when the velocity is negative is calculated as a positive value.

Distance is a scalar quantity and therefore has no direction.

$d_D = ½(2 \text{ m/s})·(1 \text{ s}) + ½(2 \text{ m/s})·(1 \text{ s}) = 2 \text{ m}$

$d_{total} = 4 \text{ m} + 6 \text{ m} + 8 \text{ m} + 2 \text{ m} = 20 \text{ m}$

If the question asked to find the displacement, the area under the curve would be calculated as negative, and the answer would be 18 m.

41. C is correct.

The two bullets have different velocities when hitting the water but only experience the force due to gravity.

Thus, the acceleration due to gravity is the same for each bullet.

42. A is correct.

$v_f = v_i + at$

$v_f = 0 + (2.5\text{ m/s}^2)\cdot(9\text{ s})$

$v_f = 22.5\text{ m/s}$

43. D is correct.

The equation for *impulse* is used for contact between two objects over a specified time:

$F\Delta t = m\Delta v$

$ma\Delta t = m(v_f - v_i)$

cancel *m* from each side of the expression

$a\Delta t = (v_f - v_i)$

$a = (v_f - v_i) / \Delta t$

$a = (-2v - v) / (0.45\text{ s})$

$a = (-3v) / (0.45\text{ s})$

$a = (-6.7\text{ s}^{-1})v$

Ratio $a : v = -6.7\text{ s}^{-1} : 1$

44. B is correct.

The time for the round trip is 4 s.

The weight reaches the top of its path in ½ time:

$½(4\text{ s}) = 2\text{ s}$

where,

$v = 0$

$a = \Delta v / t$ for the first half of the trip

$a = (v_f - v_i) / t$

$a = (0 - 3.2\text{ m/s}) / 2\text{ s}$

$a = -1.6\text{ m/s}^2$

$|a| = 1.6\text{ m/s}^2$

Acceleration is a vector, and the negative direction only indicates direction.

45. A is correct.

$\Delta v = a\Delta t$

$\Delta v = (0.3\ \text{m/s}^2)\cdot(3\ \text{s})$

$\Delta v = 0.9\ \text{m/s}$

46. D is correct.

Velocity, displacement, and acceleration are vectors. Mass is not a vector quantity.

47. B is correct.

$d = d_0 + (v_i^2 + v_f^2) / 2a$

$d = 64\ \text{m} + (0\ \text{m/s} + 60\ \text{m/s})^2 / 2(9.8\ \text{m/s}^2)$

$d = 64\ \text{m} + (3{,}600\ \text{m}^2/\text{s}^2) / (19.6\ \text{m/s}^2)$

$d = 64\ \text{m} + 184\ \text{m}$

$d = 248\ \text{m}$

48. C is correct.

$a = (v_f^2 + v_i^2) / 2d$

$a = [(60\ \text{m/s})^2 + (0\ \text{m/s})^2] / [2(64\ \text{m})]$

$a = (3{,}600\ \text{m}^2/\text{s}^2) / 128\ \text{m}$

$a = 28\ \text{m/s}^2$

49. D is correct.

Expression for the time interval during constant acceleration upward:

$d = ½at^2$

Solving for acceleration:

$a = (v_f^2 + v_i^2) / 2d$

$a = [(60\ \text{m/s})^2 + (0\ \text{m/s})^2] / [2(64\ \text{m})]$

$a = (3{,}600\ \text{m}^2/\text{s}^2) / (128\ \text{m})$

$a = 28.1\ \text{m/s}^2$

Solving for time:

$t^2 = 2d / a$

$t^2 = 2(64\ \text{m}) / 28.1\ \text{m/s}^2$

$t^2 = 4.5\ \text{s}^2$

$t = 2.1\ \text{s}$

50. C is correct.

$d = (v_i^2 + v_f^2) / 2a$

where $v_i = 0$

$d = v_f^2 / 2a$

For half the final velocity:

$d_2 = (v_f / 2)^2 / 2a$

$d_2 = \frac{1}{4}v_f^2 / 2a$

$d_2 = \frac{1}{4}d$

51. A is correct.

$v_{average} = \Delta d / \Delta t$

52. A is correct.

Use an equation that relates *v*, *d* and *t*:

$d = vt$

$v = d / t$

If *v* increases by a factor of 3, then *t* decreases by a factor of 3.

Another method to solve this problem:

$d = vt$, t = original time and t_N = new time

$d = 3vt_N$

$vt = d = 3vt_N$

$vt = 3vt_N$

$t = 3t_N$

$t / 3 = t_N$

Thus, if *v* increases by a factor of 3, then the original time decreases by a factor of 3.

53. B is correct.

$v_f = v_i + at$

$t = (v_f - v_i) / a$

Since the ball is thrown straight up, its initial speed upward equals its final speed downward (just before hitting the ground): Therefore:

$v_f = -v_i$

$t = [39 \text{ m/s} - (-39 \text{ m/s})] / 9.8 \text{ m/s}^2$ *continued…*

$t = (78 \text{ m/s}) / 9.8 \text{ m/s}^2$

$t = 8 \text{ s}$

54. D is correct.

Since the speed is changing, the velocity is changing, and therefore there *is* an acceleration.

Since the speed *decreases*, acceleration must be *in the reverse direction* (i.e., opposite to the direction of travel).

Since the particle is moving to the right, the acceleration vector points to the left.

If the speed *increases*, acceleration is in the *same* direction as the direction of travel, and the acceleration vector points to the right.

55. A is correct.

The only force that Larry applies to the package is the normal force due to his hand.

There is no horizontal force as the package moves with constant velocity.

The normal force due to his hand points upward.

The displacement of the package is horizontal:

$W = Fd \cos \theta$

where θ is the angle between the force and the displacement

$\theta = 90°$

Since $\cos 90° = 0$,

$W = 0 \text{ J}$

56. A is correct.

The *slope of a tangent line* on a velocity *vs.* time graph is the acceleration at that time point.

This is equivalent to taking the derivative of the velocity with respect to time to find the instantaneous acceleration.

57. C is correct.

Since the car is initially traveling North, let North be the positive direction and South be the negative direction:

$a = (v_f - v_i) / t$

$a = (14.1 \text{ m/s} - 17.7 \text{ m/s}) / 12 \text{ s}$

$a = (-3.6 \text{ m/s}) / 12 \text{ s}$

$a = -0.3 \text{ m/s}^2$

$a = 0.3 \text{ m/s}^2$ South

58. C is correct.

Speed is represented by the magnitude of the slope of a position *vs.* time plot.

A steeper slope equates to a higher speed.

59. B is correct.

If the object has not reached terminal velocity, it accelerates but at an ever-decreasing rate until the terminal velocity is reached.

60. D is correct.

Approach the problem by finding the distance traveled in each of the three segments.

$$d_1 = ½a_1\Delta t_1^2$$

$$d_1 = (0.5)\cdot(2\ \text{m/s}^2)\cdot(10\ \text{s})^2$$

$$d_1 = 100\ \text{m}$$

The second segment:

$$d_2 = v_2\Delta t_2$$

where $\Delta t_2 = 10$ s,

the duration of interval 2 and v_2 is the speed during interval 2, which is the speed at the end of interval 1

$$v_2 = v_{1f}$$

$$v_{1f} = a_1\Delta t_1$$

$$v_2 = a_1\Delta t_1$$

$$v_2 = (2\ \text{m/s})\cdot(10\ \text{s})$$

$$v_2 = 20\ \text{m/s}$$

So:

$$d_2 = (20\ \text{m/s})\cdot(10\ \text{s}) = 200\text{m}$$

Next, the third segment:

$$d_3 = (v^2_{3f} - v^2_{3i})\ /\ 2a_3$$

$$d_3 = [(0\ \text{m/s})^2 - (20\ \text{m/s})^2]\ /\ 2(-2\ \text{m/s}^2)$$

$$d_3 = 100\ \text{m}$$

The total distance traveled is the sum of d_1, d_2 and d_3:

$$d = 100\ \text{m} + 200\ \text{m} + 100\ \text{m}$$

$$d = 400\ \text{m}$$

Notes for active learning

Force and Motion – Detailed Explanations

1. B is correct.

The tension of the string keeps the weight traveling in a circular path; otherwise, it would move linearly on a tangent path to the circle. Without the string, there are no horizontal forces on the weight and no horizontal acceleration. The horizontal motion of the weight is in a straight line at a constant speed.

2. D is correct.

The vertical force on the garment bag from the left side of the clothesline is:

$$T_{y,\text{left}} = T\cos\theta$$

Similarly, for the right side:

$$T_{y,\text{right}} = T\cos\theta$$

where $T = 10$ N (tension) and $\theta = 60°$

Since the garment bag is at rest, its acceleration is zero. Therefore, according to Newton's second law:

$$T_{y,\text{left}} + T_{y,\text{right}} - mg = 0 = 2T(\cos\theta) - mg$$

Or: $$2T(\cos\theta) = mg$$

$$m = 2T(\cos\theta) / \text{g}$$

$$m = 2(10\text{ N})\cdot(\cos 60°) / (10.0\text{ m/s}^2)$$

$$m = 2(10\text{ N})\cdot(0.5) / (10.0\text{ m/s}^2)$$

$$m = 1\text{ kg}$$

3. A is correct.

An object's inertia is its resistance to change in motion. The milk carton has enough inertia to overcome the force of static friction.

4. C is correct.

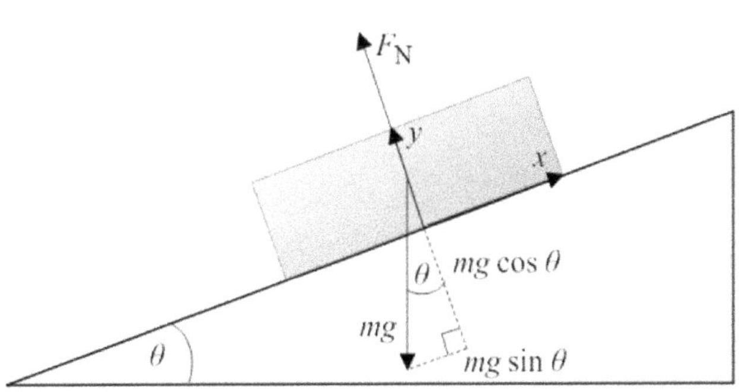

$$(F_{\text{net}})_y = (F_N)_y - (F_g)_y$$

The car is not moving up or down, so $a_y = 0$:

$$(F_{\text{net}})_y = 0$$

$$0 = (F_N)_y - (F_g)_y$$

$$F_N = (F_g)_y$$

$$F_N = F_g\cos\theta$$

$$F_N = mg\cos\theta$$

The normal force is a force that is perpendicular to the plane of contact (the slope).

5. B is correct.

$F = ma$

$F = (27 \text{ kg})\cdot(1.7 \text{ m/s}^2)$

$F = 46 \text{ N}$

6. D is correct.

The mass on the table causes a tension force in the string that acts against the force of gravity.

7. A is correct.

Although the net force acting on the object is decreasing with time and the magnitude of the object's acceleration is decreasing, there exists a positive acceleration.

Therefore, the object's speed continues to increase.

8. D is correct.

An object moving at constant velocity experiences zero net force.

9. A is correct.

The sine of an angle is equal to the opposite side over the hypotenuse:

$\sin \theta = \text{opposite / hypotenuse}$

$\sin \theta = h / L$

$h = L \sin \theta$

10. B is correct.

The force of the table on the book, the normal force (F_N), from Newton's Third Law of Motion, which states that there is an equal and opposite reaction for every action.

A book sitting on the table experiences a force from the table equal to the book's weight:

$\text{W} = mg$

$F_N = \text{W}$

$F_N = mg$

$F_N = (2 \text{ kg})\cdot(10 \text{ m/s}^2)$

$F_N = 20 \text{ N}$

11. A is correct.

$a = (v_f - v_i) / t$

$a = (3.5 \text{ m/s} - 1.5 \text{ m/s}) / (3 \text{ s})$

$a = (2 \text{ m/s}) / (3 \text{ s})$

$a = 0.67 \text{ m/s}^2$

12. C is correct.

An object with uniform circular motion (i.e., constant angular velocity) only experiences centripetal acceleration directed toward the center of the circle.

13. B is correct.

$F = ma$, so zero force means zero acceleration in any direction.

14. C is correct.

$F = ma$

$a = F / m$

$a = 9 \text{ N} / 9 \text{ kg}$

$a = 1 \text{ m/s}^2$

15. B is correct.

The only force acting on a projectile in motion is the force due to gravity.

Since that force acts downward, there is only a downward acceleration.

16. A is correct.

$F_{net} = ma$

If an object moves with constant v;

$a = 0$

So:

$F_{net} = 0$

Since gravity pulls down on the can with a force of mg:

$F_g = mg$

$F_g = (10 \text{ kg})\cdot(10 \text{ m/s}^2)$

$F_g = 100 \text{ N}$

The rope pulls *up* on the can with the same magnitude of force, so the tension is 100 N for a net force = 0.

17. D is correct.

$F = ma$

$F = (1{,}000\text{ kg})\cdot(2\text{ m/s}^2)$

$F = 2{,}000\text{ N}$

18. C is correct.

$a_{cent} = v^2 / r$

$a_{cent} = (4\text{ m/s})^2 / (4\text{ m})$

$a_{cent} = (16\text{ m}^2\text{/s}^2) / (4\text{ m})$

$a_{cent} = 4\text{ m/s}^2$

19. A is correct.

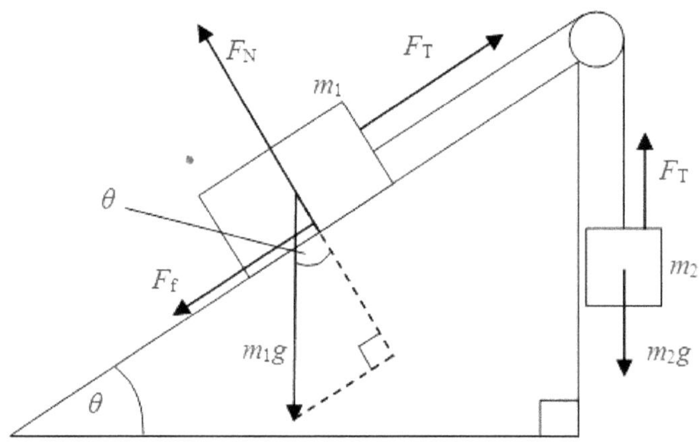

Solve for m_1:

$F_{net} = 0$

$m_2g = F_T$

$m_1g \sin \theta + F_f = F_T$

$m_1g \sin \theta + \mu_s m_1 g \cos \theta = m_2g$

cancel g from each side

$m_1(\sin \theta + \mu_s \cos \theta) = m_2$

$m_1 = m_2 / (\sin \theta + \mu_s \cos \theta)$

$m_1 = 2\text{ kg} / [\sin 20° + (0.55) \cos 20°]$

$m_1 = 2\text{ kg} / 0.86$

$m_1 = 2.3\text{ kg}$

Kinetic friction is only used when the mass is in motion.

20. B is correct.

Since the masses are identical, the force of gravity on each is the same. The force of gravity on one of the masses produces the tension force in the string, which in turn pulls on the other mass.

Since this tension force is equal to the force of gravity, there is no net force, and the objects remain at rest.

21. B is correct.

Newton's Third Law states that for every action, there is an equal and opposite reaction.

22. A is correct.

There is an equal and opposite reaction for every action according to Newton's Third Law.

23. C is correct.

If w denotes the magnitude of the box's weight, then the component of this force parallel to the inclined plane is $w \sin \theta$, where θ is the incline angle.

If θ is less than 90°, then $\sin \theta$ is less than 1.

The component of w parallel to the inclined plane is less than w.

24. B is correct.

The package experiences projectile motion upon leaving the truck, so it experiences no horizontal forces, and its initial velocity of 30 m/s remains unchanged.

25. D is correct.

f = revolutions / unit of time

The time (period) for one complete revolution is:

$T = 1 / f$

Each revolution represents a length of $2\pi r$.

Velocity is the distance traveled in one revolution over duration of one revolution (circumference over period):

$v = 2\pi r / t$

$v = 2\pi rf$

If f doubles, then v doubles.

26. A is correct.

$F = ma$

$m = F / a$

m = 4,500 N / 5 m/s^2

m = 900 kg

27. D is correct.

Newton's First Law states that every object will remain at rest or in uniform motion unless acted upon by an outside force.

Steve and the bus are in uniform constant motion until the bus stops due to sudden deceleration (the ground exerts no frictional force on Steve). There is no force acting upon Steve.

However, his inertia carries him forward because he is still in uniform motion while the bus comes to a stop.

28. D is correct.

The ball is in a state of rest, so $F_{net} = 0$

$$F_{down} = F_{up}$$

$$F_{external} + F_w = F_{buoyant}$$

$$F_{external} = F_{buoyant} - F_w$$

$$F_{external} = 8.4\text{ N} - 4.4\text{ N}$$

$F_{external} = 4$ N, in the same direction as the weight

29. A is correct.

The luggage and the train move at the same speed, so when the luggage moves forward with respect to the train, it means the train has slowed down while the luggage is continuing to move at the train's original speed.

30. D is correct.

The mass does not change by changing the object's location.

Since the object is outside of Earth's atmosphere, the object's weight is represented by the equation:

$$F_g = GmM_{Earth} / R^2$$

If the altitude is $2R_{Earth}$, then the distance from the center of the Earth is $3R_{Earth}$.

The gravitational acceleration decreases by a factor of:

$$3^2 = 9\ (g = GmM / R^2)$$

Weight decreases by a factor of 9.

New weight:

$$360\text{ N} / 9 = 40\text{ N}$$

31. C is correct.

The velocity of the rock just after its release is the same as the truck.

Once in free fall, there are no horizontal forces on the rock.

The rock's velocity remains unchanged and is equal to that of the truck.

32. A is correct.

The acceleration of Jason due to thrust is:

$$F_{net} = ma_1$$

$$ma_1 = F_{ski} - \mu_k mg$$

continued…

$$a_1 = (F_{ski} - \mu_k mg) / m$$

$$a_1 = [200 \text{ N} - (0.1)\cdot(75 \text{ kg})\cdot(9.8 \text{ m/s}^2)] / 75 \text{ kg}$$

$$a_1 = (126.5 \text{ N}) / 75 \text{ kg}$$

$$a_1 = 1.69 \text{ m/s}^2$$

The distance traveled during the acceleration stage is:

$$d_1 = \tfrac{1}{2}a_1t^2$$

$$d_1 = \tfrac{1}{2}(1.69 \text{ m/s}^2)\cdot(67 \text{ s})^2$$

$$d_1 = 3{,}793 \text{ m}$$

The distance traveled after the skis run out of fuel is:

$$d_2 = (v_f^2 - v_i^2) / 2a_2$$

a_2 is Jason's acceleration after the fuel runs out:

$$F_{net} = ma_2$$

$$ma_2 = -\mu_k mg$$

cancel m from both sides of the expression

$$a_2 = -\mu_k g$$

$$a_2 = -(0.1)\cdot(9.8 \text{ m/s}^2)$$

$$a_2 = -0.98 \text{ m/s}^2$$

The acceleration is negative since the frictional force opposes the direction of motion.

v_i is the velocity at the moment when the fuel runs out:

$$v_i = a_1t$$

$$v_i = (1.69 \text{ m/s}^2)\cdot(67 \text{ s})$$

$$v_i = 113.2 \text{ m/s}$$

Substitute a_2 and v_i into the equation for d_2:

$$d_2 = [(0 \text{ m/s})^2 - (113.2 \text{ m/s})^2] / 2(-0.98 \text{ m/s}^2)$$

$$d_2 = (-12{,}814.2 \text{ m}^2/\text{s}^2) / -1.96 \text{ m/s}^2$$

$$d_2 = 6{,}538 \text{ m}$$

The total distance Jason traveled is:

$$d_{total} = d_1 + d_2$$

$$d_{total} = 3{,}793 \text{ m} + 6{,}538 \text{ m}$$

$$d_{total} = 10{,}331 \text{ m}$$

33. D is correct.

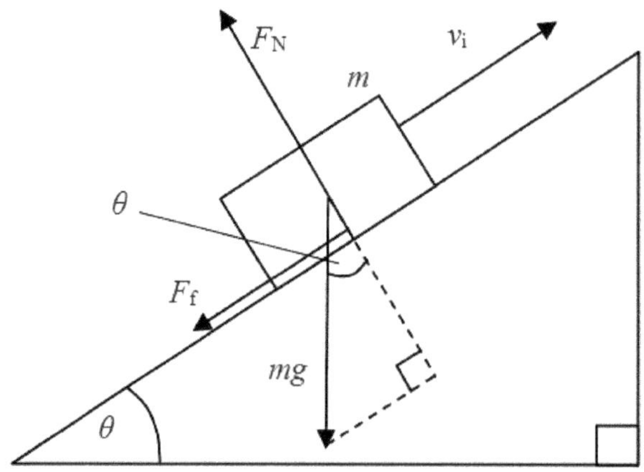

Using the force analysis:

$F_{net} = F_g + F_{fk}$

$F_g = mg \sin \theta$

$F_g = (0.2 \text{ kg})\cdot(-9.8 \text{ m/s}^2) \sin 30°$

$F_g = (0.2 \text{ kg})\cdot(-9.8 \text{ m/s}^2)\cdot(1/2)$

$F_g = -1 \text{ N}$

$F_{fk} = \mu_k F_N$

$F_{fk} = \mu_k mg \cos \theta$

$F_{fk} = (0.3)\cdot(0.2 \text{ kg})\cdot(-9.8 \text{ m/s}^2) \cos 30°$

$F_{fk} = (0.3)\cdot(0.2 \text{ kg})\cdot(-9.8 \text{ m/s}^2)\cdot(0.866)$

$F_{fk} = -0.5 \text{ N}$

$F_{net} = -1 \text{ N} + (-0.5 \text{ N})$

$F_{net} = -1.5 \text{ N}$

$a = F_{net} / m$

$a = -1.5 \text{ N} / 0.2 \text{ kg}$

$a = -7.5 \text{ m/s}^2$

The distance it travels until it reaches a velocity of 0 at its maximum height:

$d = (v_f^2 - v_i^2) / 2a$

$d = [(0 \text{ m/s})^2 - (63 \text{ m/s})^2] / 2(-7.5 \text{ m/s}^2)$

$d = (-4{,}000 \text{ m}^2/\text{s}^2) / (-15 \text{ m/s}^2)$

$d = 267 \text{ m}$

The vertical height is:

$h = d \sin \theta$

$h = (267 \text{ m}) \sin 30°$

$h = (267 \text{ m})\cdot(0.5)$

$h = 130 \text{ m}$

Using energy to solve the problem:

$\text{KE} = \text{PE} + \text{W}_f$

$½mv^2 = mgd \sin \theta + \mu_k mgd \cos \theta$

continued…

Cancel m from the expression:

$$\frac{1}{2}v^2 = gd \sin \theta + \mu_k gd \cos \theta$$

$$\frac{1}{2}v^2 = d(g \sin \theta + \mu_k g \cos \theta)$$

$$d = v^2 / [2g(\sin \theta + \mu_k \cos \theta)]$$

$$d = (63 \text{ m/s})^2 / [(2)\cdot(9.8 \text{ m/s}^2)\cdot(\sin 30° + 0.3 \times \cos 30°)]$$

$$d = 267 \text{ m}$$

$$h = d \sin \theta$$

$$h = (267 \text{ m}) \sin 30°$$

$$h = 130 \text{ m}$$

34. A is correct.

$$F = ma$$

$$a = F / m$$

$$a_1 = F / 4 \text{ kg}$$

$$a_2 = F / 10 \text{ kg}$$

$$4a_1 = 10a_2$$

$$a_1 = 2.5a_2$$

35. D is correct.

At $\theta = 17°$, the force of static friction is equal to the force due to gravity:

$$F_f = F_g$$

$$\mu_s mg \cos \theta = mg \sin \theta$$

$$\mu_s = \sin \theta / \cos \theta$$

$$\mu_s = \tan \theta$$

$$\mu_s = \tan 17°$$

$$\mu_s = 0.31$$

36. A is correct.

Newton's Third Law describes that when one object pushes on another, the second object pushes right back with the same force. Mathematically, it can be expressed as:

$$F_{\text{AonB}} = -F_{\text{BonA}}$$

continued…

The force that the truck exerts on the car is in the opposite direction to the force that the car exerts on the truck (since they push on each other), and crucially, the *magnitudes* of the two forces are the same.

This may seem counterintuitive since it is known that the car will get far more damaged than the truck.

Newton's Second Law states that the car will accelerate faster (since it is less massive than the truck). It is this extreme acceleration that causes the car to be destroyed.

Therefore, to understand this situation thoroughly, two Newton's laws must be applied:

The Third Law states that each vehicle experiences a force of the same magnitude, and

The Second Law describes why the car *responds* to that force more violently due to its smaller mass.

37. D is correct.

$$m = F / a_{\text{Earth}}$$

$$m = 20 \text{ N} / 3 \text{ m/s}^2$$

$$m = 6.67 \text{ kg}$$

$$F_{\text{Moon}} = mg_{\text{Moon}}$$

$$F_{\text{Moon}} = (6.67 \text{ kg})\cdot(1.62 \text{ m/s}^2)$$

$$F_{\text{Moon}} = 11 \text{ N}$$

38. A is correct.

If θ is the angle with respect to a horizontal line, then:

$$\theta = ½(40°)$$

$$\theta = 20°$$

Therefore, for the third force to cause equilibrium, the sum of all three forces' components must equal zero. Since F_1 and F_2 mirror each other in the y-direction:

$$F_{1y} + F_{2y} = 0$$

Therefore, for F_3 to balance forces in the y-direction, its y component must equal zero:

$$F_{1y} + F_{2y} + F_{3y} = 0$$

$$0 + F_{3y} = 0$$

$$F_{3y} = 0$$

Since the y component of F_3 is zero, the angle that F_3 makes with the horizontal is zero:

$$\theta_3 = 0°$$

continued…

The x component of F_3:

$$F_{1x} + F_{2x} + F_{3x} = 0$$

$$F_1 \cos \theta + F_2 \cos \theta + F_3 \cos \theta = 0$$

$$F_3 = -(F_2 \cos \theta_2 + F_3 \cos \theta_3)$$

$$F_3 = -[(2.3 \text{ N}) \cos 20° + (2.3 \text{ N}) \cos 20°]$$

$$F_3 = -4.3 \text{ N}$$

$$F_3 = 4.3 \text{ N to the right}$$

39. B is correct.

The solution needs an expression that connects time and mass.

Given information for F, v_1, and d:

$$a = F / m$$

$$d = v_1 t + ½at^2$$

Combine the expressions and set $v_i = 0$ m/s because initial velocity is zero:

$$d = ½at^2$$

$$a = F / m$$

$$d = ½(F / m)t^2$$

$$t^2 = 2dm / F$$

$$t = \sqrt{(2dm / F)}$$

If m increases by a factor of 4, t increases by a factor of $\sqrt{4} = 2$

40. C is correct.

$$a = (v_f^2 - v_i^2) / 2d$$

$$a = [(0 \text{ m/s})^2 - (27 \text{ m/s})^2] / 2(578 \text{ m})$$

$$a = (-729 \text{ m}^2/\text{s}^2) / 1{,}056 \text{ m}$$

$$a = -0.63 \text{ m/s}^2$$

$$F = ma$$

$$F = (1{,}100 \text{ kg})\cdot(-0.63 \text{ m/s}^2)$$

$$F = -690 \text{ N}$$

The car is decelerating, so the acceleration (and therefore the force) is negative.

41. A is correct.

Constant speed upward means no net force.

Tension = weight (equals Mg)

42. C is correct.

Weight = mg

75 N = mg

m = 75 N / 9.8 m/s^2

m = 7.65 kg

$F_{net} = F_{right} - F_{left}$

F_{net} = 50 N – 30 N

F_{net} = 20 N

$F_{net} = ma$

$a = F_{net} / m$

a = 20 N / 7.65 kg

a = 2.6 m/s^2

43. B is correct.

The string was traveling at the same velocity as the plane with respect to the ground outside.

When the plane began accelerating backward (decelerating), the string continued to move forward at its original velocity and appeared to go towards the front of the plane.

Since the string is attached to the ceiling at one end, only the bottom of the string moved.

44. C is correct.

If the object slides down the ramp with a constant speed, velocity is constant.

Acceleration and the net force = 0

$F_{net} = F_{grav\ down\ ramp} - F_{friction}$

$F_{net} = mg \sin \theta - \mu_k mg \cos \theta$

$F_{net} = 0$

$mg \sin \theta - \mu_k mg \cos \theta = 0$

$mg \sin \theta = \mu_k mg \cos \theta$

$\mu_k = \sin \theta / \cos \theta$

45. D is correct.

Each scale weighs the fish at 17 kg, so the sum of the two scales is:

$$17\text{ kg} + 17\text{ kg} = 34\text{ kg}$$

46. C is correct.

$$a = \Delta v / \Delta t$$

$$a = (v_f - v_i) / t$$

$$a = (20\text{ m/s} - 0\text{ m/s}) / (10\text{ s})$$

$$a = (20\text{ m/s}) / (10\text{ s})$$

$$a = 2\text{ m/s}^2$$

47. D is correct.

Since the object does not move, it is in equilibrium, so forces acting on it are equal and oppose the force F that Yania applies to the object.

48. A is correct.

Newton's Third Law describes that when one object pushes on another, the second object pushes right back with the same force. Mathematically, it can be expressed as:

$$F_{\text{AonB}} = -F_{\text{BonA}}$$

In this example, if one pushes on an object with force F, the object must push back on them equally (magnitude is F) and in the opposite direction (hence the negative sign).

Therefore, the force vector of the object is $-F$.

49. C is correct.

Find equal and opposite forces:

$$F_{Rx} = -F_1$$

$$F_{Rx} = -(-6.6\text{ N})$$

$$F_{Rx} = 6.6\text{ N}$$

$$F_{Ry} = -F_2$$

$$F_{Ry} = -2.2\text{ N}$$

Pythagorean Theorem ($a^2 + b^2 = c^2$) to calculate the magnitude of the resultant force:

continued…

The magnitude of F_R:

$$F_R^2 = F_{Rx}^2 + F_{Ry}^2$$

$$F_R^2 = (6.6\text{ N})^2 + (-2.2\text{ N})^2$$

$$F_R^2 = 43.6\text{ N}^2 + 4.8\text{ N}^2$$

$$F_R^2 = 48.4\text{ N}^2$$

$$F_R = 7\text{ N}$$

The direction of F_R:

$$\theta = \tan^{-1}(-2.2\text{ N} / 6.6\text{ N})$$

$$\theta = \tan^{-1}(-1 / 3)$$

$$\theta = 342°$$

The direction of F_R with respect to F_1:

$$\theta = 342° - 180°$$

$$\theta = 162° \text{ counterclockwise of } F_1$$

50. A is correct.

$$m_{Bob} = 4m_{Sarah}$$

Conservation of momentum since the system (Bob and Sarah combined) initially had a total momentum of 0, in the final state, Sarah's momentum and Bob's momentum must add to 0 (i.e., they will be the same magnitude, but opposite directions):

$$m_{Bob}v_{Bob} = m_{Sarah}v_{Sarah}$$

$$4m_{Sarah}\, v_{Bob} = m_{Sarah}v_{Sarah}$$

$$4v_{Bob} = v_{Sarah}$$

51. C is correct.

For most surfaces, the coefficient of static friction is greater than the coefficient of kinetic friction.

Thus, the force needed to overcome static friction and start the object's motion is greater than the amount of force needed to overcome kinetic friction and keep the object moving at a constant velocity.

52. A is correct.

Newton's First Law states that an object at rest tends to stay at rest, and an object in motion tends to maintain that motion unless acted upon by an unbalanced force.

This law depends on *inertia*, which is related to the object's mass.

More massive objects are more difficult to move and manipulate than less massive objects.

53. D is correct.

Neither Joe nor Bill is moving, so the net force is zero:

$F_{net} = F_{Joe} - F_T$

$0 = F_{Joe} - F_T$

$F_{Joe} = F_T$

$F_T = 200 \text{ N}$

54. C is correct.

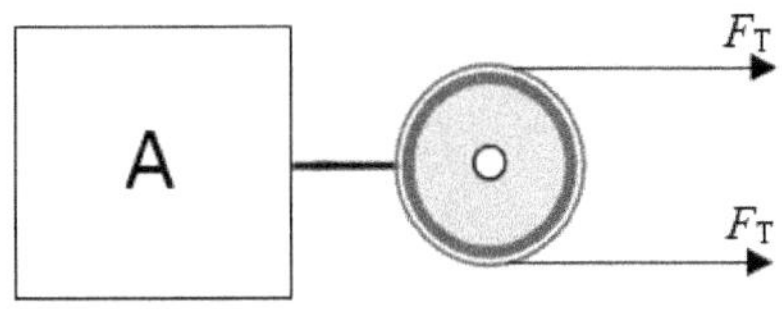

Tension in the rope is always equal to F_T.

The net force on block A to the right is:

$F_{right} = m_A a_A = 2F_T$

The net force of block B downward is:

$F_{down} = m_B a_B = m_B g - F_T$

Since block A is connected to both the pulley at the end of the table and the wall, it uses twice the rope length to travel the same distance as block B.

The distance block A moves is half that of block B, the velocity of block A is half the velocity of block B, and the acceleration of block A is half the acceleration of block B:

$a_A = a_B / 2$

$F_{right} = m_A(a_B / 2)$

$m_A(a_B / 2) = 2F_T$

$m_A a_B = 4F_T$

$F_T = ¼ m_A a_B$

$m_B a_B = m_B g - ¼ m_A a_B$

$m_B a_B + ¼ m_A a_B = m_B g$

$a_B[m_B + ¼ m_A] = m_B g$

$a_B = m_B g / [m_B + ¼ m_A]$

$a_B = (5 \text{ kg})\cdot(9.8 \text{ m/s}^2) / [5 \text{ kg} + ¼(4 \text{ kg})]$

$a_B = 49 \text{ N} / 6 \text{ kg}$

$a_B = 8.2 \text{ m/s}^2$

$a_A = a_B / 2$

$a_A = (8.2 \text{ m/s}^2) / 2$

$a_A = 4.1 \text{ m/s}^2$

55. B is correct.

The force exerted by one surface on another has a perpendicular component (i.e., normal force) and a parallel component (i.e., friction force).

The force of kinetic friction on an object acts opposite to the direction of its velocity relative to the surface.

56. D is correct.

The scale measures the force of interaction between the person and the floor, the normal force.

The question asks to find the normal force.

Use Newton's Second Law:

$$F = ma$$

where the net force is the result of the force of gravity and the normal force

$$F_{net} = F_{normal} - F_{gravity} = N - W$$

Here, W is the normal weight of the object:

$$W = mg$$

Therefore, Newton's Law becomes:

$$N - W = ma = (W / g)\ a$$

Solve for the reading of the scale, N, noting that the acceleration is negative:

$$N = W(1 + a / g)$$

$$N = (600\ \text{N})\cdot(1 + -6\ \text{m/s}^2 / 9.8\ \text{m/s}^2)$$

$$N = (600\ \text{N})\cdot(0.388)$$

$$N = 233\ \text{N}$$

57. B is correct.

If the bureau moves in a straight line at a constant speed, its velocity is constant.

Therefore, the bureau is experiencing zero acceleration and zero net force.

The force of kinetic friction equals the 30 N force that pulls the bureau.

58. D is correct.

Since the crate can only move in the horizontal direction, only consider the horizontal component of the applied force when computing the acceleration.

$F_x = F \cos \theta$

$F_x = (140 \text{ N}) \cos 30°$

$F_x = (140 \text{ N})·(0.866)$

$F_x = 121 \text{ N}$

$a = F_x / m$

$a = 121 \text{ N} / 40 \text{ kg}$

$a = 3 \text{ m/s}^2$

59. C is correct.

Vectors indicate magnitude and direction, while scalars only indicate magnitude.

60. B is correct.

The gravitational force and the direction of travel are perpendiculars:

$W = Fd \cos \theta$

$\cos \theta = 0$

61. A is correct.

Work = Force × distance

$W_{rope} = Fd_x$

$W_{rope} = Fd \cos \theta$

$d = vt$

$d = (2.5 \text{ m/s})·(4 \text{ s})$

$W_{rope} = (30 \text{ N})·(10 \text{ m}) \cos 30°$

$W_{rope} = 260 \text{ J}$

Notes for active learning

Equilibrium and Momentum – Detailed Explanations

1. A is correct.

The rate of change of angular momentum of a system is equal to the net external torque:

$\tau_{net} = \Delta L / \Delta t$

If the angular momentum is constant, then the net external torque must be zero.

2. D is correct.

If the velocity is 7 m/s down the mountain, the horizontal component v_x is:

$v_x = v \cos \theta$

$1.8 \text{ m/s} = (7 \text{ m/s}) \cos \theta$

$\cos \theta = 0.26$

$\theta \approx 75°$

3. D is correct.

The hill exerts a normal force on the sled, and this force is *perpendicular* to the surface of the hill.

There is no parallel force that the hill exerts because it is frictionless.

4. C is correct.

Assuming the water flow is tangent to the wheel, it is perpendicular to the radius vector at the point of contact.

The torque around the center of the wheel is:

$\tau = rF$

$\tau = (10 \text{ m})\cdot(300 \text{ N})$

$\tau = 3{,}000 \text{ N}\cdot\text{m}$

5. D is correct.

1 revolution = 360°

1 min = 60 s

33 rpm = 33 rev/min

(33 rev/min)·(360°/rev) = 11,880°/min

(11,880°/min)·(1 min/60 s) = 198°/s

Degrees per second is a *rate*:

rate × time = total degrees

(198°/s)·(0.32 s) ≈ 63°

6. B is correct.

momentum = mass × velocity

$p = mv$

Since *momentum is directly proportional to mass*, doubling the mass doubles the momentum.

7. D is correct.

The *total momentum before* the collision is:

$p_{total} = m_I v_I + m_{II} v_{II} + m_{III} v_{III}$

p_{before} = (1 kg)·(0.5 m/s) + (1.5 kg)·(–0.3 m/s) + (3.5 kg)·(–0.5 m/s)

p_{before} = (0.5 kg·m/s) + (–0.45 kg·m/s) + (–1.75 kg·m/s)

p_{before} = –1.7 kg·m/s

8. A is correct.

The collision of I and II does not affect the momentum of the system:

$p_{before} = p_{after}$

$p_{I\ \&\ II}$ = (1 kg)·(0.5 m/s) + (1.5 kg)·(–0.3 m/s)

$p_{I\ \&\ II}$ = (0.5 kg·m/s) – (0.45 kg·m/s)

$p_{I\ \&\ II}$ = 0.05 kg·m/s

p_{III} = (3.5 kg)·(–0.5 m/s)

p_{III} = –1.75 kg·m/s

$p_{net} = p_{I\ and\ II} + p_{III}$

p_{net} = (0.05 kg·m/s) + (–1.75 kg·m/s)

p_{net} = –1.7 kg·m/s

Momentum is always conserved.

9. B is correct.

Set initial momentum equal to the final momentum after the collisions have occurred.

$p_{before} = p_{after}$

$p_{before} = (m_I + m_{II} + m_{III})v_f$

–1.7 kg·m/s = (1 kg + 1.5 kg + 3.5 kg)v_f

v_f = (–1.7 kg·m/s) / (6 kg)

v_f = –0.28 m/s

10. C is correct.

Momentum is conserved in this system. The momentum of each car is given by mv, and the sum of the momenta before the collision must equal the sum of the momenta after the collision:

$$p_{\text{before}} = p_{\text{after}}$$

Solve for the velocity of the first car after the collision.

Each car travels in the same direction before and after the collision, so each velocity value has the same sign.

$$m_1v_{i1} + m_2v_{i2} = m_1v_{f1} + m_2v_{f2}$$

$$(480\ \text{kg})\cdot(14.4\ \text{m/s}) + (570\ \text{kg})\cdot(13.3\ \text{m/s}) = (480\ \text{kg})\cdot(v_{f2}) + (570\ \text{kg})\cdot(17.9\ \text{m/s})$$

$$(480\ \text{kg})\cdot(v_{f2}) = (480\ \text{kg})\cdot(14.4\ \text{m/s}) + (570\ \text{kg})\cdot(13.3\ \text{m/s}) - (570\ \text{kg})\cdot(17.9\ \text{m/s})$$

$$v_{f2} = [(480\ \text{kg})\cdot(14.4\ \text{m/s}) + (570\ \text{kg})\cdot(13.3\ \text{m/s}) - (570\ \text{kg})\cdot(17.9\ \text{m/s})] / (480\ \text{kg})$$

$$v_{f2} = 8.9\ \text{m/s} \approx 9\ \text{m/s}$$

11. D is correct.

Impulse is a force acting over a period of time:

$$J = F\Delta t$$

An impulse changes a system's momentum, so:

$$F\Delta t = \Delta p_{\text{system}}$$

The moving block with the lodged bullet comes to a stop when it compresses the spring, losing all momentum.

Initial velocity of the block and bullet separately can be determined by conservation of energy.

The two values of interest are the KE of the block and bullet and the PE of the spring.

$$(\text{KE} + \text{PE})_{\text{before}} = (\text{KE} + \text{PE})_{\text{after}}$$

$$½mv^2 + 0 = 0 + ½kx^2$$

x = distance of compression of the spring

k = spring constant

$$½(4\ \text{kg} + 0.008\ \text{kg})v^2 = ½(1{,}400\ \text{N/m})\cdot(0.089\ \text{m})^2$$

$$v^2 = (1{,}400\ \text{N/m})\cdot(0.089\ \text{m})^2 / (4.008\ \text{kg})$$

$$v^2 = 2.76\ \text{m}^2/\text{s}^2$$

$$v = 1.66\ \text{m/s}$$

Thus, the block with the lodged bullet hits the spring with an initial velocity of 1.66 m/s.

Since there is no friction, the block is sent in the opposite direction with the same speed of 1.66 m/s when the spring decompresses.

continued…

Calculate the momentum, with initial momentum toward the spring and final momentum away from the spring.

$\Delta p = p_{\text{final}} - p_{\text{initial}}$

$\Delta p = (4.008\ \text{kg})\cdot(-1.66\ \text{m/s}) - (4.008\ \text{kg})\cdot(1.66\ \text{m/s})$

$\Delta p = (-6.65\ \text{kg}\cdot\text{m/s}) - (6.65\ \text{kg}\cdot\text{m/s})$

$\Delta p \approx -13\ \text{kg}\cdot\text{m/s}$

$\Delta p \approx -13\ \text{N}\cdot\text{s}$

Since $F\Delta t = \Delta p$, the impulse is $-13\ \text{kg}\cdot\text{m/s} = -13\ \text{N}\cdot\text{s}$

The negative sign signifies the coordinate system chosen in this calculation: toward the spring is the positive direction, and away from the spring is the negative direction.

12. C is correct.

For a rotating body, kinetic energy is:

$K = ½\, I\, \omega^2$

Angular momentum is:

$L = I\, \omega$

Therefore:

$I = L\, /\, \omega$

Replacing this for I in the expression for kinetic energy:

$K = L^2\, /\, 2I$

Taking the ice to be frictionless, there is no external torque on the skater.

Thus, angular momentum is conserved and does not change as she brings in her arms.

The moment of inertia of a body of a given mass is smaller if its mass is more concentrated toward the rotation axis (e.g., when she draws her arms in close). Therefore, the moment of inertia of the skater decreases.

Consequently, the skater's kinetic energy increases.

13. D is correct.

The centripetal force is the net force required to maintain an object in uniform circular motion.

$F_{\text{centripetal}} = mv^2/r$

where r is the radius of the circular path

Since m is constant and r remains unchanged, the centripetal force is proportional to v^2.

$2^2 = 4$

Thus, if v is doubled, then $F_{\text{centripetal}}$ is quadrupled.

14. B is correct.

$1\ \text{J} = \text{kg·m}^2/\text{s}^2$

$p = mv = \text{kg·m/s}$

$\text{J·s/m} = (\text{kg·m}^2/\text{s}^2)\text{·(s/m)}$

$\text{J·s/m} = \text{kg·m/s}$

$\text{kg·m/s} = p$

$\text{J·s/m} = p$

15. D is correct.

Impulse is a change in momentum.

$J = \Delta p$

$J = m\Delta v$

Impulse is the product of average force and time.

$J = F\Delta t$

$F\Delta t = m\Delta v$

$ma\Delta t = m\Delta v$

Cancel m from each side of the expression:

$a\Delta t = \Delta v$

Because acceleration g is constant impulse depends only upon time and velocity.

The speed of the apple affects the impulse, as this is included in the Δv term.

Bouncing results in a change in direction; a greater change in velocity (the Δv term), so the impulse is greater.

The time of impulse changes the impulse as it is included in the Δt term.

16. D is correct.

$F\Delta t = m\Delta v$

$F = m\Delta v / \Delta t$

Choosing toward the wall as the positive direction, the initial velocity is 25 m/s, and the final is –25 m/s:

$F = m(v_f - v_i) / \Delta t$

$F = (0.8\ \text{kg})\cdot(-25\ \text{m/s} - 25\ \text{m/s}) / (0.05\ \text{s})$

$F = -800\ \text{N}$

Thus, the wall exerts an average force of 800 N on the ball in a negative direction.

From Newton's Third Law, the ball exerts a force of 800 N on the wall in the opposite direction.

17. B is correct.

$p = mv$

Sum momentum:

$p_{total} = m_1v_1 + m_2v_2 + m_3v_3$

All objects moving to the left have negative velocity.

$p_{total} = (7 \text{ kg})\cdot(6 \text{ m/s}) + (12 \text{ kg})\cdot(3 \text{ m/s}) + (4 \text{ kg})\cdot(-2 \text{ m/s})$

$p_{total} = (42 \text{ kg}\cdot\text{m/s}) + (36 \text{ kg}\cdot\text{m/s}) + (-8 \text{ kg}\cdot\text{m/s})$

$p_{total} = 70 \text{ kg}\cdot\text{m/s}$

18. D is correct.

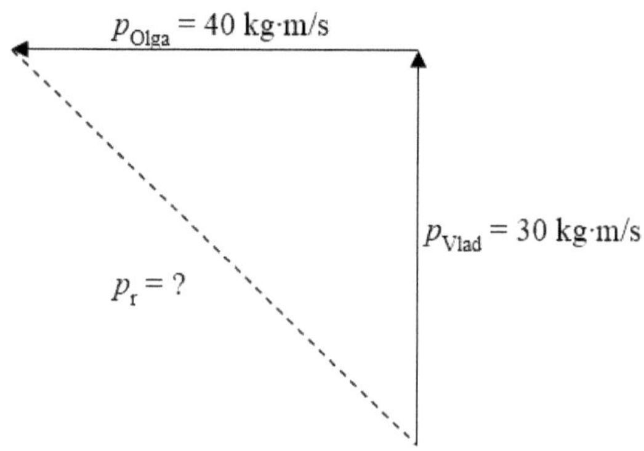

Use conservation of momentum to determine the momentum after the collision. Since they stick, treat it as a perfectly inelastic collision.

Before the collision, Vladimir's momentum is:

$(60 \text{ kg})\cdot(0.5 \text{ m/s}) = 30 \text{ kg}\cdot\text{m/s}$ pointing North

Before the collision, Olga's momentum is:

$(40 \text{ kg})\cdot(1 \text{ m/s}) = 40 \text{ kg}\cdot\text{m/s}$ pointing West

Write two expressions: one for the conservation of momentum on the y-axis (North-South) and one for the conservation of momentum on the x-axis (East-West). They do not interact since as perpendiculars.

Since Olga and Vladimir stick, the final mass is the sum of their masses.

Simply use the Pythagorean Theorem:

$a^2 + b^2 = c^2$

$(30 \text{ kg}\cdot\text{m/s})^2 + (40 \text{ kg}\cdot\text{m/s})^2 = p^2$

$900 (\text{kg}\cdot\text{m/s})^2 + 1{,}600 (\text{kg}\cdot\text{m/s})^2 = p^2$

$2{,}500(\text{kg}\cdot\text{m/s})^2 = p^2$

$p = 50 \text{ kg}\cdot\text{m/s}$

Use this to solve for velocity:

$p = mv$

$50 \text{ kg}\cdot\text{m/s} = (100 \text{ kg})v$

$v = 50 \text{ kg}\cdot\text{m/s} / 100 \text{ kg}$

$v = 0.5 \text{ m/s}$

continued…

Also, this problem can be solved algebraically:

$p_{before} = p_{after}$

$p = mv$

On the y-coordinate:

$(60\text{ kg})\cdot(0.5\text{ m/s}) = (60\text{ kg} + 40\text{ kg})v_y$

$v_y = (30\text{ kg·m/s}) / (100\text{ kg})$

$v_y = 0.3\text{ m/s}$

On the x-coordinate:

$(40\text{ kg})\cdot(1\text{ m/s}) = (60\text{ kg} + 40\text{ kg})v_x$

$v_x = (40\text{ kg·m/s}) / (100\text{ kg})$

$v_x = 0.4\text{ m/s}$

Combine these final velocity components using the Pythagorean Theorem since they are perpendicular.

$v^2 = v_x^2 + v_y^2$

$v^2 = (0.4\text{ m/s})^2 + (0.3\text{ m/s})^2$

$v = 0.5\text{ m/s}$

19. B is correct.

Use conservation of momentum to determine the momentum after the collision. Since they stick, treat it as a perfectly inelastic collision.

Before collision, Vladimir's momentum is:

$(60\text{ kg})\cdot(0.5\text{ m/s}) = 30\text{ kg·m/s}$ pointing North

Before collision, Olga's momentum is:

$(40\text{ kg})\cdot(1\text{ m/s}) = 40\text{ kg·m/s}$ pointing West

Write two expressions: one for the conservation of momentum on the y coordinate (North-South) and one for the conservation of momentum on the x coordinate (East-West).

They do not interact since they are perpendiculars. Since they stick, the final mass is the sum of their masses.

Simply use the Pythagorean Theorem:

$a^2 + b^2 = c^2$

$(30\text{ kg·m/s})^2 + (40\text{ kg·m/s})^2 = p^2$

$900(\text{kg·m/s})^2 + 1{,}600(\text{kg·m/s})^2 = p^2$

$2{,}500(\text{kg·m/s})^2 = p^2$

$p = 50\text{ kg·m/s}$

continued…

This problem can also be solved algebraically:

$p_{before} = p_{after}$

$p = mv$

On the *y*-coordinate:

$(60\text{ kg})\cdot(0.5\text{ m/s}) = (60\text{ kg} + 40\text{ kg})v_y$

$v_y = (30\text{ kg}\cdot\text{m/s}) / (100\text{ kg})$

$v_y = 0.3\text{ m/s}$

On the *x*-coordinate:

$(40\text{ kg})\cdot(1\text{ m/s}) = (60\text{ kg} + 40\text{ kg})v_x$

$v_x = (40\text{ kg}\cdot\text{m/s}) / (100\text{ kg})$

$v_x = 0.4\text{ m/s}$

Combine these final velocity components using the Pythagorean Theorem since they are perpendiculars:

$v^2 = v_x^2 + v_y^2$

$v^2 = (0.4\text{ m/s})^2 + (0.3\text{ m/s})^2$

$v = 0.5\text{ m/s}$

Use the final weight and final velocity to find the final momentum directly after the collision:

$p = mv$

$p = (60\text{ kg} + 40\text{ kg})\cdot(0.5\text{ m/s})$

$p = 50\text{ kg}\cdot\text{m/s}$

20. C is correct.

$p_0 = mv$

If *m* and *v* are doubled:

$p = (2m)\cdot(2v)$

$p = 4mv$

$p = 4p_0$

The momentum increases by a factor of 4.

21. D is correct.

Balance forces on box Q to solve for tension on box P cable:

$$m_Q a = F - T_P$$

$$T_P = F - m_Q a$$

$$0 < T_P < F$$

Thus, the tension on the cable connected to box P is less than F because it is equal to the difference between F and $m_Q a$ but is not equal because the boxes are accelerating.

22. B is correct.

At all points on a rotating body, the angular velocity is equal. The speed at different points along a rotating body is directly proportional to the radius.

$$v = \omega r$$

where v = speed, ω = angular velocity and r = radius

Thus, Melissa and her friend have different speeds due to their different radial locations.

23. A is correct.

Impulse is directly proportional to force and change in time:

$$J = F\Delta t$$

Increasing the change in time lowers the impact force while decreasing the change in time increases the force.

24. D is correct.

Angular momentum is always conserved unless a system experiences a net torque greater than zero, and this is the rotational equivalent of Newton's First Law of motion.

25. D is correct.

$$F\Delta t = m\Delta v$$

$$F = (m\Delta v) / (\Delta t)$$

$$F = (6.8\text{ kg})\cdot(-3.2\text{ m/s} - 5.4\text{ m/s}) / (2\text{ s})$$

$$F = (-58.48\text{ kg}\cdot\text{m/s}) / (2\text{ s})$$

$$F = -29.2\text{ N}$$

$$|F| = 29.2\text{ N}$$

26. A is correct.

Before collision, the total momentum of the system = 0 kg·m/s.

Momentum is conserved in the explosion.

The momentum of the moving rifle and bullet are in opposite directions:

Therefore, $p = 0$

The total momentum after the explosion = 0 kg·m/s

27. D is correct.

$$p = mv$$

Conservation of momentum:

$$p_{initial} = p_{final}$$

$$0 \text{ kg·m/s} = (0.01 \text{ kg})·(300 \text{ m/s}) + (4 \text{ kg})v_{recoil}$$

$$0 \text{ kg·m/s} = 3 \text{ kg·m/s} + (4 \text{ kg})v_{recoil}$$

$$–3 \text{ kg·m/s} = (4 \text{ kg})v_{recoil}$$

$$(–3 \text{ kg·m/s}) / (4 \text{ kg}) = v_{recoil}$$

$$v_{recoil} = –0.75 \text{ m/s}$$

Velocity is negative since the gun recoils in the opposite direction of the bullet.

28. C is correct.

Since the initial velocity only has a horizontal component, the y component of the initial velocity = 0.

Use 24 m to calculate the time the ball is in the air:

$$d_y = ½at^2$$

$$t^2 = 2d_y / a$$

$$t^2 = 2(24 \text{ m}) / (9.8 \text{ m/s}^2)$$

$$t^2 = 4.89 \text{ s}^2$$

$$t = 2.21 \text{ s}$$

Use the time in the air and the horizontal distance to calculate the horizontal speed of the ball:

$$v_x = d_x / t$$

$$v_x = (18 \text{ m}) / (2.21 \text{ s})$$

$$v_x = 8.1 \text{ m/s}$$

29. A is correct.

An object moving in a circle at constant speed is undergoing uniform circular motion.

In uniform circular motion, the acceleration is due to centripetal acceleration and points inward towards the center of a circle.

30. B is correct.

Impulse:

$J = F\Delta t$

$J = \Delta p$

where p is momentum

31. C is correct.

Conservation of energy:

$KE_i + PE_i = KE_f + PE_f$

$KE_i + PE_i = KE_f + 0$

$KE_f = ½mv_i^2 + mgh_i$

$KE_f = ½(4\ kg)·(20\ m/s)^2 + (4\ kg)·(10\ m/s^2)·(10\ m)$

$KE_f = 800\ J + 400\ J$

$KE_f = 1{,}200\ J$

32. D is correct.

The force needed to stop a car can be related to KE and work:

$KE = W$

$½mv^2 = Fd$

$F = ½mv^2 / d$

Momentum is included in the KE term.

$p = mv$

$F = ½(mv)v / d$

$F = ½(p)v / d$

If there is less stopping distance, the force increases as they are inversely proportional.

If the momentum or mass increases, the force increases as they are directly proportional.

33. C is correct.

Impulse:

$$J = F\Delta t$$

Based on Newton's Third Law, the force experienced by these two objects is equal and opposite.

Therefore, the magnitudes of impulse are the same.

34. B is correct.

Balance the counterclockwise (CCW) torque with the clockwise (CW) torque.

Let the axis of rotation be at the point where the rope attaches to the bar.

This placement causes the torque from the rope to be zero since the lever arm is zero.

$$\Sigma \tau : \tau_1 - \tau_2 = 0$$

$$\tau_1 = \tau_2$$

The CCW torque due to the weight of the 6 kg mass:

$$\tau = r_1 F_1$$

$$r_1 F_1 = (x)\cdot(6\ \text{kg})\cdot(9.8\ \text{m/s}^2)$$

The CW torque due to the weight of the 30 kg mass:

$$r_2 F_2 = (5\ \text{m} - x)\cdot(30\ \text{kg})\cdot(9.8\ \text{m/s}^2)$$

Set the two expressions equal

$$(9.8\ \text{m/s}^2)\cdot(x)\cdot(6\ \text{kg}) = (5\ \text{m} - x)\cdot(30\ \text{kg})\cdot(9.8\ \text{m/s}^2)$$

Cancel g and kg from each side of the equation:

$$6x = 30(5\ \text{m} - x)$$

$$6x = 150\ \text{m} - 30x$$

$$36x = 150\ \text{m}$$

$$x = 4.2\ \text{m}$$

35. D is correct.

If the block is at rest, then the force of static friction equals the force of gravity at angle θ.

$$F_f = mg \sin \theta$$

36. C is correct.

$F_{net} = 0$ is necessary to maintain a constant velocity.

If 45 N must be exerted on the block to maintain a constant velocity, the force of kinetic friction against the block equals 45 N.

For a horizontal surface and no other vertical forces acting, the normal force on the block equals its weight.

$$N = mg$$

$$F_{friction} = \mu_k N$$

$$F_{friction} = \mu_k mg$$

$$\mu_k = (F_{friction}) / mg$$

$$\mu_k = (45\ \text{N}) / [(30\ \text{kg})\cdot(10\ \text{m/s}^2)]$$

$$\mu_k = 0.15$$

37. B is correct.

Newton's Second Law:

$$F = ma$$

The impulse-momentum relationship can be derived by multiplying Δt on each side:

$$F\Delta t = ma\Delta t$$

$$F\Delta t = m\Delta v$$

$$J = m\Delta v$$

Thus, the impulse is equal to the change in momentum.

38. C is correct.

Force X acts perpendicular to the short arm of the rectangle; this is the lever arm.

$$\tau = rF$$

$$\tau = (0.5\ \text{m})\cdot(15\ \text{N})$$

$$\tau = 7.5\ \text{N}\cdot\text{m}$$

Since the torque causes the plate to rotate clockwise its sign is negative.

$$\tau = -7.5\ \text{N}\cdot\text{m}$$

39. D is correct.

$\tau = rF$

Force Z acts directly at the pivot, so the lever arm equals zero.

$\tau = (0 \text{ m})\cdot(30 \text{ N})$

$\tau = 0 \text{ N}\cdot\text{m}$

40. A is correct.

$\tau = rF$

Force Y acts perpendicular to the long arm of the rectangle; this is the lever arm.

$\tau = (0.6 \text{ m})\cdot(25 \text{ N})$

$\tau = 15 \text{ N}\cdot\text{m}$

The torque is clockwise, so its sign is negative.

$\tau = -15 \text{ N}\cdot\text{m}$

41. B is correct.

The tension in the string provides the centripetal force.

$T = mv^2 / r$

$m = 50 \text{ g} = 0.05 \text{ kg}$

$T = [(0.05 \text{ kg})\cdot(20 \text{ m/s})^2] / (2 \text{ m})$

$T = [(0.05 \text{ kg})\cdot(400 \text{ m}^2/\text{s}^2)] / (2 \text{ m})$

$T = (20 \text{ kg}\cdot\text{m}^2/\text{s}^2) / (2 \text{ m})$

$T = 10 \text{ N}$

42. A is correct.

Newton's Third Law states that each force is paired with an equal and opposite reaction force.

Therefore, the small car and the truck each receive the same force.

43. C is correct.

Choose the axis of rotation at the point where the bar attaches to the wall.

Since the lever arm of the force that the wall exerts is zero, the torque at that point is zero and can be ignored.

The two other torques present arise from the weight of the bar exerting a force downward and the cable exerting force upward.

continued…

The weight of the bar acts at the center of mass, so its lever arm is 1 m.

The lever arm for the cable is 2 m since it acts the full 2 m away from the wall at the end of the bar.

Torque is the product of the length of the lever arm and the component of force perpendicular to the arm.

The torque applied by the wire is:

$F_T l \sin\theta$

The sum of torques = 0 since the bar is in rotational equilibrium.

Let the torque of the cable be positive, and the torque of the weight be negative.

$(F_T \sin 30°)\cdot(2\text{ m}) - (10\text{ kg})\cdot(10\text{ m/s}^2)\cdot(1\text{ m}) = 0$

$F_T = [(10\text{ kg})\cdot(10\text{ m/s}^2)\cdot(1\text{ m})] / [(2\text{ m})\cdot(\sin 30°)]$

$F_T = [(10\text{ kg})\cdot(10\text{ m/s}^2)\cdot(1\text{ m})] / [(2\text{ m})\cdot(0.5)]$

$F_T = 100\text{ N}$

44. B is correct.

Momentum is defined as:

$p = mv$

$m_A = 2m_B$

$p_A = 2m_Bv$

$p_B = m_Bv$

$p_A = 2p_B$

If both objects reach the ground at the same time, they have equal velocities.

Because A is twice the mass, it has twice the momentum as object B.

45. D is correct.

Use conservation of momentum to make equations for momenta along the x-axis and the y-axis.

Since the mass ratio is 1 : 4, one car has a mass of m, and the other has a mass of $4m$.

The entangled cars after the collision have a combined mass of $5m$.

Let the car of mass m be traveling in the positive x-direction, and the car of mass $4m$ be traveling in the positive y-direction.

The choice of directions here is arbitrary, but the angle of impact is important.

$p_{\text{initial}} = p_{\text{final}}$ for both the x- and y-axes

$p = mv$

continued…

For the x-axis:

$$m_i v_i = m_f v_{fx}$$

$$m(12 \text{ m/s}) = 5mv_x$$

Cancel m from each side of the expression:

$$12 \text{ m/s} = 5v_x$$

$$v_x = 2.4 \text{ m/s}$$

For the y-axis:

$$m_i v_i = m_f v_{fy}$$

$$4m(12 \text{ m/s}) = 5mv_y$$

Cancel m from each side of the expression:

$$4(12 \text{ m/s}) = 5v_y$$

$$v_y = 9.6 \text{ m/s}$$

The question asks for the magnitude of the final velocity, so combine the x and y components of the final velocity using the Pythagorean Theorem.

$$v^2 = (2.4 \text{ m/s})^2 + (9.6 \text{ m/s})^2$$

$$v^2 = 5.76 \text{ m}^2/\text{s}^2 + 92.16 \text{ m}^2/\text{s}^2$$

$$v = 9.9 \text{ m/s}$$

46. C is correct.

Use conservation of momentum on the horizontal plane.

Before the throw, the total momentum of the skater-ball system is zero. Thus, after the throw, the total horizontal momentum must sum to zero: the horizontal component of the ball's momentum equals the momentum of the skater moving the opposite way.

Use m_s for the skater's mass and $m_s/3$ for the ball's mass.

$$p = mv$$

$$p_{\text{skater}} = p_{\text{ball}}$$

$$m_s v_s = m_b v_b$$

$$m_s(2.9 \text{ m/s}) = (1/3)m_s v \cos 5°$$

Cancel m from each side of the expression:

$$v = (2.9 \text{ m/s})\cdot(3) / (\cos 5°)$$

$$v = (2.9 \text{ m/s})\cdot(3) / (0.996)$$

$$v = 8.73 \text{ m/s}$$

47. B is correct.

weight = mass × gravity

$W = mg$

$m = W / g$

$m = (98\ \text{N}) / (9.8\ \text{m/s}^2)$

$m = 10\ \text{kg}$

Newton's Second Law:

$F = ma$

$F = (10\ \text{kg})\cdot(10\ \text{m/s}^2)$

$F = 100\ \text{N}$

48. A is correct.

KE is constant because speed is constant. PE increases because the cart is at a greater height at point B.

The cart as a system is not isolated since the winch does work on it, so its energy is not conserved.

Conservation of energy:

PE increase of the cart = work done by the winch

49. D is correct.

The vertical component of the initial velocity:

$v_{iy} = (140\ \text{m/s})\ \sin 35°$

$v_{iy} = (140\ \text{m/s})\cdot(0.57)$

$v_{iy} = 79.8\ \text{m/s}$

The initial velocity upward, time elapsed, and acceleration due to gravity is known.

Determine the final velocity after 4 s.

$v_y = v_{iy} + at$

$v_y = 79.8\ \text{m/s} + (-9.8\ \text{m/s}^2)\cdot(4\ \text{s})$

$v_y = 41\ \text{m/s}$

50. C is correct.

impulse = force × time

$J = F\Delta t$

51. D is correct.

Conservation of momentum: the momentum of the fired bullet is equal and opposite to that of the rifle.

$$p = mv$$

$$p_{\text{before}} = p_{\text{after}}$$

$$0 = p_{\text{rifle}} + p_{\text{bullet}}$$

$$-p_{\text{rifle}} = p_{\text{bullet}}$$

$$-(2\text{ kg})v = (0.01\text{ kg})\cdot(220\text{ m/s})$$

$$v = (0.01\text{ kg})\cdot(220\text{ m/s}) / (-2\text{ kg})$$

$$v = -1.1\text{ m/s}$$

Thus, the velocity of the rifle is 1.1 m/s in the opposite direction as the bullet.

52. D is correct.

Airbags reduce the force by increasing the time of contact between the passenger and the surface.

In a collision, an impulse is experienced by a passenger:

$$J = F\Delta t$$

$$F = J / \Delta t$$

The impulse is constant, but the force experienced by the passenger is inversely related to the time of contact.

Airbags increase the time of impact and thus reduce the forces experienced by the person.

53. A is correct.

Since Force I is perpendicular to the beam, the entire force produces torque without a horizontal force component.

$$\tau = rF$$

$$\tau = (0.5\text{ m})\cdot(10\text{ N})$$

$$\tau = 5\text{ N}\cdot\text{m}$$

Because the force causes the beam to rotate clockwise against the positive counterclockwise direction, the torque sign should be negative:

$$\tau = -5\text{ N}\cdot\text{m}$$

54. D is correct.

To calculate torque, use the 35° angle.

For torque:

$\tau = rF \sin \theta$

$\tau = (1 \text{ m})\cdot(5 \text{ N}) \sin 35°$

$\tau = 2.9 \text{ N}\cdot\text{m}$

The torque is counterclockwise, so the sign is positive.

55. B is correct.

Force III acts purely in tension with the beam and has no component acting vertically against the beam.

Torque can only be calculated using a force with some component perpendicular to the length vector.

Because Force III has no perpendicular component to the length vector, torque is zero.

$\tau = rF$

$\tau = (1 \text{ m})\cdot(0 \text{ N})$

$\tau = 0 \text{ N}\cdot\text{m}$

56. C is correct.

Impulse can be written as:

$J = m\Delta v$

$J = F\Delta \text{t}$

Impulse is the change in the momentum of an object.

Because the yellow ball bounced higher, it can be concluded that its upward velocity after the collision must be higher than that of the red ball:

$\Delta v_{\text{yellow}} > \Delta v_{\text{red}}$

Thus, because the mass of both balls is the same, the yellow ball must have a greater impulse according to the impulse equation:

$m\Delta v_{\text{yellow}} > m\Delta v_{\text{red}}$

57. D is correct.

$J = F\Delta t$

$J = (4.5 \text{ N})\cdot(1.4 \text{ s})$

$J = (4.5 \text{ kg}\cdot\text{m/s}^2)\cdot(1.4 \text{ s})$

$J = 6.3 \text{ kg}\cdot\text{m/s}$

58. A is correct.

Both trucks experience the same acceleration due to gravity, so their acceleration and velocity are equal because these do not depend on mass:

$v_f = v_0 + a\Delta t$

However, their momentum is different, and the heavier truck has a larger momentum because of its larger mass.

$p = mv$

$m_H > m_L$

$p_H = m_H v$

$p_L = m_L v$

$p_H > p_L$

59. D is correct.

The time elapsed from release until a collision is:

time from release until collision = round-trip time / 2

$t = (4\text{ s}) / 2$

$t = 2\text{ s}$

The time of contact is negligible to the round-trip time, so this calculation ignores it.

Since this collision is elastic, the time from release until the collision is the same as the time from the collision until the ball reaches the same height again.

Given this time in the air, find the *velocity* of the ball immediately before impact:

$v = v_i + at$

$v = 0 + (9.8\text{ m/s}^2)\cdot(2\text{ s})$

$v = 19.6\text{ m/s}$

Find the *KE* of the ball before impact:

$KE = \frac{1}{2}mv^2$

$KE = \frac{1}{2}(0.078\text{ kg})\cdot(19.6\text{ m/s})^2$

$KE = 15\text{ J}$

The KE of the ball is stored as elastic energy during the collision and is then converted back to KE to send the ball upward in the opposite direction.

This stored elastic energy is equivalent to the KE before the collision.

60. B is correct.

Consider the system to be the set containing both carts.

The force on the initial object provides an impulse to the system of:

$$I = F\Delta t$$

An impulse causes a change in the momentum of the system:

$$I = \Delta p$$

The initial momentum of the system is zero, and momentum is conserved during the collision.

Since the two carts stick after the collision, the final momentum is:

$$p_f = v_f(m_A + m_A) = \Delta p = F\Delta t$$

Therefore:

$$v_f = F\Delta t \,/\, (m_A + m_A)$$

$$v_f = (3\text{ N})\ (2\text{ s}) \,/\, (5\text{ kg} + 10\text{ kg})$$

$$v_f = 0.4\text{ m/s}$$

Notes for active learning

Rotational Motion – Detailed Explanations

1. D is correct.

An object is rolling down an incline experiences three forces, and hence three torques.

The forces are the force of gravity acting on the center of mass of the object, the normal force between the incline and the object, and the force of friction between the incline and the object.

If the origin is taken to be the center of the object, the force of gravity provides zero torque. This can be seen by noting that the distance between the origin and the point of application of the force is zero.

$$\tau_{\text{gravity}} = F_{\text{gravity}}r = mg(0) = 0$$

Similarly, the normal force contributes zero torque because the direction of the force is directed through the origin (pivot point).

$$\tau_{normal} = F_{normal}\,\text{r}\sin\theta = F_{normal}(R)\cdot(\sin 180°) = F_{normal}\,(R)\cdot(0) = 0$$

Use a coordinate system in which the x-axis is parallel to the incline and the y-axis is perpendicular.

The object is rolling in the positive x-direction.

The dynamical equation for linear motion along the x-direction is:

$$F_{\text{net}} = ma$$

$$(mg\sin\theta - f) = ma$$

Note that the normal force is only in the y-direction and thus does not directly contribute to the acceleration in the x-direction.

The dynamical equation for rotational motion is:

$$\tau_{\text{net}} = I\alpha$$

$fR = I\alpha$ (Note that the frictional force is perpendicular to the r vector, and sin 90° = 1), where R is the radius of the object, f is the force of friction, and I is the moment of inertia.

A relation coupling these two dynamical equations is needed.

The equation of constraint imposed by the restriction that the object rolls without slipping:

$$\alpha = a / R$$

To find the linear acceleration, use the equation of constraint to eliminate α from the rotational equation by replacing it with a / R:

$$fR = I(a / R)$$

The force of friction is of no interest, so rearrange this last expression:

$$f = Ia / R^2$$

continued…

Substitute this into the linear dynamic equation from above in place of f:

$$(mg \sin \theta - Ia / R^2) = ma$$

Solving this for a:

$$a = mg \sin \theta / [m + (I / R^2)]$$

$$a = g \sin \theta / [1 + (I / mR^2)$$

The moment of inertia of any circular object can be written as NmR^2, where N is some real number different for different shapes.

For example, for a sphere,

$I = (2/5)mR^2$, so for a sphere $N = 2/5$.

So, for any rolling object, the linear acceleration is:

$a = g \sin \theta / (1 + N)$, which depends on neither the radius nor the mass of the object.

Only the shape of the object is important.

2. B is correct.

Use the conservation of energy.

The initial and final states are the sphere at the top and bottom of the ramp, respectively.

Take the zero of gravitational potential energy to be the configuration in which the sphere is at the bottom.

The potential energy at the bottom is zero.

The sphere starts from rest, so the kinetic energy at the top is zero:

$$mgh = K_{linear} + K_{rotation}$$

$$mgh = ½mv^2 + ½I\omega^2$$

For a sphere,

$$I = (2/5)mr^2$$

Because the sphere rolls without slipping,

$v = r\omega$.

Substituting these into the conservation of energy equation:

$$mgh = ½mr^2\omega^2 + ½(2/5)mr^2\omega^2$$

$$mgh = ½mr^2\omega^2 + (2/10)mr^2\omega^2$$

$$mgh = (7/10)mr^2\omega^2$$

continued…

Isolating ω:

$$\omega = \sqrt{(10gh / 7r^2)}$$

$$\omega = \sqrt{[10\cdot(9.8\ \text{m/s}^2)\cdot(5.3\ \text{m})] / [7\cdot(1.7\ \text{m})^2]}$$

$$\omega = 5.1\ \text{rad/s}$$

3. A is correct.

There is no torque on the ball during the fall; its rotational speed does not change.

Therefore, the rotational KE just before the ball hits the floor is the same as rolling on the horizontal surface.

The rotational kinetic energy when it was rolling on the surface can be calculated directly.

Recall that the moment of inertia of a solid sphere is:

$$I = 2/5mR^2$$

$$K_{rot} = ½I\omega^2$$

$$K_{rot} = ½(2/5)mR^2\omega^2$$

Since the ball is rolling without slipping,

$$\omega = v / R$$

$$K_{rot} = ½(2/5)mR^2(v / R)^2$$

$$K_{rot} = (2/10)mv^2$$

$$K_{rot} = (2/10)mv^2$$

$$K_{rot} = (2/10)\cdot(0.125\ \text{kg})\cdot(4.5\ \text{m/s})^2$$

$$K_{rot} = 0.51\ \text{J}$$

4. D is correct.

The angular momentum of an object in circular motion is:

$$L = I\omega$$

where I is the moment of inertia with respect to the center of motion and ω is the angular speed.

The moment of inertia of a point mass is:

$$I = mr^2$$

The angular momentum is then:

$$L = mr^2\omega$$

Angular speed is in rev/s.

continued…

Express in rad/s:

$$1.2 \text{ rev/s}\cdot(2\pi \text{ rad/rev}) = 7.540 \text{ rad/s}$$

Finally:

$$L = (0.38 \text{ kg})\cdot(1.3 \text{ m})^2\cdot(7.540 \text{ rad/s})$$

$$L = 4.8 \text{ kg m}^2\text{/s}$$

5. C is correct.

Conservation of angular momentum requires:

$$L_f = L_i$$

$$I_f\omega_f = I_i\omega_i$$

The final angular speed is:

$$\omega_f = \omega_i\,(I_i / I_f)$$

$$\omega_f = (3.0 \text{ rev/s})\cdot(5.0 \text{ kg}\cdot\text{m}^2) / (2.0 \text{ kg m}^2)$$

$$\omega_f = 7.5 \text{ rev/s}$$

6. C is correct.

An external torque changes the angular velocity of a system:

$\alpha = \sum\tau / I$, and hence its angular momentum.

To maintain a constant angular momentum, the sum of external torques must be zero.

7. D is correct.

For a rotating circular object:

$$\omega = v / r$$

$$\omega = v / (d / 2)$$

$$\omega = (4.0 \text{ m/s}) / [(0.60 \text{ m}) / 2]$$

$$\omega = 13.3 \text{ rad/s}$$

8. A is correct.

$$K = ½I\omega^2$$

The moment of inertia of a rod with respect to its "short axis" is:

$$I = (1/12)ml^2$$

$$K = (1/24)ml^2\omega^2$$

$$K = (1/24)\cdot(0.4500 \text{ kg})\cdot(1.20 \text{ m})^2\cdot(3.60 \text{ rad/s})^2$$

$$K = 0.350 \text{ J}$$

9. C is correct.

The moment of inertia can be found from the dynamic relation:

$$\tau = I\alpha$$

$$I = \tau / \alpha$$

where τ is the torque applied to the pulley, and α is the pulley's angular acceleration.

Torque is defined as:

$$FR \sin \theta$$

In this case, the force F is the tension force from the rope, R is the radius of the wheel, and $\theta = 90°$.

Thus, the torque is just the product of the tension of the rope and the radius of the pulley:

$$\tau = TR$$

The angular acceleration is related to the acceleration of a point on the circumference of the pulley:

$$\alpha = a / R$$

where a is the linear acceleration at the circumference, and R is the pulley's radius.

Combining these two results, the moment of inertia is:

$$I = TR^2 / a$$

If the rope does not slip on the pulley, then the rope, and hence the hanging mass, also has an acceleration a.

To continue, find the acceleration and the tension.

The tension is found by applying Newton's Second Law to the hanging mass.

There are two forces on the hanging mass; force of gravity pointing down and tension of the rope pointing up.

From Newton's Second Law (with down as the positive direction):

$$(mg - T) = ma$$

$$T = m(g - a)$$

With that, the moment of inertia becomes:

$$I = mR^2[(g - a) / a]$$

The acceleration can be found from the kinematic information given about the movement of the hanging mass.

The relation needed is:

$$\Delta y = \tfrac{1}{2}a(\Delta t)^2 + v_0(\Delta t)$$

$$a = 2\Delta y / (\Delta t)^2 = 2\cdot(10\text{ m}) / (2\text{ s})^2$$

$$a = 5.000\text{ m/s}^2$$

continued…

Calculate the moment of inertia:

$I = (14\ \text{kg})\cdot(2.0\ \text{m})^2\cdot[(9.8\ \text{m/s}^2 - 5.000\ \text{m/s}^2) / (5.000\ \text{m/s}^2)]$

$I = 53.76\ \text{kg}\cdot\text{m}^2$

$I = 53.8\ \text{kg}\cdot\text{m}^2$

10. B is correct.

The *final speed* of the string can be found if the acceleration is known:

$v_f^2 = v_i^2 + 2ad = 0 + 2ad$

$v_f = \sqrt{(2ad)}$

where d is the distance over which the acceleration occurs.

The *acceleration* of the string is related to the acceleration of the pulley:

$a = r\alpha$

The *angular acceleration* follows from the dynamical equation for the rotational motion:

$\tau = I\alpha$

The *torque* is the force applied times the radius of the pulley:

$Fr = I\alpha$

Combining these equations gives:

$a = r^2F / I$

The *final velocity* of the string is:

$v_f = \sqrt{(2r^2Fd / I)}$

$v_f = \sqrt{[2\cdot(0.125\ \text{m})^2\cdot(5.00\ \text{N})\cdot(1.25\ \text{m}) / (0.0352\ \text{kg}\cdot\text{m}^2)]}$

$v_f = 2.36\ \text{m/s}$

11. A is correct.

The angle of every point remains fixed relative to all other points.

The tangential acceleration increases as one moves away from the center ($a_t = \alpha r$).

The radial (or centripetal) acceleration also depends on the distance r from the center ($a_c = \omega^2 r$).

The only choice that does not depend on r (i.e., the same for all the points in the object) is I.

12. D is correct.

Linear velocity is related to angular velocity by $v = r\omega$.

13. C is correct.

For a rotating object:

$$K = \frac{1}{2}I\omega^2$$

The moment of inertia of a cylinder is:

$$I = \frac{1}{2}mr^2$$

Combining these:

$$K = \frac{1}{4}mr^2\omega^2$$

Solving for the angular speed:

$$\omega = \sqrt{(4K / mr^2)}$$

$$\omega = \sqrt{\{4 \cdot (3.2 \times 10^7\ \text{J}) / [(400.0\ \text{kg})\ (0.60\ \text{m})^2]\}}$$

$$\omega = 940\ \text{rad/s}$$

14. B is correct.

For a rotating object subject to a constant torque that has undergone a total angular displacement of $\Delta\theta$, the work done on the wheel is:

$$\text{W} = \tau\Delta\theta$$

Since work is the change in energy of the wheel from external forces, and since the wheel started with $E = 0$ ("from rest"), the final kinetic energy can be written as:

$$K = \tau\Delta\theta$$

By rotational kinematics:

$$\Delta\theta = \frac{1}{2}\alpha t^2$$

The equation of rotational dynamics is:

$$\tau = I\alpha,\ \text{or}\ \alpha = \tau / I$$

So:

$$\Delta\theta = t^2\tau / 2I$$

and

$$K = t^2\tau^2 / 2I$$

(Note that this expression can be developed by finding the final angular velocity and using the definition of rotational kinetic energy.)

$$K = [(8.0\ \text{s})^2 \cdot (3.0\ \text{N·m})^2] / [2 \cdot (5.0\ \text{kg·m}^2)]$$

$$K = 58\ \text{J}$$

15. D is correct.

Tangential speed depends on the distance of the point from the fixed axis, so points at different radii have different tangential speeds.

Angular speed and acceleration of a rigid object do not depend on radius and are the same for all points (each point on the object must rotate through the same angle in the same time interval, or else it would not be rigid).

16. D is correct.

The angular momentum of a rotating object is:

$$L = I\omega$$

For a cylinder, $I = ½mr^2$, giving:

$$L = ½mr^2\omega$$

$$L = (0.5)\cdot(15.0\text{ kg})\cdot(1.4\text{ m})^2\cdot(2.4\text{ rad/s})$$

$$L = 35\text{ kg m}^2\text{/s}$$

17. D is correct.

The speed of an accelerating object is related to the distance covered by the kinematic relation:

$$v_f^2 - v_i^2 = 2ad$$

In this case, the initial speed is zero, so:

$$v_f = \sqrt{(2ad)}$$

Thus, find the linear acceleration of the disk.

An object is rolling down an incline experiences three forces, and hence three torques.

The forces are the force of gravity acting on the center of mass of the object, the normal force between the incline and the object, and the force of friction between the incline and the object.

If the origin is taken to be the center of the object, the force of gravity provides zero torque.

This can be seen by noting that the distance between the origin and the point of application of the force is zero.

$$\tau_{\text{gravity}} = F_{\text{gravity}}r = mg(0) = 0$$

Similarly, the normal force contributes zero torque because the direction of the force is directed through the origin (pivot point).

$$\tau_{normal} = F_{normal}\,\text{r}\sin\theta = F_{normal}(R)\cdot(\sin 180°) = F_{normal}(R)\cdot(0) = 0$$

Use a coordinate system in which the x-axis is parallel to the incline and the y-axis is perpendicular.

The object is rolling in the positive x-direction.

continued…

The dynamical equation for linear motion along the x-direction is:

$$F_{net} = ma$$

$$(mg \sin \theta - f) = ma$$

Note that the normal force is only in the y-direction and thus does not directly contribute to the acceleration in the x-direction.

The dynamical equation for rotational motion is:

$$\tau_{net} = I\alpha$$

$fR = I\alpha$ (Note that the frictional force is perpendicular to the r vector, and $\sin 90° = 1$)

where R is the radius of the object, f is the force of friction, and I is the moment of inertia.

A relation coupling these two dynamical equations is needed.

Equation of constraint imposed by the restriction that the object rolls without slipping is:

$$\alpha = a / R$$

To find the linear acceleration, use the equation of constraint to eliminate α from the rotational equation by replacing it with a / R:

$$fR = I(a / R)$$

The force of friction is of no interest, so rearrange this last expression:

$$f = Ia / R^2$$

Substitute this into the linear dynamic equation from above in place of f:

$$(mg \sin \theta - Ia / R^2) = ma$$

Solving this for a:

$$a = mg \sin \theta / [m + (I / R^2)]$$

$$a = g \sin \theta / [1 + (I / mR^2)]$$

For a disk,

$$I = \tfrac{1}{2}mR^2$$

so:

$$a = g \sin \theta / (1 + \tfrac{1}{2}) = (2/3)g \sin \theta$$

Using this in the kinematic equation above:

$$v_f = \sqrt{(4gd \sin \theta / 3)}$$

$$v_f = \sqrt{[(4/3)\cdot(9.8 \text{ m/s}^2)\cdot(3.0 \text{ m})\cdot\sin (25°)]}$$

$$v_f = \sqrt{[(4/3)\cdot(9.8 \text{ m/s}^2)\cdot(3.0 \text{ m})\cdot(0.4226)]}$$

$$v_f = 4.1 \text{ m/s}$$

18. A is correct.

The center of the tire is moving at velocity v, but the bottom of the tire is in contact with the ground without slipping, so the speed at the bottom is 0 m/s.

Thus, with respect to the ground, the tire is *instantaneously* rotating about the point of contact with the ground, and all points in the tire have the same instantaneous angular speed.

The top of the tire is twice the distance from the ground as the center.

For the center of the tire:

$$v = r\omega$$

At the top:

$$v_{top} = (2r)\omega = 2(r\omega) = 2v$$

19. D is correct.

The string does not slip; the speed of the string is the same as the speed of a point on the circumference of the pulley.

The angular speed of the pulley (radius R) and the speed of a point on its circumference are related be:

$$\omega = v / R$$

$$\omega = (5.0 \text{ m/s}) / (0.050 \text{ m})$$

$$\omega = 100 \text{ rad/s}$$

20. D is correct.

One way of expressing the magnitude of angular momentum is:

$$L = rp \sin \theta$$

where r is the magnitude of the object's absolute position vector, p is the magnitude of the object's linear momentum, and θ is the angle between the position vector and the momentum vector.

The magnitude of the position vector is:

$$r = \sqrt{(r_x^2 + r_y^2)}$$

$$r = \sqrt{[(2.00 \text{ m})^2 + (3.10 \text{ m})^2]}$$

$$r = 3.689 \text{ m}$$

Its angle with respect to the positive x axis is:

$$\theta_r = \text{atan}(r_y / r_x)$$

$$\theta_r = \text{atan}(3.10 / 2.00)$$

$$\theta_r = 0.99783 \text{ rad} = 57.17°$$

continued…

The magnitude of the momentum vector is:

$p = mv$

$p = (1.4 \text{ kg})(4.62 \text{ m/s})$

$p = 6.468 \text{ kg·m/s}$

The angle of the momentum vector is given in the problem:

$\theta_p = 45°$

Thus, the angle between the two vectors is:

$\theta = \theta_r - \theta_p$

$\theta = 57.17° - 45°$

$\theta = 12.17°$

The angular momentum is then:

$L = (3.689 \text{ m})·(6.468 \text{ kg m/s})·\sin(12.17°)$

$L = 5.0 \text{ kg·m}^2\text{/s}$

21. B is correct.

Average angular acceleration is defined by:

$\alpha_{avg} = \Delta\omega / \Delta t$

$\alpha_{avg} = |(6.3 \text{ rad/s} - 10.0 \text{ rad/s}) / (5.0 \text{ s})|$

$\alpha_{avg} = 0.74 \text{ rad/s}$

22. D is correct.

The kinematic equation for angular velocity is:

$\omega_f = \omega_i + \alpha\Delta t$

Note that the sign of angular velocity is opposite from the sign of angular acceleration; the wheel is slowing down.

For the final kinetic energy to be larger than the initial kinetic energy, the wheel must slow and continue beyond zero speed so that it gains speed in the opposite direction.

That is, the final angular velocity must be negative.

The kinetic energy of a rotating wheel is:

$K = ½I\omega^2$

continued…

The kinetic energy scales as the square of the angular speed.

To double the kinetic energy, the angular speed must increase by a factor of $\sqrt{2}$:

$$\omega_f = -\omega_i\sqrt{2}$$

Putting the result in the kinematic equation:

$$-\omega_i\sqrt{2} = \omega_i + \alpha\Delta t$$

Solving for Δt:

$$\Delta t = -(1 + \sqrt{2})\omega_i / \alpha)$$

$$\Delta t = -(1 + \sqrt{2})\cdot(26.0 \text{ rad/s}) / (-0.43 \text{ rad/s}^2)$$

$$\Delta t = 146 \text{ s}$$

23. B is correct.

Apply conservation of energy.

Take the zero of gravitational potential energy to be the configuration when the disk is at the bottom of the ramp.

Conservation of energy demands:

$$E_{\text{top}} = E_{\text{bottom}}$$

The disk is at rest at the top of the ramp, so the kinetic energy is zero.

The total energy at the top is:

$$E_{\text{top}} = K_{\text{top}} + U_{\text{top}}$$

$$E_{\text{top}} = 0 + mgh$$

At the bottom of the ramp, the gravitational potential energy is zero, and the kinetic energy is the sum of the linear and rotational kinetic energies:

$$E_{\text{bottom}} = K_{\text{bottom}} + U_{\text{bottom}}$$

$$E_{\text{bottom}} = K_{\text{linear}} + K_{\text{rotational}} + 0$$

$$E_{\text{bottom}} = \tfrac{1}{2} mv^2 + \tfrac{1}{2} I\omega^2$$

Final linear velocity is not given, but the final angular velocity is given.

Eliminate the linear velocity in favor of the angular velocity by applying the constraint to a circular object rolling without slipping.

$$v = \omega r$$

$$E_{\text{bottom}} = \tfrac{1}{2}m\omega^2r^2 + \tfrac{1}{2}I\omega^2$$

The moment of inertia of a disk is:

$$I = \tfrac{1}{2}mr^2$$

continued…

Thus:

$$E_{\text{bottom}} = \tfrac{1}{2}m\omega^2r^2 + \tfrac{1}{2}(\tfrac{1}{2}mr^2)\omega^2$$

$$E_{\text{bottom}} = \tfrac{1}{2}m\omega^2r^2 + \tfrac{1}{4}mr^2\omega^2$$

$$E_{\text{bottom}} = \tfrac{3}{4}m\omega^2r^2$$

Combining results into the expression for conservation of energy:

$$mgh = \tfrac{3}{4}m\omega^2r^2$$

or

$$h = 3\omega^2r^2 / 4g$$

Note that the mass has canceled, a common occurrence in mechanics problems. The chance of error is reduced by proceeding algebraically (rather than plugging in numbers in the beginning), by which mass can be canceled.

Finally, note that diameter is given, not the radius.

$$r = d / 2 = 1.6 \text{ m}$$

$$h = [3(4.27 \text{ rad/s})^2 \cdot (1.60 \text{ m})^2] / [4(9.8 \text{ m/s}^2)]$$

$$h = 3.57 \text{ m}$$

24. A is correct.

The direction of angular velocity is taken by convention by applying the right-hand rule to the rotation.

In the case of a wheel of a forward-moving bicycle, that direction is to the left of the rider.

25. A is correct.

Angular acceleration can be found if linear acceleration is calculated from:

$$\alpha = a / R$$

The linear acceleration follows from the kinematic relationship:

$$v_f^2 - v_i^2 = 2ad$$

In this case, v_f is zero.

$$-v_i^2 = 2ad$$

Only absolute value of the acceleration is needed, so drop the minus sign and solve for a:

$$a = v_i^2 / 2d$$

The angular acceleration is:

$$\alpha = v_i^2 / 2dR$$

$$\alpha = (8.4 \text{ m/s})^2 / [2(115.0 \text{ m})(0.34 \text{ m})]$$

$$\alpha = 0.90 \text{ rad/s}^2$$

26. A is correct.

For a rotating object:

$$K = ½I\omega^2$$

The moment of inertia of a cylinder is:

$$I = ½mr^2$$

Combining these:

$$K = ¼mr^2\omega^2$$

The angular speed is given in rpm but needs to be in rad/s:

$$33.4 \text{ rpm}\cdot(1 \text{ min} / 60 \text{ s})\cdot(2\pi \text{ rad/rev}) = 3.498 \text{ rad/s}$$

Thus:

$$K = (0.25)\cdot(3.0 \text{ kg})\cdot(0.10 \text{ m})^2\cdot(3.489 \text{ rad/s})^2$$

$$K = 0.091 \text{ J}$$

27. A is correct.

The angular momentum of a spinning object can be written as:

$$L = I\omega$$

The moment of inertia of a long thin uniform object of length l about an axis through the center perpendicular to the long axis is:

$$I = (1/12)ml^2$$

Giving:

$$L = (1/12)ml^2\omega$$

$$L = (1/12)\cdot(0.1350 \text{ kg})\cdot(1.000 \text{ m})^2\cdot(3.5 \text{ rad/s})$$

$$L = 0.0394 \text{ kg}\cdot\text{m}^2\text{/s}$$

28. C is correct.

The period of a rotating object can be expressed as:

$$T = 2\pi / \omega$$

ω can be extracted from the definition of centripetal force:

$$F = mv^2 / r$$

Combining this with the relationship

$v = \omega r$ gives:

$$F = mr\omega^2$$

continued…

Thus:

$\omega = \sqrt{(F / mr)}$

Then:

$T = 2\pi\sqrt{(mr / F)}$

$T = 2\pi\sqrt{[(23.0 \text{ kg})(1.3 \text{ m}) / (51.0 \text{ N})]}$

$T = 4.8 \text{ s}$

29. C is correct.

Angular acceleration is:

$\alpha = \Delta\omega / \Delta t$

$\alpha = (38.0 \text{ rad/s} - 0.00 \text{ rad/s}) / (10.0 \text{ s})$

$\alpha = 3.80 \text{ rad/s}$

The other information given in the question is not needed.

30. C is correct.

The magnitude of *torque* can be expressed as:

$\tau = rF \sin \theta$

where r is the distance from the origin (taken here to be the pivot point) to the point of application of the force F, and θ is the angle between the position vector of the point of application and the force vector.

$\tau = (0.63 \text{ m})\cdot(17.0 \text{ N})\cdot\sin(45°)$

$\tau = 7.6 \text{ N m}$

31. D is correct.

Use the conservation of energy.

Take the zero of potential energy to be the configuration in which the sphere is at the bottom. The potential energy at the bottom is zero.

The disk starts from rest, so the kinetic energy at the top is zero:

$mgh = K_{\text{linear}} + K_{\text{rotation}}$

$mgh = \frac{1}{2}mv^2 + \frac{1}{2}I\omega^2$

For a disk,

$I = \frac{1}{2}mr^2$

Because the disk rolls without slipping,

$v = r\omega$

continued…

Substituting these into the conservation of energy equation:

$$mgh = ½m(r^2\omega^2) + ½(½mr^2)\omega^2$$

$$mgh = ½mr^2\omega^2 + ¼mr^2\omega^2$$

$$mgh = ¾mr^2\omega^2$$

Isolating h:

$$h = 3r^2\omega^2 / 4g$$

$$h = [3\cdot(1.60\text{ m})^2\cdot(4.27\text{ rad/s})^2] / [4\cdot(9.8\text{ m/s}^2)]$$

$$h = 3.57\text{ m}$$

32. D is correct.

The moment of inertia is proportional to the square of the distance of an object from the center of rotation.

As Paul moves toward the center, the moment of inertia decreases.

Angular momentum,

$L = I\omega$ is conserved.

As I decreases, ω, the angular speed increases to compensate.

33. B is correct.

Converting units:

210.0 rpm·(1 min / 60 s)·(2π rad/revolution) = 22.0 rad/s

34. C is correct.

Use the conservation of energy:

$$E_f = E_i$$

$$KE_f + PE_f = KE_i + PE_i$$

Take the zero of potential energy at the initial height so that PE_i is zero.

The kinetic energy of a rolling object is the sum of the KE of translation plus the KE of rotation:

$$KE = ½mv^2 + ½I\omega^2$$

If the object rolls without slipping, as is the case here, then $v = r\omega$, where R is the radius of the sphere.

The moment of inertia of a sphere is:

$$I = (2/5)m\omega^2$$

continued…

With these, the kinetic energy of a sphere that rolls without slipping:

$$KE = ½mv^2 + ½(2 / 5)mr^2(v / r)^2$$

$$KE = ½mv^2 + (1 / 5)mv^2$$

$$KE = (7 / 10)mv^2$$

Conservation of energy becomes:

$$(7 / 10)mv_f^2 + mgh = (7/10)mv_i^2$$

Canceling the mass and solving for the final speed:

$$v_f = \sqrt{[v_i^2 - (10 / 7)gh]}$$

The height, h, is related to the distance traveled and the angle of incline:

$$h = d \sin \theta$$

So:

$$v_f = \sqrt{[v_i^2 - (10 / 7)gd \sin \theta]}$$

$$v_f = \sqrt{[(5.5 \text{ m/s})^2 - (10/7)\cdot(9.8 \text{ m/s}^2)\cdot(3.0 \text{ m})\cdot\sin (25°)]}$$

$$v_f = \sqrt{[(5.5 \text{ m/s})^2 - (10/7)\cdot(9.8 \text{ m/s}^2)\cdot(3.0 \text{ m})\cdot(0.4226)]}$$

$$v_f = 3.5 \text{ m/s}$$

35. D is correct.

The dynamical relation for rotational motion is:

$$\tau = I\alpha$$

$$I = \tau / \alpha$$

where τ is the torque applied to the wheel, and α is the wheel's angular acceleration.

The torque is the product of the force and the radius of the pulley:

$$\tau = FR$$

Note that $\sin \theta = 1$, since $\theta = 90°$ [the problem mentions the force is applied tangentially].

So:

$$I = \tau / \alpha$$

$$I = FR / \alpha$$

$$I = (16.88 \text{ N})\cdot(0.340 \text{ m}) / (1.20 \text{ rad/s}^2)$$

$$I = 4.78 \text{ kg}\cdot\text{m}^2$$

36. B is correct.

The energy required to bring a rotating object to rest is:

$$E = \tfrac{1}{2}I\omega_0^2$$

The moment of inertia of a cylinder is:

$$I = \tfrac{1}{2}mr^2$$

So, the energy needed to stop the object is:

$$E = \tfrac{1}{4}mr^2\omega_0^2$$

Solving for the mass:

$$m = 4E / r^2\omega_0^2$$

The angular speed in rpm is given; convert to rad/s:

$$500.0 \text{ rpm}\cdot(1 \text{ min} / 60 \text{ sec})\cdot(2\pi \text{ rad} / 1 \text{ rev}) = 52.36 \text{ rad/s}$$

The mass of the object is:

$$m = 4(3900 \text{ J}) / (1.2 \text{ m})^2\cdot(52.36 \text{ rad/s})^2$$

$$m = 4.0 \text{ kg}$$

37. C is correct.

The torque at P_2 due to the weight of the billboard is:

$$\tau = r_{perp}F = r_{perp}mg$$

where r_{perp} is the perpendicular distance between P_2 and the line of application of the force.

Here, the weight of the billboard acts as if all of the mass is concentrated at the center of mass, which is the center of the billboard.

Thus, the line of application of the force of gravity is P_1P_3, the vertical line that passes through P_1 and P_3.

The length r_{perp} is the length of the horizontal line segment between P_2 and the line of application of the force (P_1P_3).

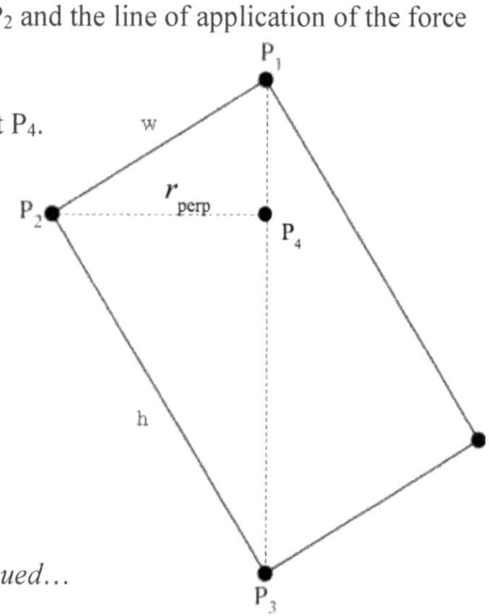

This line segment is perpendicular to P_1P_3 and intersects P_1P_3 at point P_4.

To find r_{perp}, note that the triangle $P_1P_2P_3$ is like the triangle formed by $P_1P_4P_2$.

The ratios of the lengths of corresponding sides of similar triangles are equal.

Apply this by forming the ratio of the length of the long side to the length of the hypotenuse of these two triangles:

$$h / \sqrt{(h^2 + w^2)} = r_{perp} / w$$

$$r_{perp} = hw / \sqrt{(h^2 + w^2)}$$

continued…

Giving an expression for torque:

$$\tau = mg[hw / \sqrt{(h^2 + w^2)}]$$

$$\tau = (5.0 \text{ kg})\cdot(9.8 \text{ m/s}^2)\cdot(0.20 \text{ m})\cdot(0.11 \text{ m}) / \sqrt{[(0.20 \text{ m})^2 + (0.11 \text{ m})^2]}$$

$$\tau = 4.7 \text{ N m}$$

38. B is correct.

An object is rolling down an incline experiences three forces, and hence three torques.

The forces are the force of gravity acting on the center of mass of the object, the normal force between the incline and the object, and the force of friction between the incline and the object.

If the origin is taken to be the center of the object, the force of gravity provides zero torque.

This can be seen by noting that the distance between the origin and the point of application of the force is zero.

$$\tau_{\text{gravity}} = F_{\text{gravity}}r = mg(0) = 0$$

Similarly, the normal force contributes zero torque because the direction of the force is directed through the origin (pivot point):

$$\tau_{normal} = F_{normal}\,\mathrm{r}\sin\theta = F_{normal}(R)\cdot(\sin 180°) = F_{normal}(R)\cdot(0) = 0$$

Use a coordinate system in which the x-axis is parallel to the incline and the y-axis is perpendicular.

The object is rolling in the positive x-direction.

The dynamical equation for linear motion along the x-direction is:

$$F_{\text{net}} = ma$$

$$(mg \sin\theta - f) = ma$$

Note that the normal force is only in the y-direction and thus does not directly contribute to the acceleration in the x-direction.

The dynamical equation for rotational motion is:

$$\tau_{\text{net}} = I\alpha$$

$$fR = I\alpha$$

Note that the frictional force is perpendicular to the r vector, and sin 90° = 1,

where R is the radius of the object, f is the force of friction, and I is the moment of inertia.

A relation coupling these two dynamical equations is necessary.

The equation of constraint imposed by the restriction that the object rolls without slipping:

$$\alpha = a / R$$

To find the linear acceleration, use the equation of constraint to eliminate α from the rotational equation by replacing it with a / R:

continued…

$fR = I(a / R)$

The force of friction is of no interest, so rearrange this last expression:

$f = Ia / R^2$

Substitute this into the linear dynamic equation from above in place of f:

$(mg \sin \theta - Ia / R^2) = ma$

Solving this for a:

$a = mg \sin \theta / [m + (I / R^2)]$

$a = g \sin \theta / [1 + (I / mR^2)]$

For a sphere,

$I = (2/5)mR^2$

so:

$a_{\text{sphere}} = g \sin \theta / [1 + (2/5)]$

$a_{\text{sphere}} = (5/7)g \sin \theta$

$a_{\text{sphere}} = (0.714)g \sin \theta$

For the disk,

$I = ½ mR^2$

so:

$a_{\text{disk}} = g \sin \theta / (1 + ½)$

$a_{\text{disk}} = (2/3)g \sin \theta$

$a_{\text{disk}} = (0.667)g \sin \theta$

For the hoop,

$I = mR^2$

so:

$a_{\text{hoop}} = g \sin \theta / (1 + 1)$

$a_{\text{hoop}} = (1/2)g \sin \theta$

$a_{\text{hoop}} = (0.500)g \sin \theta$

The object with the largest acceleration will reach the bottom first.

The order is sphere, disk, hoop.

39. B is correct.

The kinetic energy of a rotation object is:

$K = ½I\omega^2$

The moment of inertia is given in SI units, but the angular speed is in rpm, which is not an SI unit.

Convert:

96.0 rpm·(1 m/ 60 s)·(2π rad/rev) = 10.05 rad/s

$K = (0.5)\cdot(6.0 \times 10^{-3}\ \text{kg}\cdot\text{m}^2)\cdot(10.05\ \text{rad/s})^2$

$K = 0.30$ J

40. D is correct.

If a wheel of radius r rolls without slipping on the pavement, the relationship between angular speed ω and translational speed is:

$v = r\omega$

$\omega = v / r$

$\omega = (6.00\ \text{m/s}) / (0.120\ \text{m})$

$\omega = 50.00$ rad/s

Converting this to rpm:

$\omega = (50.00\ \text{rad/s})\cdot(1\ \text{rev} / 2\pi\ \text{rad})\cdot(60\ \text{s} / 1\ \text{min})$

$\omega = 477.5$ rpm

$\omega = 478$ rpm

41. C is correct.

Her moment of inertia does not remain constant because the radial position of her hands is changing.

Angular momentum remains constant because there is no torque on her, assuming the ice is frictionless.

Her kinetic energy changes. To visualize this, notice that her hands initially execute uniform circular motion, and hence there is no tangential force. If this condition is maintained, her kinetic energy is constant.

However, *as she pulls her hands in,* they are no longer in uniform circular motion.

During this time, there will be a tangential component of force, and work will be done.

Another way to think about it is to recognize that because L is conserved,

$L_0 = L_f$

$I_0\omega_0 = I_f\omega_f$

I will decrease by some amount, and ω will multiply by that same amount, e.g., I halves and ω doubles. When examining $KE = ½I\omega^2$, I has gone down by half, but ω has doubled. Since KE depends on ω squared, the change to ω has a greater impact on KE. KE increases.

42. D is correct.

There are two perpendicular components to Tanya's acceleration.

Centripetal acceleration:

$$a_c = r\omega^2$$

and tangential acceleration:

$$a_t = r\alpha$$

Since these two acceleration components are perpendiculars, find the magnitude of the total linear acceleration:

$$a = \sqrt{(a_c^2 + a_t^2)}$$

$$a = \sqrt{[(r\omega^2)^2 + (r\alpha)^2]}$$

$$a = r\sqrt{(\omega^4 + \alpha^2)}$$

$$a = (4.65 \text{ m})\cdot\sqrt{[(1.25 \text{ rad/s})^4 + (0.745 \text{ rad/s}^2)^2]}$$

$$a = 8.05 \text{ m/s}^2$$

43. C is correct.

For a rotating object:

$$\Delta\theta = \omega\Delta t$$

The angular speed is given in rpm, but needs to be in deg/s:

$$33.0 \text{ rpm}\cdot(1 \text{ min} / 60 \text{ s})\cdot(360 \text{ deg} / \text{rev}) = 198.0 \text{ deg/s}$$

$$\Delta\theta = (198.0 \text{ deg/s})\cdot(0.32 \text{ s})$$

$$\Delta\theta = 63°$$

44. C is correct.

The question asks to find the *magnitude* of the force of the floor on the bottom of the ladder.

Two forces are acting on the bottom of the ladder: the normal force of the floor, N, which acts in the vertical direction, and the force of friction, f, which acts horizontally.

These two forces are directed in mutually perpendicular directions, so the magnitude of the net force of the floor on the bottom of the ladder is:

$$F_{\text{net, bottom}} = \sqrt{(N^2 + f^2)}$$

The values of N and f can be found by applying the laws of rotational and linear static equilibrium.

Assume the ladder is static (not moving), but this is not a valid assumption until it is known whether the force of friction exceeds the limit imposed by static friction. This will be verified at the end.

continued…

If the ladder is in equilibrium, the sum of the forces in both the x- and y-directions is zero, and the sum of the torques is zero:

$$\sum F_x = 0; \sum F_y = 0; \sum \tau = 0$$

The force equations are straightforward to fill in:

$$\sum F_x = F_w - f = 0$$

$$\sum F_y = N - Mg - mg = 0$$

where F_w is the normal force from the wall on the top of the ladder, M is the mass of the hanging block, and m is the mass of the ladder.

Since the wall is "smooth," there is no frictional force from the wall on the ladder.

The $\sum F_y$ equation has only one unknown quantity and can be solved for N:

$$N = (M + m)g = (80\text{kg} + 50\text{kg})(9.8\text{m/s}^2) = 1{,}274 \text{ N}$$

For the torque equation, choose the origin to be the point where the top of the ladder touches the wall. With that choice, the torque due to the normal force of the wall is zero, and the torque due to the hanging block is also zero (since these forces are exerted right at the origin).

There are three non-zero torques.

> One is the torque due to gravity on the ladder that acts as if all of the mass of the ladder is concentrated at the center of mass.
>
> The second torque is due to the normal force of the floor on the bottom of the ladder.
>
> The third torque is the force of friction acting on the bottom of the ladder.

Using the general formula for torque,

$$\tau = Fr \sin \phi,$$

where F is the force, r is the distance from the origin to the point where the force is being applied, and ϕ is the angle between the force vector and the r vector (which points from the origin to the point where the force is being applied):

$$\sum \tau = 0 + 0 - mg(L / 2) \sin \beta - fL \sin \alpha + NL \sin \beta = 0$$

where m is the mass of the ladder and α and β are the angles in the figure shown.

Take counterclockwise torques to be positive and clockwise torques to be negative.

Use trigonometry to find α and β:

$$\alpha = \sin^{-1}(h / L) = \sin^{-1}(3.7 / 5) = 47.73°$$

$$\beta = 90° - \alpha = 42.27°$$

There is only one unknown quantity in the torque equation. Divide by L and rearrange.

continued...

Solve for f:

$$f \sin \alpha = N \sin \beta - ½mg \sin \beta$$

$$f = [(N - ½mg) \cdot \sin \beta] / \sin \alpha$$

$$f = [1274 \text{ N} - ½(50 \text{ kg}) \cdot (9.8 \text{ m/s}^2) \cdot (\sin 42.27°)] / [\sin 47.73°]$$

$$f = 935.3 \text{ N}$$

Calculate the magnitude of the force on the bottom of the latter due to the floor:

$$F_{\text{net, bottom}} = \sqrt{(N^2 + f^2)}$$

$$F_{\text{net, bottom}} = \sqrt{[(1{,}274 \text{ N})^2 + (935.3 \text{ N})^2]}$$

$$F_{\text{net, bottom}} = 1{,}580 \text{ N}$$

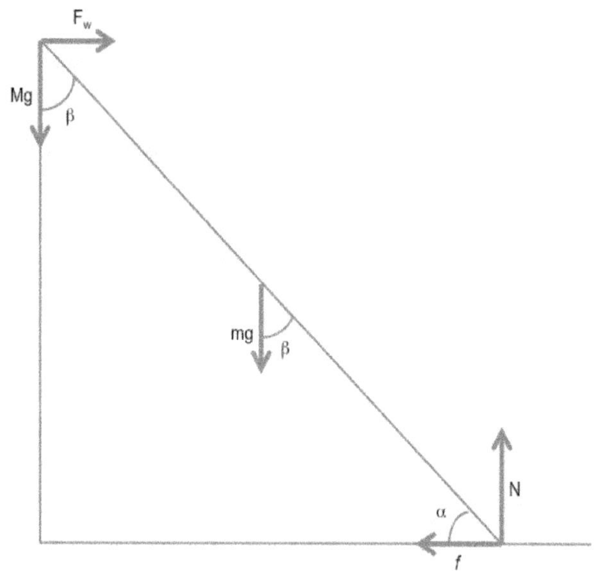

Finally, check if the force of friction is lower than the limit imposed by static friction.

$$f_{\max} = \mu_s N = (0.750) \cdot (1{,}274 \text{ N}) = 955 \text{ N}$$

The calculated value for the force of static friction is 935.3 N, lower than the limit imposed by static friction,

So, the assumption that the system is in static equilibrium is valid.

If this number had been lower than the value of f, the static frictional force would *not* be able to hold the ladder stationary.

45. B is correct.

The maximum displacement occurs when the acceleration stops the forward motion and the wheel reverses direction.

Since.

$$\theta_0 = 0, \Delta\theta = \theta - 0 = \theta.$$

The kinematic relation is:

$$\omega_f^2 - \omega_i^2 = 2\alpha\theta$$

The final angular speed is zero at the instant that the wheel changes direction.

$$-\omega_i^2 = 2\alpha\theta$$

Thus, the angular displacement at that instant is:

$$\theta = -(\omega_i^2 / 2\alpha)$$

$$\theta = -(29.0 \text{ rad/s})^2 / [2 \cdot (-0.52 \text{ rad/s}^2)]$$

$$\theta = 809 \text{ rad}$$

46. C is correct.

Determine the direction of the angular momentum for a rotating object with the *right-hand rule*.

Take the fingers of your right hand and curl them in the direction that the object is rotating and stick your thumb out perpendicular to your fingers (as in a "thumbs-up" or "thumbs down" signal). The direction of your thumb points in the direction of the angular momentum – in this case, down.

For a mathematical approach, use the definition:

$L = r \times p$ applied to a point on the outer edge of the object.

Taking a point on the disk along the positive x-axis, the position vector r is in the positive x-direction, and the momentum vector p is positive z-direction.

The cross-product of these two vectors points down in the negative y-direction.

47. B is correct.

Consider this formula for rotational kinematics:

$$\omega_f = \omega_i + \alpha\Delta t$$

Solve for Δt:

$$\Delta t = (\omega_f - \omega_i) / \alpha$$

$$\Delta t = (33.3 \text{ rad/s} - 15.0 \text{ rad/s}) / 3.45 \text{ rad/s}^2$$

$$\Delta t = 5.30 \text{ s}$$

48. D is correct.

Use the conservation of energy.

Note that in the initial state, the ball is at rest, so the initial kinetic energy is zero.

Take the zero of gravitational potential energy to be the height of the ball at its final position.

The final kinetic energy has a translational part and a rotational part:

$$E_f = E_i$$

$$KE_f = PE_i$$

$$KE_{\text{translation}} + KE_{\text{rotation}} = PE_i$$

$$½mv^2 + ½I\omega^2 = mgh$$

The ball is rolling without slipping, so the relationship between translational and rotational velocity is $v = r\omega$.

Then, the above becomes:

$$½mr^2\omega^2 + ½(2 / 5)mr^2\omega^2 = mgh$$

$$(7 / 10)mr^2\omega^2 = mgh$$

continued…

Solving for the angular speed:

$$\omega = \sqrt{(10gh / 7r^2)}$$

$$\omega = \sqrt{\{[10\cdot(9.8\text{ m/s}^2)\cdot(5.3\text{ m})] / [7\cdot(1.7\text{ m})^2]\}}$$

$$\omega = 5.1\text{ rad/s}$$

49. B is correct.

There are three rotating bodies.

The total angular momentum is the sum of their angular momenta.

$$L_{\text{total}} = L_1 + L_2 + L_3$$

The angular momentum of a rotating object is:

$$L = I\omega$$

The flywheels are identical, so they each have the same rotational inertia I, so:

$$L = I(\omega_1 + \omega_2 + \omega_3)$$

The moment of inertia of the flywheels is:

$$I = ½mr^2$$

$$I = 0.5(65.0\text{ kg})\cdot(1.47\text{ m})^2$$

$$I = 70.23\text{ kg m}^2$$

Thus:

$$L = (70.23\text{ kg m}^2)\cdot(3.83\text{ rad/s} + 3.83\text{ rad/s} – 3.42\text{ rad/s})$$

$$L = 298\text{ kg m}^2\text{/s}$$

50. D is correct.

It seems that there is not enough information because neither the masses nor the radii of the sphere and cylinder are known. However, various parameters often cancel.

Try to find the acceleration of the two objects.

The object with the larger acceleration will reach the bottom first.

An object is rolling down an incline experiences three forces, and hence three torques.

The forces are the force of gravity acting on the center of mass of the object, the normal force between the incline and the object, and the force of friction between the incline and the object.

If the origin is taken to be the center of the object, the force of gravity provides zero torque.

This can be seen by noting that the distance between the origin and the point of application of the force is zero.

continued…

$$\tau_{gravity} = F_{gravity}r$$

$$F_{gravity}r = mg(0)$$

$$mg(0) = 0$$

Similarly, the normal force contributes zero torque because the direction of the force is directed through the origin (pivot point):

$$\tau_{normal} = F_{normal}\,\mathrm{r}\sin\theta$$

$$F_{normal}\,\mathrm{r}\sin\theta = F_{normal}(R)\cdot(\sin 180°)$$

$$F_{normal}(R)\cdot(\sin 180°) = F_{normal}\,(R)\cdot(0)$$

$$F_{normal}\,(R)\cdot(0) = 0$$

Use a coordinate system in which the x-axis is parallel to the incline and the y-axis is perpendicular. The object is rolling in the positive x-direction.

The dynamical equation for linear motion along the x-direction is:

$$F_{net} = ma$$

$$(mg\sin\theta - f) = ma$$

Note that the normal force is only in the y-direction and thus does not directly contribute to the acceleration in the x-direction.

The dynamical equation for rotational motion is:

$$\tau_{net} = I\alpha$$

$$fR = I\alpha$$

Note that the frictional force is perpendicular to the r vector, and $\sin 90° = 1$,

where R is the radius of the object, f is the force of friction, and I is the moment of inertia.

A relation is needed to couple these two dynamical equations. This is the equation of constraint imposed by the restriction that the object rolls without slipping:

$$\alpha = a / R$$

To find the linear acceleration, use the equation of constraint to eliminate α from the rotational equation by replacing it with a / R:

$$fR = I(a / R)$$

The force of friction is of no interest, so rearrange this last expression:

$$f = Ia / R^2$$

Substitute this into the linear dynamic equation from above in place of f:

$$mg\sin\theta - (Ia / R^2) = ma$$

continued…

Solving this for a:

$$a = mg \sin \theta / [(m + (I / R^2)]$$

$$a = g\sin \theta / [1 + (I / mR^2)]$$

For a sphere,

$$I = (2/5)mR^2$$

So:

$$a_{sphere} = g \sin \theta / (1 + 2/5)$$

$$a_{sphere} = (5/7)g \sin \theta$$

For the cylinder,

$$I = ½mR^2$$

So:

$$a_{cylinder} = g \sin \theta / (1 + ½)$$

$$a_{cylinder} = (2/3)g \sin \theta$$

Since 5/7 > 2/3, the acceleration of the sphere is greater than the acceleration of the cylinder so that the sphere will reach the bottom first.

Interestingly, neither the mass nor the size of the sphere or cylinder enters into the result.

Indeed, both the mass and radius cancel.

Since neither the masses nor the radii were given in the statement of the problem, it would not be possible to solve this problem by brute force numerical calculation.

51. D is correct.

$$813.0 \text{ rpm} \cdot (1 \text{ min} / 60 \text{ s})\cdot(2\pi \text{ rad/rev}) = 85.14 \text{ rad/s}$$

52. B is correct.

The angular speed changes according to the kinematic relation:

$$\Delta\omega = \alpha\Delta t$$

$$\Delta t = \Delta\omega / \alpha$$

$$\Delta t = (0 \text{ rad/s} - 96.0 \text{ rad/s}) / (-1.5 \text{ rad/s}^2)$$

The angular acceleration is negative because the wheel is slowing down, and the initial ω is positive.

$$\Delta t = 64.0 \text{ s}$$

53. D is correct.

Apply the law of the conservation of energy.

There is no potential energy in this situation – all of the energy is kinetic (rotational kinetic energy in the initial state, translational kinetic energy in the final state).

$$E_f = E_i$$

$$K_f = K_i$$

$$\tfrac{1}{2}m_{\text{car}}v^2 = \tfrac{1}{2}I\omega^2$$

The moment of inertia of a disk is:

$$\tfrac{1}{2}mR^2,$$

where m is the mass of the disk, and R is its radius.

Then:

$$\tfrac{1}{2}m_{\text{car}}v^2 = \tfrac{1}{2}(\tfrac{1}{2}m_{\text{wheel}}R^2)\omega^2$$

Solving for velocity:

$$v = R\omega\sqrt{(m_{\text{wheel}} / 2m_{\text{car}})}$$

The angular speed in rev/s must be converted to rad/s:

$$200.0 \text{ rev/s}\cdot(2\pi \text{ rad/rev}) = 1{,}256.6 \text{ rad/s}$$

Finally:

$$v = (0.50 \text{ m})\cdot(1{,}256.6 \text{ rad/s})\cdot\sqrt{[370.0 \text{ kg} / (2\cdot 1{,}500.0 \text{ kg})]}$$

$$v = 221 \text{ m/s}$$

54. D is correct.

Since the motor spins at constant speed, there is no tangential component for the linear acceleration.

The linear acceleration is just the centripetal acceleration:

$$a = r\omega^2$$

Angular speed is in rpm, but needs to be in rad/s:

$$2695.0 \text{ rpm}\cdot(1 \text{ min}/60 \text{ s})\cdot(2\pi \text{ rad/rev}) = 282.22 \text{ rad/s}$$

$$a = (0.07165 \text{ m})\cdot(282.22 \text{ rad/s})^2$$

$$a = 5{,}707 \text{ m/s}^2$$

55. A is correct.

The kinetic energy stored in a rotating object is:

$$K = ½\, I\omega^2$$

With this, the angular speed as a function of energy is:

$$\omega^2 = 2K / I$$

The moment of inertia of a disk is:

$$I = ½mr^2$$

Giving:

$$\omega^2 = 4K / mr^2$$

$$\omega = \sqrt{(4K / mr^2)}$$

$$\omega = \sqrt{[4\cdot(3.2 \times 10^7\ \text{J})] / [(400.0\ \text{kg})\cdot(0.60\ \text{m})^2]}$$

$$\omega = 943\ \text{rad/s}$$

56. B is correct.

An object having a moment of inertia I rotating with an angular speed of ω has angular momentum:

$$L = I\omega$$

The moment of inertia of a solid right circular cylinder about the axis indicated is $½mR^2$:

$$L = (½mR^2)\omega$$

$$L = 0.5(15.0\ \text{kg})\cdot(1.4\ \text{m})^2\cdot(2.7\ \text{rad/s})$$

$$L = 39.69\ \text{kg m}^2/\text{s}$$

$$L = 40\ \text{kg m}^2/\text{s}$$

57. B is correct.

The angular speed changes according to the kinematic relation:

$$\Delta\omega = \alpha\Delta t$$

$$\Delta t = \Delta\omega / \alpha$$

The torque is given, and the moment of inertia of a right circular cylinder is:

$$½mR^2$$

Find the angular acceleration from the dynamic relation:

$$\tau = I\alpha$$

$$\alpha = \tau / I$$

continued…

$\tau / I = \tau / (\frac{1}{2}mR^2)$

$\alpha = 2\tau / mR^2$

Combining the two results:

$\Delta t = mR^2\Delta\omega / 2\tau$

$\Delta t = (10.0 \text{ kg})\cdot(3.00 \text{ m})^2\cdot(8.13 \text{ rad/s}) / [2(110.0 \text{ N m})]$

$\Delta t = 3.33 \text{ s}$

58. C is correct.

The gravitational force is the only influence making the satellite move in a circular path.

The centripetal acceleration in orbit is the gravitational acceleration at that orbital radius.

$v^2 / r = g'$

$v = \sqrt{(g'r)}$

$v = \sqrt{[(2.3 \text{ m/s}^2)\cdot(34{,}000 \text{ m})]}$

$v = 280 \text{ m/s}$

Notes for active learning

Work and Energy – Detailed Explanations

1. D is correct.

The final velocity in projectile motion is related to the maximum height of the projectile through the conservation of energy:

KE = PE

$\frac{1}{2}mv^2 = mgh$

When the stone thrown straight up passes its starting point on its way back down, its downward speed equals its initial upward velocity (2D motion).

The stone thrown straight downward contains the same magnitude of initial velocity as the stone thrown upward, and thus the stone thrown upward and the stone thrown downward have the same final speed.

A stone thrown horizontally (or, for example, a stone thrown at 45°) does not achieve the same height h as a stone thrown straight up, so it has a smaller final vertical velocity.

2. B is correct.

Work = force × displacement × cos θ

$W = Fd \cos \theta$

where θ is the angle between the vectors F and d

$W = (5\ N)\cdot(10\ m) \cos 45°$

$W = (50\ J)\cdot(0.7)$

$W = 35\ J$

3. A is correct.

$KE = \frac{1}{2}mv^2$

KE is influenced by mass and velocity.

Since velocity is squared, its influence on KE is greater than the influence of mass.

4. B is correct.

Work = force × displacement × cos θ

$W = Fd \cos \theta$

$\cos 90° = 0$

$W = 0$

Since the force of gravity acts perpendicular to the distance traveled by the ball, the force due to gravity does no work in moving the ball.

5. C is correct.

$KE = ½mv^2$

$KE = ½(5\text{ kg})\cdot(2\text{ m/s})^2$

$KE = 10\text{ J}$

6. A is correct.

$W = Fd\cos\theta$

$\cos\theta = 1$

$F = W / d$

$F = (360\text{ J}) / (8\text{ m})$

$F = 45\text{ N}$

$F = ma$

$m = F / a$

$m = (45\text{ N}) / (10\text{ m/s}^2)$

$m = 4.5\text{ kg}$

7. D is correct.

On a displacement (x) *vs.* force (F) graph, the displacement is the y-axis, and the force is the x-axis.

The slope is x / F (in units of m/N), which is the reciprocal of the spring constant k, measured in N/m.

8. C is correct.

Work done by a spring equation:

$W = ½kx^2$

$W = ½(22\text{ N/m})\cdot(3\text{ m})^2$

$W = 99\text{ J}$

9. A is correct.

The force of gravity points down.

When the ball is moving upwards, its displacement is opposite of that of the force of gravity, and therefore the work done by gravity is negative.

On the way down, the direction of displacement is the same as that of the force of gravity, and therefore the work done by gravity is positive.

10. B is correct.

Work done by gravity is an object's change in gravitational PE.

$W = –PE$

$A_1 = 400$ J

By the work-energy theorem,

$W = KE$

$B_1 = 400$ J

11. D is correct.

Work is calculated as the product of force and displacement parallel to the direction of the applied force:

$W = Fd \cos \theta$

where some component of d is in the direction of the force

12. B is correct.

Work only depends on force and distance:

$W = Fd \cos \theta$

Power = W / t is the amount of work done in a unit of time.

13. A is correct.

The area under the curve on a graph is the product of the values of $y \times x$.

Here, the y value is force, and the x value is distance:

$Fd = W$

14. C is correct.

This is the conservation of energy. The only force acting on the cat is gravity.

$KE = PE_g$

$KE = mgh$

$KE = (3 \text{ kg})\cdot(10 \text{ m/s}^2)\cdot(4 \text{ m})$

$KE = 120$ J

15. B is correct.

Although the book is stationary with respect to the plank, the plank is applying a force to the book, causing it to accelerate in the direction of the force. Since the displacement of the point of application of the force is in the same direction as the force, the work done is positive.

Choice D is not correct because work is a scalar and has no direction.

16. D is correct.

$W = Fd$

$d = W / F$

$d = (350\text{ J}) / (900\text{ N})$

$d = 0.39\text{ m}$

17. A is correct.

Conservation of energy between kinetic energy and potential energy:

$KE = PE$

$KE = ½mv^2$ and $PE = mgh$

Set the equations equal:

$½mv^2 = mgh$

cancel m from each side

$½v^2 = gh$

h is only dependent on the initial v, equal between both objects, so the two objects rise to the same height.

18. A is correct.

Work = Power × time

$P_1 = W / t$

$P_2 = (3\text{ W}) / (1/3\ t)$

$P_2 = 3(3/1)\cdot(W / t)$

$P_2 = 9(W / t)$

$P_2 = 9(P_1)$

19. D is correct.

Conservation of energy:

$KE = PE$

$KE = mgh$

$W = mg$

$KE = Wh$

$KE = (450\text{ N})\cdot(9\text{ m})$

$KE = 4{,}050\text{ J}$

20. A is correct.

$F_1 = -kx_1$

Solve for the spring constant k:

$k = F / x_1$

$k = (160 \text{ N}) / (0.23 \text{ m})$

$k = 696 \text{ N/m}$

$F_2 = -kx_2$

$F_2 = (696 \text{ N/m})\cdot(0.34 \text{ m})$

$F_2 = 237 \text{ N}$

21. B is correct.

There is a frictional force since the net force = 0

The mule pulls in the same direction as the direction of travel, so $\cos \theta = 1$

$\text{W} = Fd \cos \theta$

$d = v\Delta t$

$\text{W} = Fv\Delta t$

22. D is correct.

$\text{W} = Fd \cos \theta$

$F_T = \text{W} / (d \times \cos \theta)$

$F_T = (540 \text{ J}) / (18 \text{ m} \times \cos 32°)$

$F_T = (540 \text{ J}) / (18 \text{ m} \times 0.848)$

$F_T = 35 \text{ N}$

23. B is correct.

The spring force balances the gravitational force on the mass.

Therefore:

$F_g = -kx$

$mg = -kx$

By adding an extra 120 grams, the mass is doubled:

$(2m)g = -kx$

continued…

Since the weight mg and the spring constant k are constant, only x changes.

Thus, after the addition of 120 g, x doubles:

$PE_1 = ½kx^2$

$PE_2 = ½k(2x)^2$

$PE_2 = ½k(4x^2)$

$PE_2 = 4(½kx^2)$

The potential energy increases by a factor of 4.

24. C is correct.

In each case, the car's energy is reduced to zero by the work done by the frictional force:

$KE + (-W) = 0$

$KE = W$

Each car starts with kinetic energy:

$KE = (½)mv^2$

The initial speed is the same for each car, so the Ferrari has the most KE due to the differences in mass.

Reducing the Ferrari's energy to zero requires the most work.

25. D is correct.

The hammer does work on the nail as it drives it into the wood.

The amount of work done is equal to the amount of kinetic energy lost by the hammer:

$\Delta KE = \Delta W$

26. A is correct.

The only force doing work is the road's friction, so the work done by the road's friction is the total work.

This work equals the change in KE.

$W = \Delta KE$

$W = KE_f - KE_i$

$W = ½mv_2^2 - ½mv_1^2$

$W = 0 - [½(1{,}500\text{ kg})\cdot(25\text{ m/s})^2]$

$W = -4.7 \times 10^5\text{ J}$

27. D is correct.

$KE = ½mv^2$

$KE_{car} = ½(1{,}000\ kg)·(4.72\ m/s)^2$

$KE_{car} = 11{,}139\ J$

Calculate the KE of the 2,000 kg truck with 20 times the KE:

$KE_{truck} = KE_{car} \times 20$

$KE_{truck} = (11{,}139\ J) \times 20$

$KE_{truck} = 222.7\ kJ$

Calculate the speed of the 2,000 kg truck:

$KE = ½mv^2$

$v^2 = 2KE / m$

$v^2 = 2(222.7\ kJ) / (2{,}000\ kg)$

$v_{truck} = \sqrt{[2(222.7\ kJ) / (2{,}000\ kg)]}$

$v_{truck} = 14.9\ m/s$

28. C is correct.

Gravity and the normal force are balanced, vertical forces.

Since the car is slowing (i.e., accelerating backward), there is a net force backward due to friction (i.e., braking).

Newton's First Law of Motion states that the car keeps moving forward in the absence of any forces.

29. B is correct.

Energy is always conserved, so the work needed to lift the piano is 0.15 m is equal to the work needed to pull the rope 1 m:

$W_1 = W_2$

$F_1d_1 = F_2d_2$

$F_1d_1 / d_2 = F_2$

$F_2 = (6{,}000\ N)·(0.15\ m) / 1\ m$

$F_2 = 900\ N$

30. C is correct.

The area under the curve on a graph is the product of the values of $y \times x$.

Here, the y value is force, and the x value is distance:

$Fd = W$

31. B is correct.

The vast majority of the Earth's energy comes from the sun, which produces radiation that penetrates the Earth's atmosphere.

Likewise, radiation is emitted from the Earth's atmosphere.

32. C is correct.

$$W = Fd$$

$$W = \Delta KE$$

$$F \times d = ½mv^2$$

If v is doubled:

$$F \times d_2 = ½m(2v)^2$$

$$F \times d_2 = ½m(4v^2)$$

$$F \times d_2 = 4(½mv^2)$$

For equations to remain equal, d_2 must be 4 times d.

33. D is correct.

$$\text{Work} = \text{Power} \times \text{time}$$

$$P = W / t$$

$$W = Fd$$

$$P = (Fd) / t$$

$$P = [(2{,}000\ \text{N})\cdot(320\ \text{m})] / (60\ \text{s})$$

$$P = 10{,}667\ \text{W} = 10.7\ \text{kW}$$

34. A is correct.

Solution using the principle of conservation of energy

Assuming the system consists of the barbell alone, the force of gravity and the force of the hands raising the barbell are both external forces.

Since the system contains only a single object, potential energy is not defined.

The net power expended is:

$$P_{net} = W_{ext} / \Delta t$$

Conservation of energy requires:

$$W_{ext} = \Delta KE$$

$$W_{ext} = ½m(v_f^2 - v_i^2)$$

continued…

For constant acceleration situations:

$$(v_f + v_i) / 2 = v_{average} = \Delta y / \Delta t$$

$$(v_f + 0.0 \text{ m/s}) / 2 = 3.0 \text{ m} / 3.0 \text{ s}$$

$$v_f = 2.0 \text{ m/s}$$

Therefore:

$$W_{ext} = \frac{1}{2}(25 \text{ kg})\cdot(2.0 \text{ m/s})^2$$

$$W_{ext} = 50.0 \text{ J}$$

The net power expended is:

$$P_{net} = 50.0 \text{ J} / 3.0 \text{ s} = 17 \text{ W}$$

$$P_{net} = 17 \text{ W}$$

Solution using work

The power expended in raising the barbell is:

$$P_{net} = W_{net} / \Delta t$$

The net work is defined as:

$$W_{net} = F_{net}\Delta y$$

By Newton's Second law:

$$F_{net} = ma$$

Find the acceleration:

$$\Delta y = \frac{1}{2}a\Delta t^2$$

$$a = (2)\cdot(3.0 \text{ m}) / (3.0 \text{ s})^2$$

$$a = 0.67 \text{ m/s}^2$$

The net force on the barbell is:

$$F_{net} = (25 \text{ kg})\cdot(0.67 \text{ m/s}^2)$$

$$F_{net} = (50 / 3) \text{ N}$$

The net work is:

$$W_{net} = F_{net}\Delta y$$

$$W_{net} = [(50 / 3) \text{ N}] \cdot (3.0 \text{ m})$$

$$W_{net} = 50.0 \text{ J}$$

The net power expended:

$$P_{net} = 50.0 \text{ J} / 3.0 \text{ s}$$

$$P_{net} = 17 \text{ W}$$

35. B is correct.

The bag was never lifted off the ground and moved horizontally at a constant velocity.

$F = 0$

$W = Fd$

$W = 0$ J

Because there is no acceleration, the force is zero, and thus the work is zero.

36. B is correct.

Using energy conservation to solve the problem:

$W = |\Delta KE|$

$Fd = |\tfrac{1}{2}m(v_f^2 - v_0^2)|$

$d = |m(v_f^2 - v_0^2) / 2F|$

$d = |(1{,}000\ \text{kg})\cdot[(22\ \text{m/s})^2 - (30\ \text{m/s})^2] / (2)\cdot(9{,}600\ \text{N})|$

$d = |(1{,}000\ \text{kg})\cdot(484\ \text{m}^2/\text{s}^2 - 900\ \text{m}^2/\text{s}^2) / 19{,}200\ \text{N}|$

$d = 22$ m

Kinematic approach:

$F = ma$

$a = F / m$

$a = (9{,}600\ \text{N}) / (1{,}000\ \text{kg})$

$a = 9.6\ \text{m/s}^2$

$v_f^2 = v_0^2 + 2a\Delta d$

$(v_f^2 - v_0^2) / 2a = \Delta d$

Note that acceleration is negative due to it acting opposite the velocity.

$\Delta d = [(22\ \text{m/s})^2 - (30\ \text{m/s})^2] / 2(-9.6\ \text{m/s}^2)$

$\Delta d = (484\ \text{m}^2/\text{s}^2 - 900\ \text{m}^2/\text{s}^2) / (-19.2\ \text{m/s}^2)$

$\Delta d = (-416\ \text{m}^2/\text{s}^2) / (-19.2\ \text{m/s}^2)$

$\Delta d = 21.7\ \text{m} \approx 22\ \text{m}$

37. C is correct.

W = 100 J

Work = Power × time

P = W / t

P = 100 J / 50 s

P = 2 W

38. D is correct.

All of the original potential energy (with respect to the bottom of the cliff) is converted into kinetic energy.

$mgh = \frac{1}{2} mv_f^2$

Therefore:

$v_f = \sqrt{2gh}$

$v_f = \sqrt{(2)\cdot(10\ \text{m/s}^2)\cdot(58\ \text{m})}$

v_f = 34 m/s

Kinematic approach:

$v_f^2 = v_0^2 + 2a\Delta x$

$v_f^2 = 0 + 2a\Delta x$

$v_f = \sqrt{2a\Delta x}$

$v_f = \sqrt{[2(10\ \text{m/s}^2)\cdot(58\ \text{m})]}$

$v_f = \sqrt{(1{,}160\ \text{m}^2/\text{s}^2)}$

v_f = 34 m/s

39. A is correct.

PE = mgh

If height and gravity are constant, then potential energy is directly proportional to mass.

If the second stone has four times the mass of the first, it must have four times the potential energy of the first.

$m_2 = 4m_1$

$PE_2 = 4PE_1$

Therefore, the second stone has four times the potential energy.

40. B is correct.

W = Fd

W = mgh, work done by gravity

continued…

$W = (1.3\ kg)\cdot(10\ m/s^2)\cdot(6\ m)$

$W = 78\ J$

41. A is correct.

Potential energy is associated with the relative positions of pairs of objects, regardless of their state of motion.

Kinetic energy is the energy associated with the motion of single particles, regardless of their location.

42. A is correct.

$F_{spring} = F_{centripetal}$

$F_{spring} = kx$

$kx = 15\ N$

$x = (15\ N) / (65\ N/m)$

$x = 0.23\ m$

$PE_{spring} = \frac{1}{2}kx^2$

$PE_{spring} = \frac{1}{2}(65\ N/m)\cdot(0.23\ m)^2$

$PE_{spring} = 1.7\ J$

43. C is correct.

total time = (3.5 h/day)·(7 days)·(5 weeks)

total time = 122.5 h

cost = (8.16 cents/kW·h)·(122.5 h)·(0.12 kW)

cost = 120 cents = $1.20

44. B is correct.

$x = 5.1\ m \times (\cos 32°)$

$x = 4.33\ m$

$h = 5.1\ m - 4.33\ m$

$h = 0.775\ m$

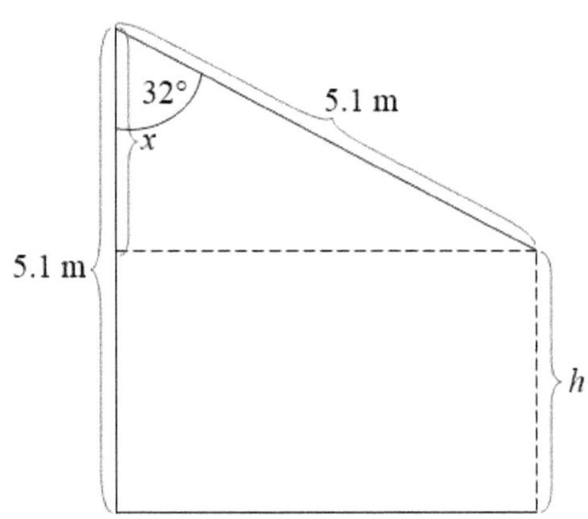

$W = Fd$

$W = mg \times h$

$m = W / gh$

$m = (120\ J) / (9.8\ m/s^2)\cdot(0.775\ m)$

$m = 15.8\ kg$

45. D is correct.

Potential energy of spring:

$PE_i + W = PE_f$

$\frac{1}{2}\, k\, x_i^2 + 111\text{J} = \frac{1}{2}\, k\, x_f^2$

$111\text{J} = \frac{1}{2}\, k\, (x_f^2 - x_i^2)$

$111\text{J} = \frac{1}{2}\, k\, [(2.9\text{m})^2 - (1.4\text{m})^2]$

$111\text{J} = \frac{1}{2}\, k\, [(8.41\text{m}^2) - (1.96\text{m}^2)]$

$111\text{J} = \frac{1}{2}\, k\, (6.45\text{m}^2)$

$k = 2(111\text{ J}) / (6.45\text{ m}^2)$

$k = 34\text{ N/m}$

Unit check:

$\text{J} = \text{kg}\cdot\text{m}^2/\text{s}^2$

$\text{J/m}^2 = (\text{kg}\cdot\text{m}^2/\text{s}^2)\cdot(1/\text{m}^2)$

$\text{J/m}^2 = (\text{kg/s}^2)$

$\text{N/m} = (\text{kg}\cdot\text{m/s}^2)\cdot(1/\text{m})$

$\text{N/m} = (\text{kg/s}^2)$

46. D is correct.

Potential energy, kinetic energy and work are measured in joules:

$\text{J} = \text{kg}\cdot\text{m}^2/\text{s}^2$

$\text{KE} = \frac{1}{2}mv^2 = \text{kg}(\text{m/s})^2 = \text{J}$

$\text{PE} = mgh$

$\text{PE} = \text{kg}(\text{m/s}^2)\cdot(\text{m}) = \text{J}$

$\text{W} = Fd = \text{J}$

47. A is correct.

Potential energy of spring:

$$PE = ½kx^2$$

Kinetic energy of mass:

$$KE = ½mv^2$$

Set equal and rearrange:

$$½kx^2 = ½mv^2$$

cancel ½ from each side of the expression

$$kx^2 = mv^2$$

$$x^2 = (mv^2) / k$$

$$x^2 = (m / k)v^2$$

Since m / k is provided:

$$x^2 = (0.038 \text{ kg·m/N})·(18 \text{ m/s})^2$$

$$x^2 = 12.3 \text{ m}^2$$

$$x = \sqrt{12.3} \text{ m}$$

$$x = 3.5 \text{ m}$$

48. A is correct.

$$m_t = 2m_c$$

$$v_t = 2v_c$$

KE of the truck:

$$KE_t = ½m_tv_t^2$$

Replace mass and velocity of the truck with the equivalent mass and velocity of the car:

$$KE_t = ½(2m_c)·(2v_c)^2$$

$$KE_t = ½(2m_c)·(4v_c^2)$$

$$KE_t = ½(8m_cv_c^2)$$

The truck has 8 times the kinetic energy of the car.

49. C is correct.

When a car stops, the KE is equal to the work done by the force of friction from the brakes.

Through friction, the KE is transformed into heat.

50. B is correct.

When the block comes to rest at the end of the spring, the upward force of the spring balances the downward force of gravity.

$F = kx$

$mg = kx$

$x = mg / k$

$x = (30\text{ kg})\cdot(10\text{ m/s}^2) / 900\text{ N/m}$

$x = 0.33\text{ m}$

51. D is correct.

$\text{KE} = \frac{1}{2}mv^2$

$\text{KE} = \frac{1}{2}(0.33\text{ kg})\cdot(40\text{ m/s})^2$

$\text{KE} = 264\text{ J}$

52. C is correct.

Work is the area under a force *vs.* position graph.

$\text{area} = Fd = \text{W}$

The area of the triangle as the object moves from 0 to 4 m:

$\text{A} = \frac{1}{2}bh$

$\text{A} = \frac{1}{2}(4\text{ m})\cdot(10\text{ N})$

$\text{A} = 20\text{ J}$

$\text{W} = 20\text{ J}$

53. C is correct.

$\text{KE} = \text{PE}$

$\frac{1}{2}mv^2 = mgh$

$v^2 / 2g = h$

If v is doubled:

$h_\text{B} = v_\text{B}^2 / 2g$

$v_\text{J} = 2v_\text{B}$

$(2v_\text{B})^2 / 2g = h_\text{J}$

$4(v_\text{B}^2 / 2g) = h_\text{J}$

$4h_\text{B} = h_\text{J}$ James's ball travels 4 times higher than Bob's ball.

54. B is correct.

Hooke's Law is given as:

$$F = -kx$$

The negative is only by convention to demonstrate that the spring force is a restoring force.

Graph B is correct because force is linearly increasing with increasing distance.

All other graphs are constant or exponential.

55. C is correct.

A decrease in the KE for the rocket causes a gain in its gravitational PE, the transfer of heat, or a combination.

The rocket loses some KE due to air resistance (friction).

Some of the rocket's KE converts to heat, which causes the air temperature surrounding the rocket to increase.

Therefore, the average KE of the air molecules increases.

56. D is correct.

Kinetic energy is given as:

$$\text{KE}_1 = \tfrac{1}{2}mv^2$$

$$\text{KE}_2 = \tfrac{1}{2}m(4v)^2$$

$$\text{KE}_2 = \tfrac{1}{2}m(16v^2)$$

Increasing the velocity by a factor of 4 increases the KE by a factor of 16.

57. C is correct.

The total energy of the system is conserved.

A relationship between the initial compression of the spring and the final speed of the mass can thus be found.

$$E_i = E_f$$

$$KE_i + PE_i = KE_f + PE_f$$

Initially, the spring is compressed and has PE, and the mass is at rest, so the initial KE is zero.

At the end, the spring is uncompressed, and the mass is moving, so the final PE is zero, and the mass has KE.

$$PE_i = KE_f$$

$$\tfrac{1}{2}kx^2 = \tfrac{1}{2}mv^2$$

$$kx^2 = mv^2$$

$$x\sqrt{k} = v\sqrt{m}$$

continued…

The velocity and the compression distance of the spring are directly proportional.

If the spring is compressed by four times the original distance, then the velocity is four times the original.

$x_2 = 4x_1$

$v_2 = 4v_1$

58. C is correct.

Force: $F = ma$ (N)

Work: $W = Fd$ (N·m)

Power: $P = W / t$ (N·m/s)

59. A is correct.

$W_{net} = \Delta KE$

$\Delta KE = KE_f - KE_i$

$\Delta KE + KE_i = KE_f$

60. D is correct.

$v = (70\ \text{km/h})\cdot(1{,}000\ \text{m/km})\cdot(1\ \text{h}/60\ \text{min})\cdot(1\ \text{min}/60\ \text{s})$

$v = 19.4\ \text{m/s}$

Force acting against the car:

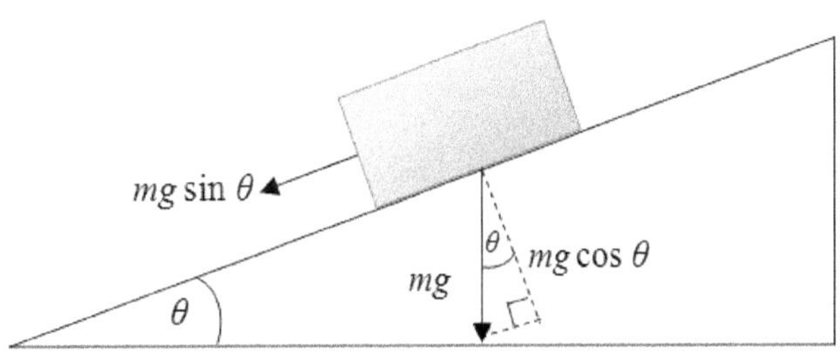

$F = mg \sin \theta$

$F = (1{,}320\ \text{kg})\cdot(9.8\ \text{m/s}^2) \sin 5°$

$\sin 5° = 0.0872$, round to 0.09

$F = (1{,}320\ \text{kg})\cdot(9.8\ \text{m/s}^2)\cdot(0.09)$

$F = 1{,}164\ \text{N}$

$\text{N} = \text{kg}\cdot\text{m/s}^2$

Rate of energy is power:

$\text{Watts} = \text{kg}\cdot\text{m}^2/\text{s}^3$

Multiply velocity by the downward force:

$P = Fv$

$P = (1{,}164\ \text{N})\cdot(19.4\ \text{m/s})$

$P = 22.6\ \text{kW}$

Notes for active learning

Periodic Motion – Detailed Explanations

1. B is correct.

Frequency measures the number of cycles per second a wave experiences, independent of the wave's amplitude.

2. D is correct.

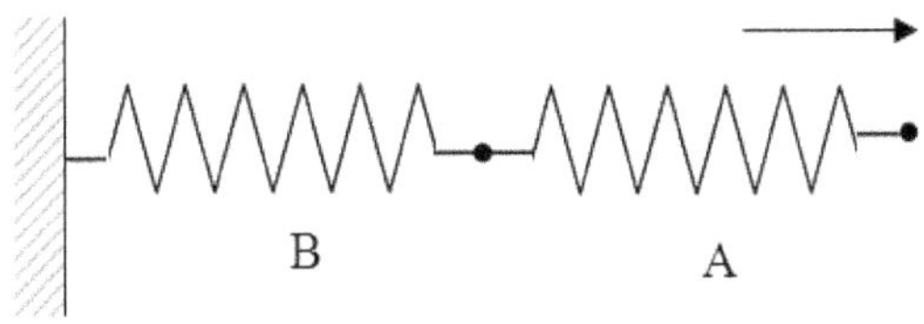

Hooke's Law:

$F = kx$

It is known that the force on each spring must be equal if they are in static equilibrium, therefore:

$F_A = F_B$

Therefore, the expression can be written as:

$k_A L_A = k_B L_B$

Solve for the spring constant of spring B:

$k_B = (k_A L_A) / L_B$

3. D is correct.

In a longitudinal wave, particles of a material are displaced parallel to the direction of the wave.

4. C is correct.

speed = wavelength × frequency

$v = \lambda f$

$v = (0.25 \text{ m})\cdot(1{,}680 \text{ Hz})$

$v = 420 \text{ m/s}$

5. A is correct.

$E_{stored} = \text{PE}$

$\text{PE} = ½kA^2$

Stored energy is potential energy.

In a simple harmonic motion (e.g., a spring), the potential energy is:

$\text{PE} = ½kx^2$ or $½kA^2$

where k is a constant and A (or x) is the distance from equilibrium

A is the amplitude of a wave in simple harmonic motion (SHM).

6. D is correct.

The spring oscillates around its new equilibrium position (3 cm below the equilibrium position with no mass hanging) with period $T = 2\pi\sqrt{(m / k)}$ since it's a mass-spring system undergoing simple harmonic motion.

To find k, consider how much the spring stretched when the mass was hung from it. Since the spring found a new equilibrium point 3 cm below its natural length, the upwards force from the spring (F_s) must balance the downwards gravitational force (F_g) at that displacement:

$$|F_s| = |F_g|$$

$$kd = mg$$

$$k\ (0.03\ \text{m}) = (11\ \text{kg})\cdot(9.8\ \text{m/s}^2)$$

$$k = 3593\ \text{N/m}$$

Now, solve for T:

$$T = 2\pi\sqrt{(m / k)}$$

$$T = 2\pi\sqrt{(11\ \text{kg} / 3593\ \text{N/m})}$$

$$T = 0.35\ \text{s}$$

The frequency is the reciprocal of the period:

$$f = 1 / T$$

$$f = 1 / (0.35\ \text{s})$$

$$f = 2.9\ \text{Hz}$$

7. C is correct.

The period is the reciprocal of the frequency:

$$T = 1 / f$$

8. D is correct.

The *period of a pendulum*:

$$T = 2\pi\sqrt{(L / g)}$$

The period only depends on the pendulum's length and gravity.

In an elevator, the apparent force of gravity only changes if the elevator is accelerating in either direction.

9. A is correct.

The *period* is the reciprocal of the frequency:

$$T = 1 / f$$

$$T = 1 / 100\ \text{Hz}$$

$$T = 0.01\ \text{s}$$

10. B is correct.

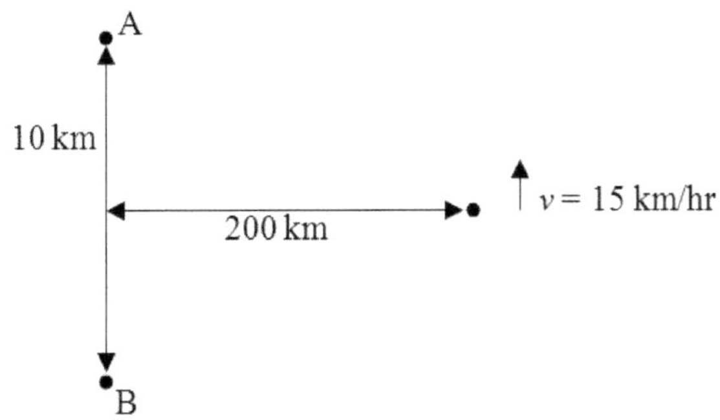

Convert v to m/s:

$v = (15\text{ km}/1\text{ h})\cdot(1\text{ h}/60\text{ min})\cdot(1\text{ min}/60\text{ s})\cdot(10^3\text{ m}/1\text{ km})$

$v = 4.2\text{ m/s}$

Convert frequency to λ:

$\lambda = c / f$

$\lambda = (3 \times 10^8\text{ m/s}) / (4.7 \times 10^6\text{ Hz})$

$\lambda = 63.8\text{ m}$

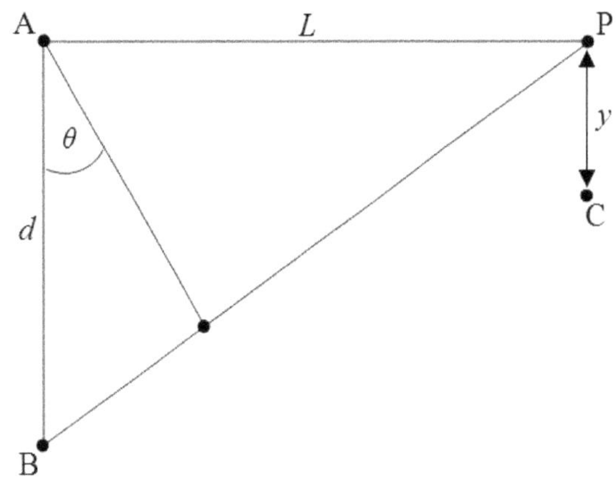

According to *Young's Equation*:

$\lambda = yd / mL$

where m = 0, 1, 2, 3, 4…

Solve for y by rearranging to isolate y:

$y = \lambda L m / d$

y = distance traveled by the ship:

$y = vt$

Since the first signal came at the point of maximum intensity, m = 0 at that time, at the next maximum m = 1.

Therefore:

$t = L\lambda / vd$

$t = (200{,}000\text{ m})\cdot(63.8\text{ m}) / (4.2\text{ m/s})\cdot(10{,}000\text{ m})$

$t = 304\text{ s}$

Convert time from seconds to minutes:

$t = (304\text{ s})\cdot(1\text{ min}/60\text{ s})$

$t = 5.06\text{ min} \approx 5.1\text{ min}$

For m values greater than 1, the calculated times are beyond the answer choices, so 5.1 min is the answer.

11. D is correct.

The *tension in the rope* is given by the equation:

$$T = (mv^2) / L$$

where v is the velocity of the wave and L is the length of the rope

Substituting:

$$v = L / t$$

$$T = [m(L / t)^2] / L$$

$$T = mL / t^2$$

$$T^2 = mL / T$$

$$t = \sqrt{(mL / T)}$$

$$t = \sqrt{[(2.31 \text{ kg})\cdot(10.4 \text{ m}) / 74.4 \text{ N}]}$$

$$t = \sqrt{(0.323 \text{ s}^2)}$$

$$t = 0.57 \text{ s}$$

12. A is correct.

$$\omega_A = 2\omega_B$$

$$\omega_B = \sqrt{g} / l_B$$

Therefore:

$$l_B = g / \omega^2{}_B$$

Similarly, for A:

$$l_A = g / \omega^2{}_A$$

$$l_A = g / (2\omega_B)^2$$

$$l_A = \tfrac{1}{4}g / \omega^2{}_B$$

$$l_A = \tfrac{1}{4}l_B$$

13. B is correct.

$$F = -kx$$

Since the motion is simple harmonic, the restoring force is proportional to displacement.

Therefore, if the displacement is 5 times greater, then so is the restoring force.

14. C is correct.

Period = (60 s) / (10 oscillations)

T = 6 s

The period is the time for one oscillation.

If 10 oscillations take 60 s, then one oscillation takes 6 s.

15. A is correct.

Conservation of Energy:

total ME = ΔKE + ΔPE = constant

$\frac{1}{2}mv^2 + \frac{1}{2}kx^2$ = constant

16. B is correct.

A displacement from the position of maximum elongation to maximum compression represents *half* a cycle.

If it takes 1 s, then the time required for a complete cycle is 2 s.

$f = 1 / T$

$f = 1 / 2$ s

$f = 0.5$ Hz

17. C is correct.

Sound waves are longitudinal waves.

18. B is correct.

speed = wavelength × frequency

speed = wavelength / period

$v = \lambda / T$

$\lambda = vT$

λ = (362 m/s)·(0.004 s)

λ = 1.5 m

19. A is correct.

$a = -A\omega^2 \cos(\omega t)$

where A is the amplitude or displacement from the resting position

20. D is correct.

The acceleration of a simple harmonic oscillation is:

$$a = -A\omega^2 \cos(\omega t)$$

Its maximum occurs when cos (ωt) is equal to 1

$$a_{max} = -\omega^2 x$$

If ω is doubled:

$$a = -(2\omega)^2 x$$

$$a = -4\omega^2 x$$

The maximum value of acceleration changes by a factor of 4.

21. B is correct.

Resonant frequency of a spring and mass system in any orientation:

$$\omega = \sqrt{(k / m)}$$

$$f = \omega / 2\pi$$

$$T = 1 / f$$

$$T = 2\pi\sqrt{(m / k)}$$

Period of a spring does not depend on gravity.

The period remains constant because only mass and the spring constant affect the period.

22. C is correct.

$$v = \lambda f$$

$$\lambda = v / f$$

An increase in v and a decrease in f must increase λ.

23. B is correct.

Frequency is the measure of oscillations or vibrations per second.

frequency = 60 vibrations in 1 s

frequency = 60 Hz

speed = 30 m / 1 s

speed = 30 m/s

24. A is correct.

$T = (mv^2) / L$

$m = TL / v^2$

$m = (60\text{ N})\cdot(16\text{ m}) / (40\text{ m/s})^2$

$m = (960\text{ N}\cdot\text{m}) / (1{,}600\text{ m}^2/\text{s}^2)$

$m = 0.6\text{ kg}$

25. D is correct.

Amplitude is independent of frequency.

26. C is correct.

$f = \#\text{ cycles / time}$

$f = 60\text{ drips} / 40\text{ s}$

$f = 1.5\text{ Hz}$

27. D is correct.

Transverse waves are characterized by their crests and valleys, caused by the particles of the wave traveling "up and down" with respect to the lateral movement of the wave.

The particles in longitudinal waves travel parallel to the direction of the wave.

28. B is correct.

The velocity *vs.* time graph shows that at $t = 0$, the particle's velocity is positive, and the speed is increasing.

When speed increases, velocity and acceleration point in the same direction.

Therefore, the acceleration is non-zero and positive.

Only graph B displays a positive acceleration at $t = 0$.

29. C is correct.

The *speed of a wave* is determined by the characteristics of the medium (and the type of wave).

Speed is *independent* of amplitude.

30. A is correct.

$f = 1 / \text{period}$

$f = \#\text{ cycles / second}$

$f = 1\text{ cycle} / 2\text{ s}$

$f = ½\text{ Hz}$

31. B is correct.

$f = v / \lambda$

$\lambda = v / f$

$\lambda = (340 \text{ m/s}) / (2{,}100 \text{ Hz})$

$\lambda = 0.16 \text{ m}$

32. D is correct.

Period (T) $= 2\pi\sqrt{(L / g)}$

The period is independent of the mass.

33. A is correct.

$v = \omega x$

$\omega = v / x$

$\omega = (15 \text{ m/s}) / (2.5 \text{ m})$

$\omega = 6.0 \text{ rad/s}$

34. D is correct.

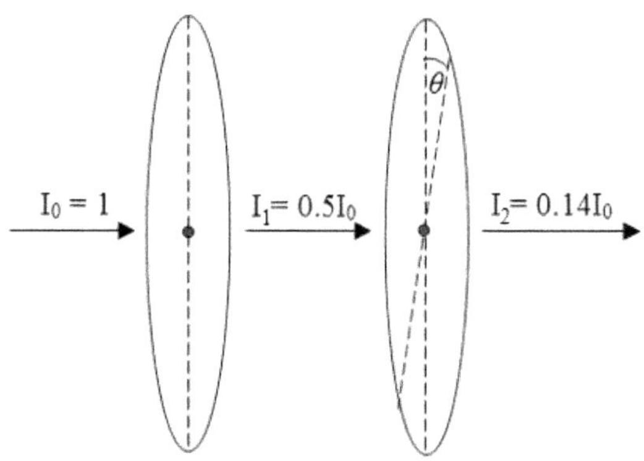

Unpolarized light on a polarizer reduces the intensity by ½.

$I = (½)I_0$

After that, the light is further reduced in intensity by the second filter.

Law of Malus:

$I = I_0 \cos^2 \theta$

$(0.14\ I_0) = (0.5\ I_0) \cos^2 \theta$

$0.28 = \cos^2 \theta$

$\cos^{-1} \sqrt{(0.28)} = \theta$

$\theta = 58°$

35. C is correct.

At a maximum distance from equilibrium, the energy in the system is potential energy, and the speed is zero.

Therefore, kinetic energy is zero.

Since there is no kinetic energy, the mass has no velocity.

36. D is correct.

$v = \lambda f$

$f = v / \lambda$

$f = (240\text{ m/s}) / (0.1\text{ m})$

$f = 2{,}400\text{ Hz}$

37. B is correct.

In a transverse wave, the vibrations of particles are perpendicular to the direction of travel of the wave.

Transverse waves have crests and troughs that move along the wave.

In a longitudinal wave, the vibrations of particles are parallel to the direction of travel of the wave.

Longitudinal waves have compressions and rarefactions that move along the wave.

38. C is correct.

$v = \sqrt{(T / \mu)}$

where μ is the linear density of the wire

$T = v^2\mu$

$\mu = \rho A$

where A is the cross-sectional area of the wire and equals πr^2

$\mu = (2{,}700\text{ kg/m}^3)\pi(4.6 \times 10^{-3}\text{ m})^2$

$\mu = 0.18\text{ kg/m}$

$T = (36\text{ m/s})^2\cdot(0.18\text{ kg/m})$

$T = 233\text{ N}$

39. D is correct.

Refraction is the change in the direction of a wave caused by the change in the wave's speed.

Examples of waves include sound waves and light waves.

Refraction is often observed when a wave passes from one medium to a different medium (e.g., from air to water and vice versa).

40. C is correct.

f = # cycles / second

f = 2 cycles / 1 s

f = 2 Hz

41. A is correct.

Pitch is how the brain perceives frequency. The pitch becomes higher as frequency increases.

42. C is correct.

The KE is maximum when the spring is neither stretched nor compressed.

If the object is bobbing, KE is maximum at the midpoint between fully stretched and fully compressed because all the spring's energy is KE rather than a mix of KE and PE.

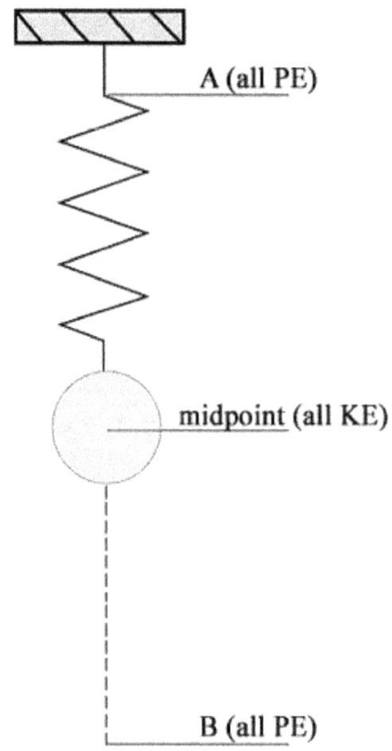

43. B is correct.

Torque = $rF \sin \theta$

$F = ma$

substitute mg for F

$\tau = rmg \sin \theta$

$\tau = (1 \text{ m})\cdot(0.5 \text{ kg})\cdot(10 \text{ m/s}^2) \sin 60°$

$\tau = (5 \text{ kg}\cdot\text{m}^2/\text{s}^2) \times 0.87$

$\tau = 4.4 \text{ N}\cdot\text{m}$

44. D is correct.

The Doppler effect can be observed in all types of waves.

45. A is correct.

$v = \sqrt{(T / \mu)}$

where μ is the linear density of the wire

$F_T = ma$

$F_T = (2{,}500 \text{ kg})\cdot(10 \text{ m/s}^2)$

$F_T = 25{,}000 \text{ N}$

$v = \sqrt{(25{,}000 \text{ N} / 0.65 \text{ kg/m})}$

$v = 196 \text{ m/s}$

The weight of the wire can be assumed to be negligible compared to the cement block.

46. B is correct.

$f = ½π[√(g / L)]$

Frequency is independent of mass

47. A is correct.

$T = 2π√(L / g)]$

$T = 2π√(3.3 \text{ m} / 10 \text{ m/s}^2)$

$T = 3.6 \text{ s}$

48. C is correct.

$f = (1/2π)√(k / m)$

If k increases by a factor of 2, then f increases by a factor of √2 (or 1.41).

Increasing by a factor of 1.41 or 41%

49. D is correct.

In a simple harmonic motion, the acceleration is greatest at the ends of motions (points A and D), where velocity is zero.

Velocity is greatest at the nadir, where acceleration is equal to zero (point C).

50. A is correct.

At the lowest point, the KE is at a maximum, and the PE is at a minimum.

The loss of gravitational PE equals the gain in KE:

$mgh = ½mv^2$

cancel m from each side of the expression

$gh = ½v^2$

$(10 \text{ m/s}^2)·(10 \text{ m}) = ½v^2$

$(100 \text{ m}^2/\text{s}^2) = ½v^2$

$200 \text{ m}^2/\text{s}^2 = v^2$

$v = 14 \text{ m/s}$

51. A is correct.

Pitch is a psychophysical phenomenon when the sensation of a frequency is commonly referred to as the pitch of a sound.

A perception of high-pitch sound corresponds to a high-frequency sound wave, and a low-pitch sound corresponds to a low-frequency sound wave.

Amplitude plays no role, and speed is constant.

52. C is correct.

Because wind is blowing in the reference frame of both the train and the observer, it is not considered.

$f_{observed} = [v_{sound} / (v_{sound} - v_{source})]f_{source}$

$f_{observed} = [340 \text{ m/s} / (340 \text{ m/s} - 50 \text{ m/s})]\cdot 500 \text{ Hz}$

$f_{observed} = 586 \text{ Hz}$

$\lambda = v / f$

$\lambda = 340 \text{ m/s} / 586 \text{ Hz}$

$\lambda = 0.58 \text{ m}$

53. B is correct.

$\text{PE} = \frac{1}{2}kx^2$

Doubling the amplitude x increases PE by a factor of 4.

54. C is correct.

The elastic modulus is given by:

E = tensional strength / extensional strain

$E = \sigma / \varepsilon$

55. D is correct.

Resonance is when one system transfers energy to another at that system's resonant frequency (natural frequency).

Resonance is a forced vibration that produces the highest amplitude response for a given force amplitude.

56. A is correct.

Period of a pendulum:

$$T_P = 2\pi\sqrt{(L / g)}$$

Period of a spring:

$$T_S = 2\pi\sqrt{(m / k)}$$

The period of a spring does not depend on gravity and is unaffected.

57. B is correct.

At the top of its arc, the pendulum comes to rest momentarily; the KE and the velocity equal zero.

Since its height above the bottom of its arc is at a maximum at this point, its (angular) displacement from the vertical equilibrium position is at a maximum.

The pendulum constantly experiences the forces of gravity and tension and is therefore continuously accelerating.

58. D is correct.

The Doppler effect is the observed change in frequency when a sound source is in motion relative to an observer (away or towards).

If the sound source moves with the observer, there is no relative motion between the two, and the Doppler effect does not occur.

59. C is correct.

The amplitude of a wave is the magnitude of its oscillation from its equilibrium point.

60. A is correct.

$$f = \sqrt{(k / m)}$$

An increase in m causes a decrease in f.

Notes for active learning

Fluids and Solids – Detailed Explanations

1. C is correct.

Refer to the unknown liquid as "A" and the oil as "O."

$$\rho_A h_A g = \rho_O h_O g$$

cancel g from both sides of the expression

$$\rho_A h_A = \rho_O h_O$$

$$h_A = 5 \text{ cm}$$

$$h_O = 20 \text{ cm}$$

$$h_A = \tfrac{1}{4} h_O$$

$$\rho_A(\tfrac{1}{4})h_O = \rho_O h_O$$

$$\rho_A = 4\rho_O$$

$$\rho_A = 4(850 \text{ kg/m}^3)$$

$$\rho_A = 3{,}400 \text{ kg/m}^3$$

2. D is correct.

$$P = \rho_{oil} \times V_{oil} \times g \,/\, (A_{tube})$$

$$P = [\rho_o \pi (r_{tube})^2 \times hg] \,/\, \pi (r_{tube})^2$$

cancel $\pi(r_{tube})^2$ from the numerator and denominator

$$P = \rho_o g h$$

$$P = (850 \text{ kg/m}^3)\cdot(9.8 \text{ m/s}^2)\cdot(0.2 \text{ m})$$

$$P = 1{,}666 \text{ Pa}$$

3. A is correct.

$$m_{oil} = \rho_{oil} V_{oil}$$

$$V = \pi r^2 h$$

$$m_{oil} = \rho_{oil} \pi r^2 h$$

$$m_{oil} = \pi(850 \text{ kg/m}^3)\cdot(0.02 \text{ m})^2 \times (0.2 \text{ m})$$

$$m_{oil} = 0.21 \text{ kg}$$

$$m_{oil} = 210 \text{ g}$$

4. A is correct.

Gauge pressure is the pressure experienced by an object referenced at atmospheric pressure.

When the block is lowered, its gauge pressure increases according to:

$$P_G = \rho gh$$

At $t = 0$, the block just enters the water and $h = 0$ so $P_G = 0$.

As time passes, the height of the block below the water increases linearly, so P_G increases linearly.

5. D is correct.

Using *Bernoulli's principle* and assuming the opening of the tank is so large that the initial velocity is essentially zero:

$$\rho gh = \tfrac{1}{2}\rho v^2$$

cancel ρ from both sides of the expression

$$gh = \tfrac{1}{2}v^2$$

$$v^2 = 2gh$$

$$v^2 = 2\cdot(9.8 \text{ m/s}^2)\cdot(0.8 \text{ m})$$

$$v^2 = 15.68 \text{ m}^2/\text{s}^2$$

$$v = 3.96 \text{ m/s} \approx 4 \text{ m/s}$$

Note: the diameter is not used to solve the problem.

6. C is correct.

The *ideal gas law* is:

$$PV = nRT$$

Keeping nRT constant:

$$\text{If } P_{final} = \tfrac{1}{2}P_{initial}$$

$$V_{final} = \tfrac{1}{2}V_{initial}$$

However, in an isothermal process, there is no change in internal energy.

Therefore, because energy must be conserved:

$$\Delta U = 0$$

7. B is correct.

Hooke's Law for a spring:

$$F = kx$$

Solve for k:

$$k = F / x$$

$$k = (9.5 \text{ kg})\cdot(9.8 \text{ m/s}^2) / (0.004 \text{ m})$$

$$k = 23{,}275 \text{ N/m}$$

$$k = 2.3 \times 10^4 \text{ N/m}$$

8. D is correct.

The object sinks when the buoyant force is less than the weight of the object.

Since the buoyant force is equal to the weight of the displaced fluid, an object sinks precisely when the weight of the fluid it displaces is less than the weight of the object itself.

9. A is correct.

$$P = \rho gh$$

$$P = (10^3 \text{ kg/m}^3)\cdot(9.8 \text{ m/s}^2)\cdot(100 \text{ m})$$

$$P = 9.8 \times 10^5 \text{ N/m}^2$$

10. C is correct.

Absolute pressure = gauge pressure + atmospheric pressure

$$P_{abs} = P_G + P_{atm}$$

$$P_{abs} = \rho gh + P_{atm}$$

Atmospheric pressure is added to the total pressure at the bottom of a volume of liquid.

Therefore, if the atmospheric pressure increases, absolute pressure increases by the same amount.

11. A is correct.

Surface tension increases as temperature decreases.

Generally, the *cohesive forces* maintaining surface tension decrease as molecular thermal activity increases.

12. A is correct.

A force per unit area causes the stretch of a wire.

If force is constant, an increase in weight increases the area related to d^2.

$$E = \sigma / \mathcal{E}$$

where E = Young's modulus, σ = stress (F / A) and $\mathcal{E}$ = strain (ΔL / L)

$$E = (F / A) / (\Delta L / L)$$

$$E = FL / A\Delta L$$

Relate force and area:

$$F = (E)\cdot(\Delta L)\cdot(A) / (L)$$

$$F = (E)\cdot(\Delta L)\cdot(\pi / 4 \times d^2) / (L)$$

The force is directly proportional to the ΔL and is directly proportional to d^2.

13. A is correct.

The buoyant force upward must balance the weight downward.

Buoyant force = weight of the volume of water displaced

$$F_B = W_{object}$$

$$\rho Vg = W_{object}$$

$$W_{object} = 60 \text{ N}$$

$$W_{object} = (\rho_{water})\cdot(V_{water})\cdot(g)$$

$$60 \text{ N} = (1{,}000 \text{ kg/m}^3)\cdot(V_{water})\cdot(10 \text{ m/s}^2)$$

$$V_{water} = 60 \text{ N} / (1{,}000 \text{ kg/m}^3)\cdot(10 \text{ m/s}^2)$$

$$V_{water} = 0.006 \text{ m}^3$$

14. D is correct.

Volume flow rate:

$$Q = vA$$

$$Q = (2.5 \text{ m/s})\pi r^2$$

$$Q = (2.5 \text{ m/s})\cdot(0.015 \text{ m})^2\pi$$

$$Q = 1.8 \times 10^{-3} \text{ m}^3\text{/s}$$

15. C is correct.

Force equation for the cork that is not accelerating:

$$F_B - mg = 0$$

Let m be the mass and V be the volume of the cork.

Replace:

$$m = \rho V$$

$$F_B = (\rho_{water})\cdot(V_{disp})\cdot(g)$$

$$(\rho_{water})\cdot(V_{disp})\cdot(g) = (\rho_{cork})\cdot(V)\cdot(g)$$

$$V_{disp} = \tfrac{3}{4}V$$

$$\rho_{water}(\tfrac{3}{4}Vg) = (\rho_{cork})\cdot(V)\cdot(g)$$

cancel g and V from both sides of the expression

$$\tfrac{3}{4}\rho_{water} = \rho_{cork}$$

$$\rho_{cork} / \rho_{water} = \tfrac{3}{4} = 0.75$$

16. D is correct.

volume = mass / density

$$V = (600\text{ g}) / (0.93\text{ g/cm}^3)$$

$$V = 645\text{ cm}^3$$

17. C is correct.

For monatomic gases:

$$U = 3/2k_BT$$

where U is the average KE per molecule and k_B is the Boltzmann constant

18. B is correct.

The object weighs 150 N less while immersed because the buoyant force supports 150 N of total weight.

Since the object is totally submerged, the volume of water displaced equals the volume of the object.

$$F_B = 150\text{ N}$$

$$F_B = \rho_{water} \times V_{water} \times g$$

$$V = F_B / \rho g$$

$$V = (150\text{ N}) / (1{,}000\text{ kg/m}^3)\cdot(10\text{ m/s}^2)$$

$$V = 0.015\text{ m}^3$$

19. C is correct.

$F_B / \rho_w = (m_c g) / \rho_c$

$F_B = (\rho_w m_c g) / \rho_c$

$F_B = [(1\ g/cm^3)\cdot(0.03\ kg)\cdot(9.8\ m/s^2)] / (8.9\ g/cm^3)$

$F_B = 0.033\ N$

$m_{total} = m_w + (F_B / g)$

$m_{total} = (0.14\ kg) + [(0.033\ N) / (9.8\ m/s^2)]$

$m_{total} = 0.143\ kg = 143\ g$

20. D is correct.

$v_1 A_1 = v_2 A_2$

$v_2 = v_1 A_1 / A_2$

$v_2 = [v_1(\pi/4)d_1^2] / (\pi/4)d_2^2$

cancel ($\pi/4$) from the numerator and denominator

$v_2 = v_1(d_1^2 / d_2^2)$

$v_2 = (1\ m/s)\cdot[(6\ cm)^2 / (3\ cm)^2]$

$v_2 = 4\ m/s$

21. C is correct.

Static fluid pressure:

$P = \rho g h$

The pressure is only dependent on gravity (g), the height (h) of the fluid above the object, and fluid density (ρ).

Pressure does depend on the depth of the object but does not depend on the surface area of the object.

Both objects are submerged to the same depth, so the fluid pressure is equal.

Note that the buoyant force on the blocks is NOT equal, but the pressure (force / area) is equal.

22. C is correct.

Surface tension force acts as the product of surface tension and the total length of contact.

$$F = \text{A}L$$

For a piece of thread, the length of contact is l as shown:

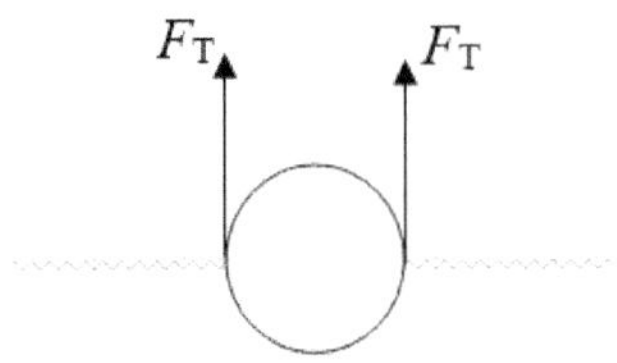

cross-sectional view

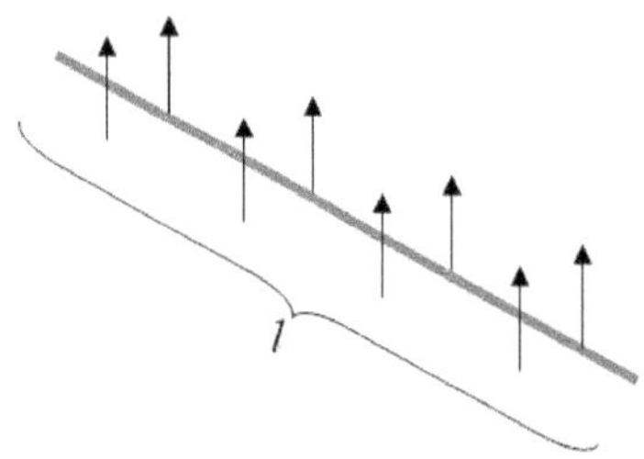

$L = 2L$ because the force acts on both sides of the thread.

Thus, for a thread rectangle, the total contact length is the total length times two.

$$L = 2(l + w + l + w)$$

$$F_{max} = 2\text{A}(l + w + l + w)$$

$$F_{max} = 2\text{A}(2l + 2w)$$

$$F_{max} = 4\text{A}(l + w)$$

23. D is correct.

Gauge pressure is the measure of pressure with respect to atmospheric pressure.

So if the pressure inside the tire is equal to the air pressure outside, the gauge reads zero.

24. B is correct.

The pressure is measured in force per unit area, which is the force divided by the area.

25. D is correct.

The shear stress is the force per unit area and has units of N/m^2.

26. B is correct.

Because the reservoir area is assumed to be virtually infinite, the velocity of the flow at the top of the tank is assumed to be zero.

Using Bernoulli's equation, find the speed through the 3 cm pipe:

$$(\tfrac{1}{2}\rho v^2 + \rho gh)_{out} = (\tfrac{1}{2}\rho v^2 + \rho gh)_{in}$$

$$\tfrac{1}{2}\rho v^2 = \rho gh$$

cancel ρ from both sides of the expression:

$$\tfrac{1}{2}v^2 = gh$$

continued…

$v^2 = 2gh$

$v = \sqrt{(2gh)}$

$v = \sqrt{[2(9.8 \text{ m/s}^2)\cdot(4 \text{ m})]}$

$v_{3cm} = 8.9 \text{ m/s}$

To find the speed through the 5 cm pipe use the continuity equation:

$A_{3cm}v_{3cm} = A_{5cm}v_{5cm}$

$(\pi / 4)\cdot(3 \text{ cm})^2\cdot(8.9 \text{ m/s}) = (\pi / 4)\cdot(5 \text{ cm})^2\cdot(v_{5cm})$

$v_{5cm} = 3.2 \text{ m/s}$

27. D is correct.

The ideal gas law is:

$PV = nRT$

where n, R and T are constants

P is pressure, V is volume, n is the number of particles, R is the ideal gas law constant, and T is temperature.

If P → 3P, then V→ (1/3)V to maintain a constant temperature.

28. D is correct.

$F = PA + F_{cover}$

$P = \rho gh$

$F = \rho ghA + F_{cover}$

$F = [(1{,}000 \text{ kg/m}^3)\cdot(10 \text{ m/s}^2)\cdot(1 \text{ m})\cdot(1 \text{ m}^2)] + 1{,}500 \text{ N}$

$F = 11{,}500 \text{ N}$

29. B is correct.

The buoyant force on a totally submerged object is independent of its depth below the surface (since any increase in the water's density is ignored).

The buoyant force on the ball at a depth of 4 m is 20 N, the same as the buoyant force at 1 m.

When it sits at the bottom of the pool, the two upward forces (i.e., the buoyant force F_B and the normal force F_N) must balance the downward force of gravity.

$F_B + F_N = F_g$

$(20 \text{ N}) + F_N = 80 \text{ N}$

$F_N = 80 \text{ N} - 20 \text{ N}$

$F_N = 60 \text{ N}$

30. B is correct.

$V_{\text{Fluid Displaced}} = \frac{1}{2}V_{\text{block}}$

Buoyant force:

$F_B = \rho g V$

$\rho_F g V_F = \rho_B g V_B$

cancel g from both sides of the expression

$\rho_F V_F = \rho_B V_B$

$\rho_F(\frac{1}{2}V_B) = \rho_B V_B$

cancel V_B from both sides of the expression

$\frac{1}{2}\rho_F = \rho_B$

$\rho_F = (1.6)\rho_{\text{water}}$

$\rho_B = (1.6)\cdot(10^3\ \text{kg/m}^3)\cdot(\frac{1}{2})$

$\rho_B = 800\ \text{kg/m}^3$

31. C is correct.

Terminal velocity:

$v_t = \sqrt{(2mg / c\rho A)}$

where c is the coefficient of air drag, ρ is the density of air, and A is the projected area (πr^2)

If the masses of the three balls are equal, then the velocity of each ball is:

$r = R$:

$A = \pi r^2$

$v_1 = \sqrt{(2mg / c\rho A)}$

$r = 2R$:

$A_1 = \pi(2r)^2$

$A_1 = 4\pi r^2$

$A_1 = 4A$

$v_2 = \sqrt{(2mg / c\rho 4A)}$

$v_2 = (\frac{1}{2})\sqrt{(2mg / c\rho A)}$

$v_2 = (\frac{1}{2})v_1$

$r = 3R$:

$A_2 = \pi(3r)^2$

continued…

$A_2 = 9\pi r^2$

$A_2 = 9A$

$v_3 = \sqrt{(2mg / c\rho 9A)}$

$v_3 = (1/3)\sqrt{(2mg / c\rho A)}$

$v_3 = (1/3)v_1$

As the ball becomes larger, its terminal velocity decreases, and therefore time until impact increases.

velocity = distance / time

32. A is correct.

The pressure due to the density of a fluid surrounding an object submerged at depth d below the surface is:

$P = \rho gd$

Since the distance that the objects are below the surface is not specified, the conclusion is that object B experiences less fluid pressure than object A.

This difference is because object B is higher off the floor of the container.

Thus, its depth is less than object A.

33. B is correct.

$A_1v_1 = A_2v_2$

$A = \pi r^2$

$A_2 = \pi(2r)^2$

$A_2 = 4\pi r^2$

$A_2 = 4A_1$

If r is doubled, then area is increased by 4 times

$A_1(14 \text{ m/s}) = (4A_1)v_2$

$v_2 = (A_1 \times 14 \text{ m/s}) / (4 \times A_1)$

$v_2 = (14 \text{ m/s}) / (4)$

$v_2 = 3.5 \text{ m/s}$

Use Bernoulli's equation to find resulting pressure:

$P_1 + ½\rho v_1^2 = P_2 + ½\rho v_2^2$

$(3.5 \times 10^4 \text{ Pa}) + ½(1{,}000 \text{ kg/m}^3)\cdot(14 \text{ m/s})^2 = P_2 + ½(1{,}000 \text{ kg/m}^3)\cdot(3.5 \text{ m/s})^2$

$P_2 = (13.3 \times 10^4 \text{ Pa}) – (6.1 \times 10^3 \text{ Pa})$

$P_2 = 12.7 \times 10^4 \text{ Pa}$

34. D is correct.

The pressure due to the atmosphere is equal to its weight per unit area.

At an altitude of 2 km, there is less atmosphere pushing down than at Earth's surface.

Therefore, atmospheric pressure decreases with increasing altitude.

35. C is correct.

The buoyant force:

$F_B = \rho_{air}V_{disp}g$, and V_{disp} is the volume of the man m / ρ_{man}

$F_B = (\rho_{air} / \rho_{man})mg$

$F_B = [(1.2 \times 10^{-3}\ g/cm^3) / (1\ g/cm^3)]\cdot(80\ kg)\cdot(9.8\ m/s^2)$

$F_B = 0.94\ N$

36. D is correct.

Graham's law: the rate at which gas diffuses is inversely proportional to the square root of its density.

37. B is correct.

If the metal rod is thought of like a spring, then Hooke's law applies:

$F = -kx$

The force is proportional to the distance that the spring is stretched.

The proportional limit is reached when the "spring" begins to deform.

The elastic limit is the point when the deformation becomes permanent.

The breaking point equals the fracture point and is when the rod snaps in two.

The elastic modulus is stress (σ) / strain (slope of the elastic region)

38. D is correct.

$A = \pi r^2$

$A_Tv_T = A_Pv_P$

$v_T = v_PA_P / A_T$

Ratio of the diameter^2 is equal to the ratio of area:

$v_T = v_P(d_1 / d_2)^2$

$v_T = (0.03\ m/s)\cdot[(0.12\ m) / (0.002\ m)]^2$

$v_T = 108\ m/s$

39. B is correct.

Specific gravity:

$$\rho_{object} / \rho_{water}$$

Archimedes' principle:

$$F = \rho g V$$

$$\rho_{water} = F / g(0.9V)$$

$$\rho_{object} = F / gV$$

$$\rho_{object} / \rho_{water} = (F / gV) / [F / g(0.9V)]$$

$$\rho_{object} / \rho_{water} = 0.9$$

V_{water} is 0.9V because 90% of the object is in the water, so 90% of the object's volume equals water displaced.

40. D is correct.

The factors considered are length, density, radius, pressure difference, and viscosity.

The *continuity equation* does not apply here because it relates velocity and radius to the flow rate.

Bernoulli's equation does not apply because it relates density and velocity.

The *Hagen-Poiseuille equation* is needed because it includes terms except for density and therefore is the most applicable to this question.

Volumetric flow rate (Q) is:

$$Q = \Delta P \pi r^4 / 8\eta L$$

The radius is raised to the fourth power.

A 15% change to r results in the greatest change.

41. A is correct.

A force meter provides a force, and the reading indicates what the force is.

Since the hammer is not accelerating, the force equation is:

$$F_{meter} + F_B - m_h g = 0$$

$$m_h g = (0.68 \text{ kg})\cdot(10 \text{ m/s}^2)$$

$$m_h g = 6.8 \text{ N}$$

The displaced volume is the volume of the hammer:

$$V_{disp} = m_h / \rho_{steel}$$

$$V_{disp} = (680 \text{ g}) / (7.9 \text{ g/cm}^3)$$

continued…

$V_{disp} = 86\ cm^3$

$F_B = \rho_{water} \times V_{disp} \times g$

$F_B = (1 \times 10^{-3}\ kg/cm^3)\cdot(86\ cm^3)\cdot(10\ m/s^2)$

$F_B = 0.86\ N$

$F_{meter} = m_h g - F_B$

$F_{meter} = 6.8\ N - 0.86\ N$

$F_{meter} = 5.9\ N$

42. D is correct.

Pascal's Principle states that pressure is transmitted undiminished in an enclosed static fluid.

43. B is correct.

The normal force (N) exerted by the seafloor is the net force between the weight of the submarine and the buoyant force:

$F_N = F_{net}$

$F_{net} = mg - F_B$

$F_{net} = mg - \rho g V$

$F_{net} = mg - W_{water}$

$F_N = mg - W_{water}$

44. C is correct.

$P = \rho g h$

Because the bottom of the brick is at a lower depth than the rest of the brick, it will experience the highest pressure.

45. A is correct.

Mass flow rate:

$\dot{m}$ = cross-sectional area × density × velocity

$\dot{m} = A_C \rho v$

$\dot{m} = (7\ m)\cdot(14\ m)\cdot(10^3\ kg/m^3)\cdot(3\ m/s)$

$\dot{m} = 2.9 \times 10^5\ kg/s$

46. C is correct.

Since the object is motionless:

$a = 0$

$F_{net} = 0$

The magnitude of the buoyant force upward = weight downward:

$F = W$

$F = mg$

$F = (3\text{ kg})\cdot(10\text{ m/s}^2)$

$F = 30\text{ N}$

47. A is correct.

$P = P_{atm} + \rho gh$

$P = (1.01 \times 10^5\text{ Pa}) + (10^3\text{ kg/m}^3)\cdot(10\text{ m/s}^2)\cdot(6\text{ m})$

$P = (1.01 \times 10^5\text{ Pa}) + (0.6 \times 10^5\text{ Pa})$

$P = 1.6 \times 10^5\text{ Pa}$

48. B is correct.

density = mass / volume

49. A is correct.

The bulk modulus is defined as how much material is compressed under a given external pressure:

$B = \Delta P / (\Delta V / V)$

Most solids and liquids compress slightly under external pressure.

However, gases have the greatest change in volume and, thus, the lowest value of B.

50. D is correct.

Young's Modulus is expressed as:

$E = \sigma\text{ (stress)} / \varepsilon\text{ (strain)}$

$E = (F / A) / (\Delta L / L)$

$E = (FL) / (\Delta L A)$

Solve for ΔL:

$\Delta L = FL / EA$

$\Delta L = [(8\text{ kg})\cdot(9.8\text{ m/s}^2)\cdot(2.7\text{ m})] / [(20 \times 10^{10}\text{ N/m}^2)(\pi / 4)\cdot(8 \times 10^{-4}\text{ m})^2]$

$\Delta L = 0.0021\text{ m} = 2.1\text{ mm}$

51. A is correct.

Young's Modulus is expressed as:

$E = \sigma$ (stress) / ε (strain)

$E = (F / A) / (\Delta L / L)$

$E = (FL) / (\Delta L A)$

Solve for ΔL and double the diameter:

$\Delta L = FL / EA$

$\Delta L = [(8\text{ kg})\cdot(9.8\text{ m/s}^2)\cdot(2.7\text{ m})] / [(20 \times 10^{10}\text{ N/m}^2)(\pi / 4)\cdot(2 \times 8 \times 10^{-4}\text{ m})^2]$

$\Delta L = 0.000526\text{ m}$

$\Delta L = 0.5\text{ mm}$

52. B is correct.

Absolute pressure is measured relative to absolute zero pressure (i.e., a perfect vacuum), and gauge pressure is measured relative to atmospheric pressure.

If the atmospheric pressure increases by ΔP, the absolute pressure increases by ΔP, but the gauge pressure does not change.

Absolute pressure at an arbitrary depth h in the lake:

$P_{abs} = P_{atm} + \rho_{water}gh$

Gauge pressure at an arbitrary depth h in the lake:

$P_{gauge} = \rho_{water}gh$

53. D is correct.

Strain is defined and calculated by:

$E = \Delta L / L$

54. D is correct.

By Poiseuille's Law:

$v = \Delta P r^2 / 8\eta L$

$\eta = \Delta P r^2 / 8L v_{effective}$

$\eta = (970\text{ Pa})\cdot(0.0021\text{ m})^2 / 8\cdot(1.8\text{ m/s})\cdot(0.19\text{ m})$

$\eta = 0.0016\text{ N}\cdot\text{s/m}^2$

55. C is correct.

Bernoulli's equation:

$$P_1 + ½\rho_1 v_1^2 + \rho_1 g h_1 = P_2 + ½\rho_2 v_2^2 + \rho_2 g h_2$$

There is no height difference, so the equation reduces to:

$$P_1 + ½\rho_1 v_1^2 = P_2 + ½\rho_2 v_2^2$$

If the flow of air across the wing tip is $v_{1,}$ then:

$$v_1 > v_2$$

Since the air that flows across the top has a higher velocity, as it travels a larger distance (curved surface of the top) over the same period, then:

$$½\rho_1 v_1^2 > ½\rho_2 v_2^2$$

To keep both sides equal:

$$P_1 < P_2$$

$$\Delta P = (P_2 - P_1)$$

Thus, the lower portion of the wing experiences greater pressure and therefore lifts the wing.

Notes for active learning

Notes for active learning

Electrostatics and Magnetism – Detailed Explanations

1. C is correct.

Since charge is quantized, the charge Q must be a whole number (n) times the charge on a single electron:

Charge = # electrons × electron charge

$Q = n(e^-)$

$n = Q / e^-$

$n = (-1\text{ C}) / (-1.6 \times 10^{-19}\text{ C})$

$n = 6.25 \times 10^{18} \approx 6.3 \times 10^{18}$ electrons

2. D is correct.

In Gaus's Law, the area is a vector perpendicular to the plane.

Only the component of the electric field strength parallel is used.

Gaus's Law:

$\Phi = EA \cos \theta$

where Φ is electric flux (scalar), E is electric field strength, and A is area vector

Solve:

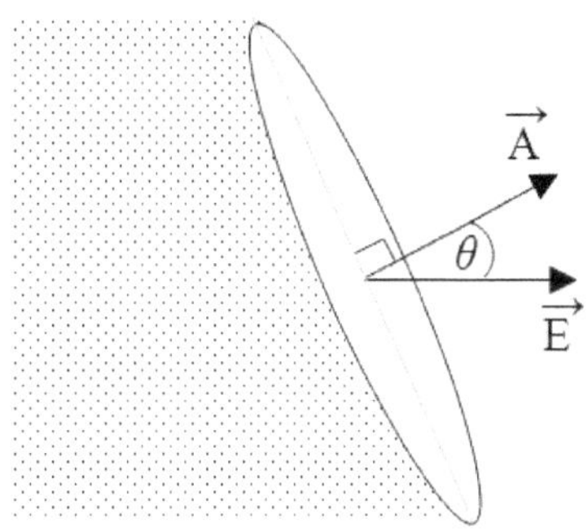

$\Phi = EA \cos (\pi / 6)$

$A = \pi r^2$

$A = \pi D^2 / 4$

$\Phi = (740\text{ N/C})\cdot(\pi / 4)\cdot(1\text{ m})^2 \cos (\pi / 6)$

$\Phi = 160\pi\text{ N}\cdot\text{m}^2/\text{C}$

For calculation, use radians mode, not degree mode.

3. A is correct.

The magnitude of the negative charge's electric field:

$|E_2| = kQ_2 / d_2^2$

$|E_2| = (9 \times 10^9\ \text{N·m}^2/\text{C}^2)·[(–1.3 \times 10^{-9}\ \text{C}) / (+10^{-3}\ \text{m})^2]$

$|E_2| = 1.17 \times 10^7$ N/C to the left

The magnitude of the positive charge's electric field:

$|E_1| = kQ_1 / d_1^2$

$|E_1| = (9 \times 10^9\ \text{N·m}^2/\text{C}^2)·[(1.3 \times 10^{-9}\ \text{C}) / (2 \times 10^{-3}\ \text{m})^2]$

$|E_1| = 2.9 \times 10^6$ N/C to the right

$\Delta E = E_2 - E_1$

$\Delta E = (1.17 \times 10^7\ \text{N/C}) - (2.9 \times 10^6\ \text{N/C})$

$\Delta E = 8.8 \times 10^6$ N/C, to the left

4. C is correct.

Coulomb's law:

$F_1 = kQ_1Q_2 / r^2$

If r is increased by a factor of 4:

$F_e = kQ_1Q_2 / (4r)^2$

$F_e = kQ_1Q_2 / (16r^2)$

$F_e = (1/16)kQ_1Q_2 / r^2$

$F_e = (1/16)F_1$

As the distance increases by a factor of 4, the force decreases by a factor of $4^2 = 16$.

5. B is correct.

Calculate the distance between two charges using the Pythagorean Theorem:

$r^2 = (1\ \text{nm})^2 + (4\ \text{nm})^2$

$r^2 = 17\ \text{nm}^2$

$r = 4.1$ nm

Coulomb's Law, which describes the repulsive force between two particles, is given as:

$F = kQ_1Q_2 / r^2$

$F = [(9 \times 10^9\ \text{N·m}^2/\text{C}^2)·(1.6 \times 10^{-19}\ \text{C})·(1.6 \times 10^{-19}\ \text{C})] / (4.1 \times 10^{-9}\ \text{m})^2$

$F = 1.4 \times 10^{-11}$ N

6. A is correct.

7. D is correct.

Use a coordinate system where a repulsive force is positive, and an attractive force is in the negative direction.

Gravitational force: F_g

$$F_g = -Gm_1m_2 / r^2$$

$$F_g = -[(6.673 \times 10^{-11}\ \text{N·m}^2/\text{kg}^2)·(54{,}000\ \text{kg})·(51{,}000\ \text{kg})] / (180\ \text{m})^2$$

$$F_g = -0.18\ \text{N·m}^2 / (32{,}400\ \text{m}^2)$$

$$F_g = -5.7 \times 10^{-6}\ \text{N}$$

Electrostatic force: F_e

$$F_e = kQ_1Q_2 / r^2$$

$$F_e = [(9 \times 10^9\ \text{N·m}^2/\text{C}^2)·(-15 \times 10^{-6}\ \text{C})·(-11 \times 10^{-6}\ \text{C})] / (180\ \text{m})^2$$

$$F_e = (1.49\ \text{N·m}^2) / (32{,}400\ \text{m}^2)$$

$$F_e = 4.6 \times 10^{-5}\ \text{N}$$

Net force:

$$F_{\text{net}} = F_g + F_e$$

$$F_{\text{net}} = (-5.7 \times 10^{-6}\ \text{N}) + (4.6 \times 10^{-5}\ \text{N})$$

$$F_{\text{net}} = 4 \times 10^{-5}\ \text{N}$$

F_{net} is positive, which means there is a net repulsive force on the asteroids.

The repulsive electrostatic force between them is stronger than the attractive gravitational force.

8. B is correct.

Newton's Third Law states that for every force, there is an equal and opposite reaction force.

This applies to electrostatic forces.

Coulomb's Law:

$$F_1 = kQ_1Q_2 / r^2$$

$$F_2 = kQ_1Q_2 / r^2$$

$$F_1 = F_2$$

9. D is correct.

Forces balance to yield:

$F_{\text{electric}} = F_{\text{gravitation}}$

$F_{\text{electric}} = mg$

The values for an electric field are provided.

$F_{\text{electric}} = QE$

$F_{\text{electric}} - F_{\text{gravitation}} = 0$

$QE - mg = 0$

where Q is the charge on the ball

$QE = mg$

$Q = mg / E$

$Q = (0.008\text{ kg})\cdot(9.8\text{ m/s}^2) / (3.5 \times 10^4\text{ N/C})$

$Q = -2.2 \times 10^{-6}\text{ C}$

If the electric field points down, then a positive charge experiences a downward force.

The charge must be negative, so the electric force balances gravity.

10. A is correct.

$a = qE / m$

The electron moves against the electric field in the upward direction, so its acceleration:

$a_e = qE / m_e$

The proton moves with the electric field, which is down, so:

$a_p = qE / m_p$

However, the masses considered are small to where the gravity component is negligible.

$m_p / m_e = (1.67 \times 10^{-27}\text{ kg}) / (9.11 \times 10^{-31}\text{ kg})$

$m_p / m_e = 1{,}830$

The mass of an electron is about 1,830 times smaller than the mass of a proton.

$(1{,}830)m_e = m_p$

$a_p = qE / (1{,}830)m_e$

$a_p = a_e / (1{,}830)$

$a_e = 1{,}830a_p$

11. D is correct.

Calculate the strength of the field at point P due to only one charge:

$$E = kQ / r^2$$

$$E_1 = (9 \times 10^9 \text{ N·m}^2/\text{C}^2)·[(2.3 \times 10^{-11} \text{ C}) / (5 \times 10^{-3} \text{ m})^2]$$

$$E_1 = 8.3 \times 10^3 \text{ N/C}$$

Both electric field vectors point toward the negative charge, so the magnitude of each field at point P is doubled:

$$E_T = 2E_1$$

$$E_T = 2(8.3 \times 10^3 \text{ N/C})$$

$$E_T = 1.7 \times 10^4 \text{ N/C}$$

12. A is correct.

charge = # electrons × electron charge

$$Q = ne^-$$

$$n = Q / e^-$$

$$n = (-10 \times 10^{-6} \text{ C}) / (-1.6 \times 10^{-19} \text{ C})$$

$$n = 6.3 \times 10^{13} \text{ electrons}$$

13. C is correct.

Coulomb's law:

$$F_e = kQ_1Q_2 / r^2$$

If the separation is halved, then r decreases by ½:

$$F_2 = kq_1q_2 / (½r)^2$$

$$F_2 = 4(kq_1q_2 / r^2)$$

$$F_2 = 4F_e$$

14. D is correct.

Coulomb's law:

$$F = kQ_1Q_2 / r^2$$

Doubling the charges and distance:

$$F = [k(2Q_1)·(2Q_2)] / (2r)^2$$

$$F = [4k(Q_1)·(Q_2)] / (4r^2)$$

$$F = (4/4)[kQ_1Q_2 / (r^2)]$$

$F = kQ_1Q_2 / r^2$, remains the same

15. D is correct.

Like charges repel each other.

From Newton's Third Law, the magnitude of the force experienced by each charge is equal.

16. A is correct.

Coulomb's law:

$$F = kQ_1Q_2 / r^2$$

The Coulomb force between opposite charges is attractive.

Since the strength of the force is inversely proportional to the square of the separation distance (r^2), the force decreases as the charges are pulled apart.

17. B is correct.

Currents occur in a conducting circuit element when there is a potential difference between the two ends of the element, and hence an established electric field.

18. D is correct.

Voltage is related to the number of coils in a wire.

More coils yield a higher voltage.

Turns ratio:

$$V_s / V_p = n_s / n_p$$

In this example:

$$n_s < n_p$$

Therefore,

$$V_s < V_p$$

Because the secondary voltage (V_s) is lower than the primary voltage (V_p), the transformer is a step-down.

19. B is correct.

Coulomb's Law:

$$F_1 = kQ_1Q_2 / r^2$$

$$F_2 = kQ_1Q_2 / r^2$$

$$F_1 = F_2$$

Newton's Third Law: the force exerted by one charge on the other has the same magnitude as the force the other exerts on the first.

20. C is correct.

$F_e = kQ_1Q_2 / r^2$

$F_e = (9 \times 10^9 \text{ N·m}^2/\text{C}^2)·(–1.6 \times 10^{-19} \text{ C})·(–1.6 \times 10^{-19} \text{ C}) / (0.03 \text{ m})^2$

$F_e = 2.56 \times 10^{-25} \text{ N}$

21. A is correct.

According to Lenz's Law, inserting a magnet into the coil causes the magnetic flux through the coil to change.

This produces an emf in the coil, which drives a current through the coil:

Lenz's Law:

$\text{emf} = –\text{N}\Delta\text{BA} / \Delta t$

The bulb's brightness changes with a change in the current, but it cannot be known if the bulb gets brighter or dimmer without knowing the orientation of the coil with respect to the incoming magnetic pole of the magnet.

22. C is correct.

Charge = # electrons × electron charge

$Q = ne^-$

$n = Q / e^-$

$n = (8 \times 10^{-6} \text{ C}) / (1.6 \times 10^{-19} \text{ C})$

$n = 5 \times 10^{13}$ electrons

23. A is correct.

An object with a charge can attract another object of opposite charge or a neutral charge.

Like charges cannot attract, but the type of charge does not matter otherwise.

24. B is correct.

1 amp = 1 C / s

The ampere is the unit to express the flow rate of electric charge known as current.

25. A is correct.

Initially, the current flows clockwise, but after 180° of rotation, the current reverses.

After 360° of rotation, the current will reverse itself again.

Thus, there are 2 current reverses in 1 revolution.

26. D is correct.

$W = Q\Delta V$

$V = kQ / r$

Consider the charge Q_1 to be fixed and move charge Q_2 from initial distance r_i to final distance r_f.

$W = Q_2(V_f - V_i)$

$W = Q_2[(kQ_1 / r_f) - (kQ_1 / r_i)]$

$W = kQ_1Q_2(1 / r_f - 1 / r_i)$

$W = (9 \times 10^9\ N\cdot m^2/C)\cdot(2.3 \times 10^{-8}\ C)\cdot(2.5 \times 10^{-9}\ C)\cdot[(1 / 0.01\ m) - (1 / 0.1\ m)]$

$W = 4.7 \times 10^{-5}\ J$

27. C is correct.

The electric field is oriented so that a positively-charged particle would be forced to move to the top because negatively-charged particles move to the bottom of the cube to be closer to the source of the electric field.

Therefore, the positively charged particles are forced upward and negatively charged downward, leaving the top surface positively charged.

28. C is correct.

The magnitude of the force between the center charge and each charge at a vertex is 5 N.

The net force of these two forces is directed toward the third vertex.

To determine the magnitude of the net force, calculate the magnitude of the component of each force acting in that direction:

$F_{net} = F_1 \sin(\frac{1}{2}\theta) + F_2 \sin(\frac{1}{2}\theta)$ *(see diagram)*

Since it is an equilateral triangle $\theta = 60°$,

$F_{net} = (5\ N \sin 30°) + (5\ N \sin 30°)$

$F_{net} = (5\ N)\cdot(\frac{1}{2}) + (5\ N)\cdot(\frac{1}{2})$

$F_{net} = 5\ N$

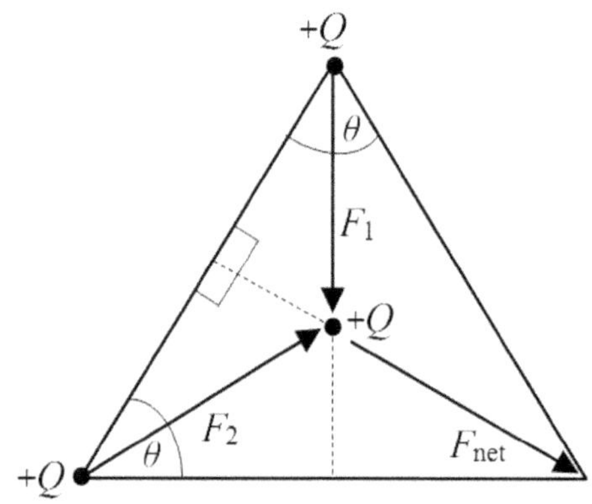

29. D is correct.

Equilibrium:

$F = kq_1q_2 / r_1^2$

$F_{\text{attractive on } q2} = F_{\text{repulsive on } q2}$

$kq_1q_2 / r_1^2 = kq_2Q / r_2^2$

cancel kq_2 from both sides of the expression

$q_1 = Qr_1^2 / r_2^2$

$q_1 = (7.5 \times 10^{-9}\text{ C})\cdot(0.2\text{ m})^2 / (0.1\text{ m})^2$

$q_1 = 30 \times 10^{-9}\text{ C}$

30. A is correct.

Gamma rays have the highest frequency on the electromagnetic spectrum, with frequencies $> 3 \times 10^{19}$ Hz.

31. A is correct.

The strength of the electrostatic field due to a single point charge is given by:

$E = kQ / r^2$ assumes that the source charge is in a vacuum

E depends on the magnitude of the source charge Q and the distance r from Q.

The sign of the source charge affects only the direction of the electrostatic field vectors.

The sign of the source charge does not affect the strength of the field.

32. B is correct.

$F = qvB$

$F = mv^2 / r$

$mv^2 / r = qvB$

cancel v from each side of the expression

$mv / r = qB$

$r = (mv) / (qB)$

If the velocity doubles, the radius doubles.

33. C is correct.

Coulomb's law:

$$F_e = kQ_1Q_2 / r^2$$

$$1 \text{ N} = kQ_1Q_2 / r^2$$

Doubling charges and keeping distance constant:

$$k(2Q_1)\cdot(2Q_2) / r^2 = 4kQ_1Q_2 / r^2$$

$$4kQ_1Q_2 / r^2 = 4F_e$$

$$4F_e = 4(1 \text{ N}) = 4 \text{ N}$$

34. D is correct.

The Na^+ ion is positively charged and attracts the oxygen atom.

Oxygen is slightly negative because it is more electronegative than the hydrogen atoms to which it is bonded.

35. C is correct.

Charge = # of electrons × electron charge

$$Q = ne^-$$

$$Q = (30)\cdot(-1.6 \times 10^{-19} \text{ C})$$

$$Q = -4.8 \times 10^{-18} \text{ C}$$

36. A is correct.

Cyclotron frequency is given as:

$$f = qB / 2\pi m$$

This expression does not consider speed.

37. B is correct.

The repulsive force between two particles is:

$$F = kQ_1Q_2 / r^2$$

As r increases, F decreases

Using $F = ma$, a also decreases

38. A is correct.

The given unit can be written $[\text{kg}\cdot\text{m}^2/\text{s}^2] / \text{C} = \text{J} / \text{C}$, which is the definition of the volt, the unit of electric potential difference.

39. C is correct.

The Coulomb is the basic unit of electrical charge in the SI unit system.

40. A is correct.

By the Law of Conservation of Charge, a charge cannot be created nor destroyed.

41. B is correct.

Coulomb's law:

$$F = kQ_1Q_2 / r^2$$

If both charges are doubled,

$$F = k(2Q_1)\cdot(2Q_2) / r^2$$

$$F = 4kQ_1Q_2 / r^2$$

F increases by a factor of 4.

42. C is correct.

Coulomb's law:

$$F = kQ_1Q_2 / r^2$$

$$Q_1 = Q_2$$

Therefore:

$$Q_1Q_2 = Q^2$$

$$F = kQ^2 / r^2$$

Rearranging:

$$Q^2 = Fr^2 / k$$

$$Q = \sqrt{(Fr^2 / k)}$$

$$Q = \sqrt{[(4\text{ N})\cdot(0.01\text{ m})^2 / (9 \times 10^9\text{ N}\cdot\text{m}^2/\text{C}^2)]}$$

$$Q = 2 \times 10^{-7}\text{ C}$$

43. A is correct.

Faraday's law states that electromotive force (emf) is equal to the rate of change of magnetic flux.

Magnetic flux is the product of the magnetic field and projected area:

$$\Phi = BA_\perp$$

where $A_\perp$ is the area of the loop projected on a plane perpendicular to the magnetic field

continued...

In this problem, B is vertical (and constant), so the projection plane is horizontal.

Therefore, find the orientation of the axis of rotation that guarantees that as the loop rotates, the projection of its area on a horizontal plane does not change with time.

Notice that if the orientation of the axis is at an arbitrary angle to the field, the emf can be made to be zero by aligning the axis of rotation with the axis of the loop (i.e., perpendicular to the loop). With this orientation, the projection of the area never changes, which is not true of other alignments to the loop.

Although the emf can be made to be zero, it is not *guaranteed* to be zero.

The only orientation of the axis that *guarantees* that the projected area is constant is the vertical orientation.

Notice that because of the high symmetry of the vertical-axis orientation, rotating the loop about a vertical axis is equivalent to changing the perspective of the viewer from one angle to another.

The answer cannot depend on the viewer's perspective.

Therefore, the projected area cannot change as the loop is rotated about the vertical axis; the emf is guaranteed to be zero.

44. C is correct.

Coulomb's law:

$$F = kQ_1Q_2 / r^2$$

When each particle has lost ½ its charge:

$$F_2 = k(\tfrac{1}{2}Q_1)\cdot(\tfrac{1}{2}Q_2) / r^2$$

$$F_2 = (\tfrac{1}{4})kQ_1Q_2 / r^2$$

$$F_2 = (\tfrac{1}{4})F$$

F decreases by a factor of ¼

45. B is correct.

The time taken for one revolution around the circular path:

$$T = 2\pi R/v$$

where R is the radius of the circle and v is the speed of the proton

If the speed is increased, the radius increases.

The relationship between speed and radius follows from the fact that the magnetic interaction provides the centripetal force:

$$mv^2/ R = qvB$$

Thus: $R = mv / qB$

If the speed is tripled, the radius triples, other things equal, the final period for a revolution is:

$$T_f = 2\pi R_f / v_f = 2\pi(3R) / 3v = 2\pi R / v = T$$

46. C is correct.

An electrostatic field shows the path that would be taken by a positively-charged particle.

As this positive particle moves closer to the negatively-charged one, the force between them increases.

Coulomb's law:

$$F = kQ_1Q_2 / r^2$$

By convention, electric field vectors point towards negative source charges.

Since electrical field strength is inversely proportional to the square of the distance from the source charge, the magnitude of the electric field progressively increases as an object moves towards the source charge.

47. D is correct.

Protons are charges, so they have an electric field.

Protons have mass, so they have a gravitational field.

Protons have an intrinsic magnetic moment, so they have a magnetic field.

48. A is correct.

$$W = Q\Delta V$$

$$V = kq / r$$

$$W = (kQq)\cdot(1 / r_2 - 1 / r_1)$$

$$W = (kQq)\cdot(1 / 2\text{ m} - 1 / 6\text{ m})$$

$$W = (kQq)\cdot(1 / 3\text{ m})$$

$$W = (9 \times 10^9\text{ N}\cdot\text{m}^2/\text{C}^2)\cdot(3.1 \times 10^{-5}\text{ C})\cdot(-10^{-6}\text{ C}) / (1 / 3\text{ m})$$

$$W = -0.093\text{ J} \approx -0.09\text{ J}$$

The negative sign indicates that the electric field does the work on charge q.

49. D is correct.

charge = # electrons × electron charge

$$Q = ne^-$$

$$n = Q / e^-$$

$$n = (-600 \times 10^{-9}\text{ C}) / (-1.6 \times 10^{-19}\text{ C})$$

$$n = 3.8 \times 10^{12}\text{ electrons}$$

50. C is correct.

The analog to N/kg would be N/C, the unit for the electric field.

51. D is correct.

All the electromagnetic waves travel through space (vacuum) at the same speed:

$$c = 3 \times 10^8 \text{ m/s}$$

52. A is correct.

As the proton of charge q moves in the direction of the electric field lines, it moves *away* from a positive charge (because field lines emanate from a positive charge).

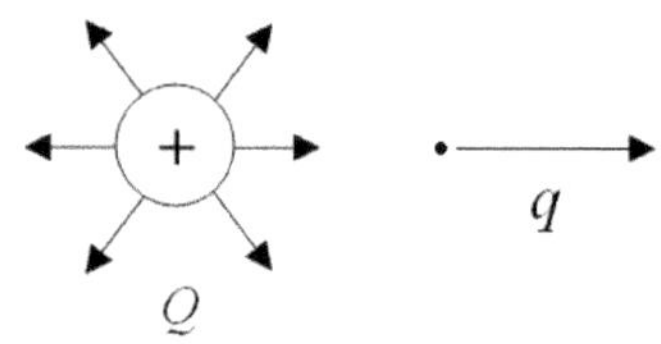

Electric Potential Energy:

$$U = kQq / r$$

As distance increases, the potential energy decreases because they are inversely proportional.

Electrical potential:

$$V = kQ / r$$

Electrical potential is inversely proportional to distance and decreases as distance increases.

53. B is correct.

1 watt = 1 J/s

54. D is correct.

An electrically polarized object has its charge separated into opposites and rearranges within distinct regions.

55. A is correct.

Because point P is symmetric about Q_1 and Q_2 and both charges have the same positive magnitude, the electric field cancels midway between the charges.

$$E = kQ / r^2$$

$$E_1 = -E_2$$

$$E_{tot} = E_1 + E_2$$

$$E_{tot} = (-E_2 + E_2)$$

$$E_{tot} = 0 \text{ N/C}$$

56. C is correct.

Force due to motion:

$F = ma$

$F = (0.001 \text{ kg})\cdot(440 \text{ m/s}^2)$

$F = 0.44 \text{ N}$

Force due to charge:

$F = kQ_1Q_2 / r^2$

$Q_1 = Q_2$

$Q_1Q_2 = Q^2$

$F = kQ^2 / r^2$

Rearranging:

$Q^2 = Fr^2 / k$

$Q = \sqrt{(Fr^2 / k)}$

$Q = \sqrt{[(0.44 \text{ N})\cdot(0.02 \text{ m})^2 / (9 \times 10^9 \text{ N}\cdot\text{m}^2/\text{C}^2)]}$

$Q = 1.4 \times 10^{-7} \text{ C} = 140 \text{ nC}$

57. D is correct.

A sphere or conduction object that acquires a net charge has the charge collect on the surface.

This is due to excess charge repelling itself and moving to the surface to increase the distance between themselves.

58. B is correct.

Coulomb's Law, which describes the repulsive force between two particles, is given as:

$F = kQ_1Q_2 / r^2$

	Force from +	Force from –
x-direction	→	→
y-direction	↑	↓

Net force = →

59. A is correct.

A charged particle only experiences a magnetic force if it moves with a perpendicular velocity component to the field.

Thus, there must not be a magnetic field, or the particle moves parallel to the field.

60. B is correct.

Coulomb's law: the strength of the electrostatic force between the two point charges.

$$F = kQ_1Q_2 / r^2$$

$$F = [(9 \times 10^9\ \text{N·m}^2/\text{C}^2)·(+3\ \text{C})·(–12\ \text{C})] / (0.5\ \text{m})^2$$

$$F = –1.3 \times 10^{12}\ \text{N}$$

F is positive to indicate an attractive force; therefore, the magnitude of the force is 1.3×10^{12} N.

Notes for active learning

Notes for active learning

Circuit Elements – Detailed Explanations

1. B is correct.

$$R = \rho L / A$$

where ρ is the resistivity of the wire material

If the length L is doubled, the resistance R is doubled.

If the radius r is doubled, the area $A = \pi r^2$ is quadrupled, and resistance R is decreased by ¼.

If these two changes are combined:

$$R_{new} = \rho(2L) / \pi(2r)^2$$

$$R_{new} = (2/4)\cdot(\rho L / \pi r^2)$$

$$R_{new} = (2/4)R = ½R$$

2. D is correct.

Internal resistance of battery is in series with resistors in circuit:

$$R_{eq} = R_1 + R_{battery}$$

where R_{eq} is equivalent resistance and R_1 is resistor connected to battery

$$V = IR_{eq}$$

$$V = I(R_1 + R_{battery})$$

$$R_{battery} = V / I - R_1$$

$$R_{battery} = (12\ V / 0.6\ A) - 6\ \Omega$$

$$R_{battery} = 14\ \Omega$$

3. C is correct.

An ohm Ω is defined as the resistance between two points of a conductor when a constant potential difference of 1 V, applied to these points, produces in the conductor a current of 1 A.

A series circuit experiences the same current through resistors regardless of their resistance.

However, the voltage across each resistor can be different.

Since the light bulbs are in series, the current through them is the same.

4. D is correct.

$V = kQ / r$

$V_B = kQ / r_B$

$V_B = (9 \times 10^9 \text{ N·m}^2/\text{C}^2)·(1 \times 10^{-6} \text{ C}) / 3.5 \text{ m}$

$V_B = 2{,}571 \text{ V}$

$V_A = kQ / r_A$

$V_A = (9 \times 10^9 \text{ N·m}^2/\text{C}^2)·(1 \times 10^{-6} \text{ C}) / 8 \text{ m}$

$V_A = 1{,}125 \text{ V}$

Potential difference:

$\Delta V = V_B - V_A$

$\Delta V = 2{,}571 \text{ V} - 1{,}125 \text{ V}$

$\Delta V = 1{,}446 \text{ V}$

5. D is correct.

The capacitance of a parallel plate capacitor demonstrates the influence of material, separation distance, and geometry in determining the overall capacitance.

$C = k\mathcal{E}_0 A / d$

where k = dielectric constant or permittivity of material between the plates, A = surface area of the conductor, and d = distance of plate separation

6. A is correct.

$E = qV$

$E = ½m(\Delta v)^2$

$qV = ½m(v_f^2 - v_i^2)$

$v_f^2 = (2qV / m) + v_i^2$

$v_f^2 = [2(1.6 \times 10^{-19} \text{ C})·(100 \text{ V}) / (1.67 \times 10^{-27} \text{ kg})] + (1.5 \times 10^5 \text{ m/s})^2$

$v_f^2 = (1.9 \times 10^{10} \text{ m}^2/\text{s}^2) + (2.3 \times 10^{10} \text{ m}^2/\text{s}^2)$

$v_f^2 = 4.2 \times 10^{10} \text{ m}^2/\text{s}^2$

$v_f = 2.04 \times 10^5 \text{ m/s} \approx 2 \times 10^5 \text{ m/s}$

7. C is correct.

power = current^2 × resistance

$P = I^2R$

Double current:

$P_2 = (2I)^2R$

$P_2 = 4(I^2R)$

$P_2 = 4P$

Power is quadrupled

8. D is correct.

A magnetic field is created only by electric charges in motion.

A stationary charged particle does not generate a magnetic field.

9. B is correct.

Combining the power equation with Ohm's law:

$P = (\Delta V)^2 / R$

where $\Delta V = 120$ V is a constant

To increase power, decrease the resistance.

A longer wire increases resistance, while a thicker wire decreases it:

$A = \pi r^2$

$R = \rho L / A$

A larger radius of the cross-sectional area means A is larger (denominator), which lowers R.

10. C is correct.

$W = kq_1q_2 / r$

$r = \Delta x$

$r = 2 \text{ mm} - (-2 \text{ mm})$

$r = 4 \text{ mm}$

$W = [(9 \times 10^9 \text{ N·m}^2/\text{C}^2)·(4 \times 10^{-6} \text{ C})·(8 \times 10^{-6} \text{ C})] / (4 \times 10^{-3} \text{ m})$

$W = (0.288 \text{ N·m}^2) / (4 \times 10^{-3} \text{ m})$

$W = 72 \text{ J}$

11. A is correct.

$V = IR$

$I = V / R$

$I = (220\ V) / (400\ \Omega)$

$I = 0.55\ A$

12. B is correct.

A parallel circuit experiences the same potential difference across each resistor.

However, the current through each resistor can be different.

13. D is correct.

$E = qV$

$E = \frac{1}{2}mv^2$

$qV = \frac{1}{2}mv^2$

$v^2 = 2qV / m$

$v^2 = [2(1.6 \times 10^{-19}\ C)\cdot(990\ V)] / (9.11 \times 10^{-31}\ kg)$

$v^2 = 3.5 \times 10^{14}\ m^2/s^2$

$v = 1.9 \times 10^7\ m/s$

14. B is correct.

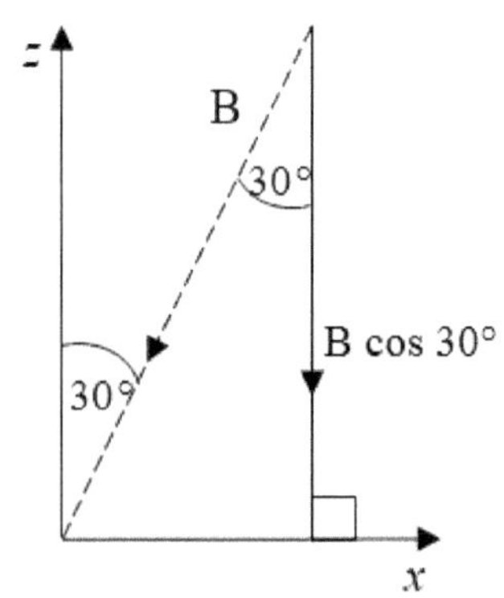

Calculate magnetic field perpendicular to loop:

$B_{Perp2} = (12\ T) \cos 30°$

$B_{Perp2} = 10.4\ T$

$B_{Perp1} = (1\ T) \cos 30°$

$B_{Perp1} = 0.87\ T$

Use Faraday's Law to calculate generated voltage:

$V = N\Delta BA / \Delta t$

$V = N\Delta B(\pi r^2) / \Delta t$

$V = [(1)\cdot(10.4\ T - 0.87\ T)\cdot(\pi(0.5\ m)^2)] / (5\ s - 0\ s)$

$V = [(1)\cdot(10.4\ T - 0.87\ T)\cdot(0.785\ m^2)] / (5\ s)$

$V = 1.5\ V$

continued...

Use Ohm's Law to calculate current:

$V = IR$

$I = V / R$

$I = (1.5\ V) / (12\ \Omega)$

$I = 0.13\ A$

15. C is correct.

Ohm's Law:

$V = IR$

$V = (10\ A)\cdot(35\ \Omega)$

$V = 350\ V$

16. A is correct.

The magnitude of the acceleration is given by:

$F = ma$

$a = F / m$

$F = qE_0$

$a = qE_0 / m$

Bare nuclei = no electrons

^1H has 1 proton, and ^4He has 2 protons and 2 neutrons

Thus, ^1H has ½ the charge and ¼ the mass of ^4He.

$a_H = q_H E_0 / m_H$

$a_{He} = q_{He} E_0 / m_{He}$

$a_H = (\frac{1}{2} q_{He}) E_0 / (\frac{1}{4} m_{He})$

$a_H = 2(q_{He} E_0 / m_{He})$

$a_H = 2a_{He}$

17. B is correct.

The current will change as the choice of lamp arrangement changes.

Since $P = V^2/R$, power increases as resistance decreases.

To rank the power in increasing order, the equivalent resistance must be ranked in decreasing order.

For arrangement B, the resistors are in series, so:

$$R_{eq} = R + R = 2R$$

For arrangement C, the resistors are in parallel, so:

$$1/R_{eq} = 1/R + 1/R$$

$$R_{eq} = R/2$$

The ranking of resistance in decreasing order is B to A to C, which is, therefore, the ranking of power in increasing order.

18. D is correct.

$$C = k\mathcal{E}_0\text{A} / d$$

where k = dielectric constant or permittivity of material between the plates, A = area and d = distance of plate separation

$$C = (2.1)\cdot(8.854 \times 10^{-12}\ \text{F/m})\cdot(0.01\ \text{m} \times 0.01\ \text{m}) / (0.001\ \text{m})$$

$$C = (1.9 \times 10^{-15}\ \text{F/m}) / (0.001\ \text{m})$$

$$C = 1.9 \times 10^{-12}\ \text{F} = 1.9\ \text{pF}$$

19. C is correct.

Resistor R_1 is connected directly across the battery.

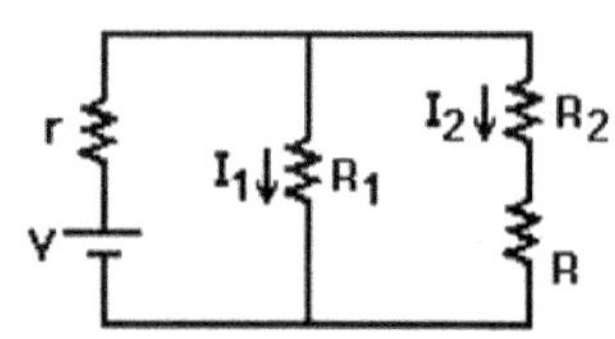

Thus, the voltage across R_1 is V and is held constant regardless of what happens in the circuit.

Similarly, the voltage across the series combination of R and R_2 is held constant at V.

Since the voltage across R_1 will always be V, the current I_1 through R_1 will be unchanged as R changes (since R_1 did not change, so $I_1 = V / R_1$ remains the same).

Since the voltage across the combination of R and R_2 will be V when R is decreased, the effective resistance of the series combination $R + R_2$ will decrease, and the current I_2 through R_2 will increase.

20. C is correct.

$V = IR$

$I = V / R$

Ohm's law states that the current between two points is directly proportional to the potential difference between the points.

21. B is correct.

Root mean square voltage equation:

$V_{rms} = V_{max} / \sqrt{2}$

$V_{rms} = 12 / \sqrt{2}$

$V_{rms} = (12 / \sqrt{2})\cdot(\sqrt{2} / \sqrt{2})$

$V_{rms} = (12\sqrt{2}) / 2$

$V_{rms} = 6\sqrt{2}$ V

22. C is correct.

By definition:

$V_{rms} = V_{max} / \sqrt{2}$

Therefore:

$V_{max} = V_{rms}\sqrt{2}$

$V_{max} = (150 \text{ V})\sqrt{2}$

$V_{max} = 212$ V

23. D is correct.

Kirchhoff's junction rule states that the sum of currents coming into a junction is the sum of currents leaving a junction.

Kirchhoff's junction rule is a statement of conservation of charge because it defines that no charge is created nor destroyed in the circuit.

24. A is correct.

The capacitance of capacitors connected in parallel is the sum of the individual capacitances:

$$C_{eq} = C_1 + C_2 + C_3 + C_4 = 4C$$

The relationship between the total charge delivered by the battery and the voltage of the battery is:

$$V = Q / C_{eq}$$

$$Q / C_{eq} = Q / 4C$$

$$V = Q / 4C$$

The charge on one capacitor is:

$$Q_1 = CV$$

$$Q_1 = C\,(Q / 4C)$$

$$Q_1 = Q / 4$$

25. B is correct.

If two conductors are connected by copper wire, each conductor will be at the same potential because current can flow through the wire and equalize the difference in potential.

26. C is correct.

Electromagnetic induction is the production of an electromotive force across a conductor.

When a changing magnetic field is brought near a coil, a voltage is generated in the coil, inducing a current.

The voltage generated can be calculated by Faraday's Law:

$$\text{emf} = -N\Delta\phi / \Delta t$$

where N = number of turns and $\Delta\phi$ = change in magnetic flux

27. A is correct.

Current is constant across resistors connected in series.

28. D is correct.

By convention, the direction of electric current is the direction that a positive charge migrates.

Therefore, current flows from the point of high potential to the point of lower potential.

29. B is correct.

$PE_e = PE_1 + PE_2 + PE_3$

$PE_e = (kQ_1Q_2) / r_1 + (kQ_2Q_3) / r_2 + (kQ_1Q_3) / r_3$

$PE_e = kQ^2 [(1 / r_1) + (1 / r_2) + (1 / r_3)]$

$r_1 = 4$ cm and $r_2 = 3$ cm are known

Use Pythagorean Theorem to find r_3:

$r_3^2 = r_1^2 + r_2^2$

$r_3^2 = (4\text{ cm})^2 + (3\text{ cm})^2$

$r_3^2 = 16\text{ cm}^2 + 9\text{ cm}^2$

$r_3^2 = 25\text{ cm}^2$

$r_3 = 5$ cm

$PE_e = (9.0 \times 10^9\text{ N·m}^2/\text{C}^2)·(3.8 \times 10^{-9}\text{ C})^2 \times [(1 / 0.04\text{ m}) + (1 / 0.03\text{ m}) + (1 / 0.05\text{ m})]$

$PE_e = (1.2 \times 10^{-7}\text{ N·m}^2)·(25\text{ m}^{-1} + 33\text{ m}^{-1} + 20\text{ m}^{-1})$

$PE_e = (1.2 \times 10^{-7}\text{ N·m}^2)·(78\text{ m}^{-1})$

$PE_e = 1.0 \times 10^{-5}\text{ J}$

30. A is correct.

The magnetic force acting on a charge q moving at velocity v in a magnetic field B is given by:

$F = qv \times B$

If q, v, and the angle between v and B are the same for both charges, then the magnitude of the force F is the same on the charges.

However, if the charges carry opposite signs, each experience oppositely-directed forces.

31. C is correct.

By convention, the direction of electric current is the direction that a positive charge migrates.

Electrons flow from regions of low potential to regions of high potential.

Electric Potential Energy:

$U = (kQq) / r$

Electric Potential:

$V = (kQ) / r$

Because the charge of an electron (q) is negative, as the electron moves opposite the electric field, it must get closer to the positive charge Q.

continued…

An increasingly negative potential energy U is produced; thus, the potential energy decreases.

Conversely, as the electron approaches Q, the electric potential V increases with less distance.

This is because the product is positive, and reducing r increases V.

32. D is correct.

Magnets provide magnetic forces.

Generators convert mechanical energy into electrical energy, turbines extract energy from fluids (e.g., air and water), and transformers transfer energy between circuits.

33. B is correct.

The potential energy of a system containing two point charges is:

$$U = kq_1q_2 / r$$

In this problem, one of the charges is positive, and the other is negative. To account for this, write:

$$q_1 = +|q_1| \text{ and } q_2 = -|q_2|$$

The potential energy can be written as:

$$U = -k\,|q_1|\,|q_2| / r$$

Moreover, the absolute value of the potential energy is:

$$|U| = k\,|q_1|\,|q_2| / r$$

All quantities are positive.

The absolute value of the potential energy is inversely proportional to the orbital radius; therefore, the absolute value of the potential energy decreases as the orbital radius increases.

34. C is correct.

$$R = (\rho L) / (\pi r^2)$$

$$R_A = (\rho L) / (\pi r^2)$$

$$R_B = [\rho(2L)] / [\pi(2r)^2]$$

$$R_B = (2/4)\cdot[(\rho L) / (\pi r^2)]$$

$$R_B = ½[(\rho L) / (\pi r^2)]$$

$$R_B = ½R_A$$

35. D is correct.

By convention, current flows from high to low potential, but it represents the flow of positive charges.

Electron flow is in the opposite direction, from low potential to high potential.

36. C is correct.

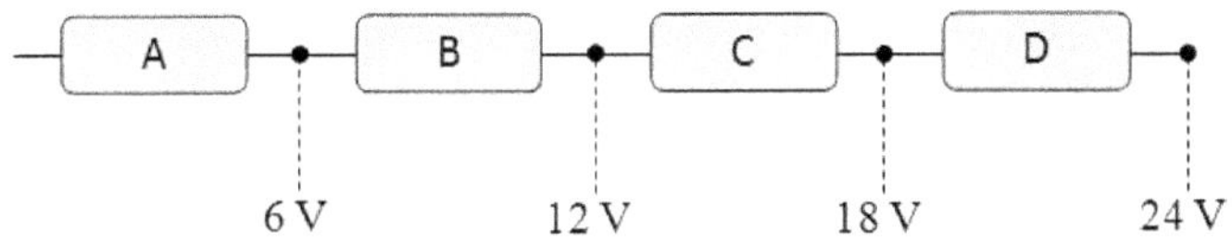

Batteries in series add voltage like resistors in series add resistance.

The resistances of the lights powered is not needed to solve the problem.

37. B is correct.

Capacitance:

$C = (k\mathcal{E}_0\text{A}) / d$

$k = (Cd) / \text{A}\mathcal{E}_0$

If capacitance increases by a factor of 4:

$k_2 = (4C)d / \text{A}\mathcal{E}_0$

$k_2 = 4(Cd / \text{A}\mathcal{E}_0)$

$k_2 = 4k$

38. C is correct.

$C = (k\mathcal{E}_0\text{A}) / d$

$C = [(1)\cdot(8.854 \times 10^{-12}\ \text{F/m})\cdot(0.4\ \text{m}^2)] / (0.04\ \text{m})$

$C = 8.854 \times 10^{-11}\ \text{F}$

$\text{V} = Q / C$

$\text{V} = (6.8 \times 10^{-10}\ \text{C}) / (8.854 \times 10^{-11}\ \text{F})$

$\text{V} = 7.7\ \text{V}$

39. A is correct.

Since force is the cross-product of velocity and magnetic field strength:

$F = qv \times B$

The force is at a maximum when v and B are perpendiculars:

$F = qvB \sin 90°$

$\sin 90° = 1$

$F = qvB$

40. D is correct.

Ampere is the unit of current, which is defined as the rate of flow of charge.

The *current* describes how much charge (in Coulombs) passes through a point every second.

So, if the current is multiplied by the number of seconds (the time interval), measure how much charge passes during that time interval.

Mathematically, the units are expressed as:

$$A = C / s$$

$$C = A \cdot s$$

41. C is correct.

$$\Delta V = \Delta E / q$$

$$\Delta V = (1 / q) \cdot (\tfrac{1}{2}mv_f^2 - \tfrac{1}{2}mv_i^2)$$

$$\Delta V = (m / 2q) \cdot (v_f^2 - v_i^2)$$

$$\Delta V = [(1.67 \times 10^{-27}\,kg) / (2) \cdot (1.6 \times 10^{-19}\,C)] \times [(3.2 \times 10^5\,m/s)^2 - (1.7 \times 10^5\,m/s)^2]$$

$$\Delta V = 384\,V$$

42. D is correct.

$$Q = VC$$

Even though the capacitors have different capacitances, the voltage across each capacitor is inversely proportional to the capacitance of that capacitor.

Like current, a charge is conserved across capacitors in series.

43. A is correct.

Calculate capacitance:

$$C = k\mathcal{E}_0 A / d$$

$$C = [(1) \cdot (8.854 \times 10^{-12}\,F/m) \cdot (0.6\,m^2)] / (0.06\,m)$$

$$C = 8.854 \times 10^{-11}\,F$$

Find potential difference:

$$C = Q / V$$

$$V = Q / C$$

$$V = (7.08 \times 10^{-10}\,C) / (8.854 \times 10^{-11}\,F)$$

$$V = 8\,V$$

44. B is correct.

"In a perfect conductor" and "in the absence of resistance" have the same meanings, and current can flow in conductors of varying resistances.

A semi-perfect conductor has resistance.

45. C is correct.

Faraday's Law: a changing magnetic environment causes a voltage to be induced in a conductor.

Metal detectors send quick magnetic pulses that cause a voltage (by Faraday's Law) and subsequent current to be induced in the conductor.

By Lenz's Law, an opposing magnetic field will then arise to counter the changing magnetic field.

The detector picks up the magnetic field and notifies the operator.

Metal detectors use Faraday's Law and Lenz's Law to detect metal objects.

46. A is correct.

$$E = q\mathrm{V}$$

$$E = (7 \times 10^{-6}\ \mathrm{C})\cdot(3.5 \times 10^{-3}\ \mathrm{V})$$

$$E = 24.5 \times 10^{-9}\ \mathrm{J}$$

$$E = 24.5\ \mathrm{nJ}$$

47. B is correct.

$$R_1 = \rho L_1 / \mathrm{A}_1$$

$$R_2 = \rho(4L_1) / \mathrm{A}_2$$

$$R_1 = R_2$$

$$\rho L_1 / \mathrm{A}_1 = \rho(4L_1) / \mathrm{A}_2$$

$$\mathrm{A}_2 = 4\mathrm{A}_1$$

$$(\pi / 4)d_2^2 = (\pi / 4)\cdot(4)d_1^2$$

$$d_2^2 = 4d_1^2$$

$$d_2 = 2d_1$$

48. D is correct.

The total resistance of a network of series resistors increases as more resistors are added.

$$\mathrm{V} = IR$$

An increase in the total resistance results in a decrease in the total current through the network.

49. A is correct.

This is a circuit with two resistors in series.

Combine the two resistors into one resistor:

$R_T = R + R_{int}$

$R_T = 0.5\ \Omega + 0.1\ \Omega$

$R_T = 0.6\ \Omega$

Ohm's law:

$V = IR$

$I = V / R$

$I = 9\ V / 0.6\ \Omega$

$I = 15\ A$

50. C is correct.

Energy stored in capacitor:

$U = ½(Q^2 / C)$

Capacitance:

$C = k\mathcal{E}_0 A / d$

$U = ½(Q^2 d) / (k\mathcal{E}_0 A)$

$Q = \sqrt{[(2U \times k\mathcal{E}_0 A) / d]}$

$Q = \sqrt{\{[(2)\cdot(10 \times 10^3\ J)\cdot(1)\cdot(8.854 \times 10^{-12}\ F/m)\cdot(2.4 \times 10^{-5}\ m^2)] / 0.0016\ m\}}$

$Q = 52\ \mu C$

51. B is correct.

Potential energy:

$U = (kQq) / r$

Electric Potential:

$V = (kQ) / r$

As r increases, the potential energy U decreases, as does the electric potential V.

Movement in the direction of the electric field is a movement *away* from a positive charge.

52. D is correct.

Electric field energy density:

$\eta_E = ½E^2 \times \mathcal{E}_0$

$\eta_E = ½(8.6 \times 10^6 \text{ V/m})^2 \cdot (8.854 \times 10^{-12} \text{ F/m})$

$\eta_E = 330 \text{ J/m}^3$

53. A is correct.

Resistance = Ohms

$\Omega = \text{V / A}$

$\Omega = [(\text{kg}\cdot\text{m}^2/\text{s}^2)/\text{C}] / [\text{C/s}]$

$\Omega = \text{kg}\cdot\text{m}^2/(\text{C}^2/\text{s})$

54. C is correct.

Ohm's Law:

$V = IR$

If V is constant, then I and R are inversely proportional.

An increase in R results in a decrease in I.

55. D is correct.

Electric Potential:

$V = kQ / r$

All other positions on the square (1, 2 or 3) are equidistant from point p, as is charge $+Q$.

Thus, a negative charge at any of these locations would have equal magnitude potential but an opposite sign because the new charge is negative.

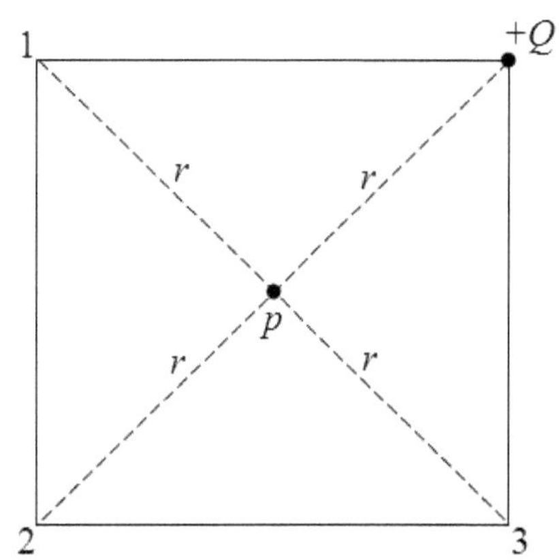

Thus: $| V | = | -V |$

$V + (-V) = 0$

56. B is correct.

Electric field energy density:

$u_E = ½E^2\varepsilon_0$

$u_E = ½(6\ N/C)^2·(8.854 \times 10^{-12}\ F/m)$

$u_E = 1.6 \times 10^{-10}\ J/m^3$

57. A is correct.

Ohm's law:

$V = IR$

Increasing V and decreasing R increases I.

58. D is correct.

$C_{Eq1} = C_2 + C_3$

$C_{Eq1} = 18\ pF + 24\ pF$

$C_{Eq1} = 42\ pF$

Voltage drops is equal in capacitors in parallel, so:

$C_{Eq1} = Q_1 / V_1$

$Q_1 = C_{Eq1} \times V_1$

$Q_1 = (42 \times 10^{-12}\ F)·(240\ V)$

$Q_1 = 1 \times 10^{-8}\ C$

Charge is equal in capacitors in series, so:

$1 / C_{Eq2} = 1 / C_{Eq1} + 1 / C_1$

$1 / C_{Eq2} = 1 / 42\ pF + 1 / 9\ pF$

$1 / C_{Eq2} = 7.4\ pF$

$1 / C_{Eq2} = V_{system} / Q_1$

$V_{system} = Q_1 / C_{Eq2}$

$V_{system} = (1 \times 10^{-8}\ C) / (7.4 \times 10^{-12}\ F)$

$V_{system} = 1{,}350\ V$

59. A is correct.

Energy stored in a capacitor:

$U = ½Q^2 / C$

$U_2 = Q^2 / 2(2C)$

$U_2 = ½Q^2 / 2C$

$U_2 = ½U$

Decreases by half.

60. C is correct.

The total resistance of a network of series resistors increases as more resistors are added to the network.

An increase in the total resistance results in a decrease in the total current through the network.

A decrease in current results in a decrease in the voltage across the original resistor:

$V = IR$

Notes for active learning

Sound – Detailed Explanations

1. B is correct.

Intensity is inversely proportional to distance (in W/m^2, not dB).

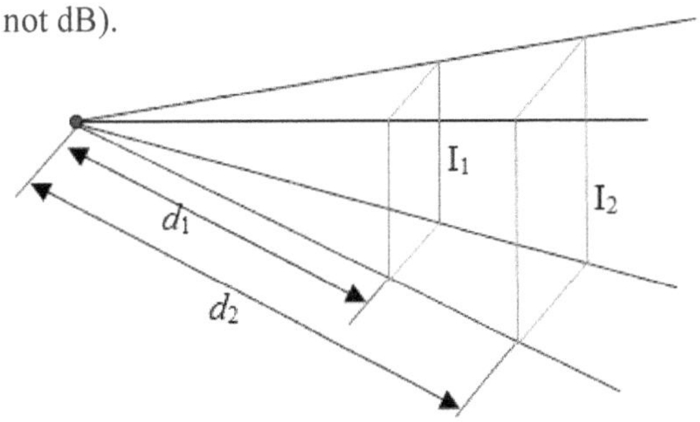

$I_2 / I_1 = (d_1 / d_2)^2$

$I_2 / I_1 = (3\text{ m} / 30\text{ m})^2$

$100\ I_2 = I_1$

The intensity is 100 times greater at 3 m than 30 m.

Intensity to decibel relationship:

$I\ (\text{dB}) = 10 \log_{10} (I / I_0)$

The intensity to dB relationship is logarithmic.

Thus, if I_1 is 100 times the original intensity then it is two times the dB intensity because:

$\log_{10} (100) = 2$

Thus, the decibel level at 3 m is:

$I\ (\text{dB}) = (2)\cdot(20\text{ dB})$

$I = 40\text{ dB}$

2. A is correct.

distance = velocity × time

$d = vt$

$t = d / v$

$t = (6{,}000\text{ m}) / (340\text{ m/s})$

$t = 18\text{ s}$

3. B is correct.

Resonance occurs when a vibrating system is driven at its resonance frequency, resulting in a relative maximum of the vibrational energy of the system.

When the force associated with the vibration exceeds the strength of the material, the glass shatters.

4. C is correct.

The third harmonic is shown in the figure below:

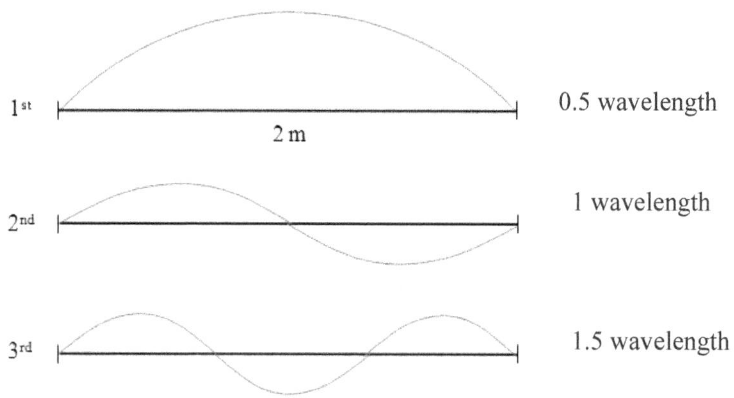

There are $(3/2)\lambda$ in the 2 m wave in the third harmonic

$L = (n / 2)\lambda$ (for n harmonic)

$L = (3 / 2)\lambda$ (for 3rd harmonic)

$L(2 / 3) = \lambda$

$\lambda = (2\text{ m})\cdot(2 / 3)$

$\lambda = 4/3\text{ m}$

5. B is correct.

High-pitched sound has a high frequency.

6. A is correct.

Snell's law:

$n_1 \sin \theta_1 = n_2 \sin \theta_2$

Solve for θ_2:

$(n_1 / n_2) \sin \theta_1 = \sin \theta_2$

$\sin \theta_1 = (n_1 / n_2) \sin \theta_2$

$\theta_2 = \sin^{-1}[(n_1 / n_2) \sin \theta_1]$

Substituting the given values:

$\theta_2 = \sin^{-1}[(1 / 1.5) \sin 60°]$

$\theta_2 = \sin^{-1}(0.67 \sin 60°)$

7. D is correct.

For a standing wave, the length and wavelength are related:

$L = (n / 2)\lambda$ (for n harmonic)

From the diagram, the wave is the 6th harmonic:

$L = (6 / 2)\lambda$

$\lambda = (2\ \text{m})\cdot(2 / 6)$

$\lambda = 0.667\ \text{m}$

$f = v / \lambda$

$f = (92\ \text{m/s}) / (0.667\ \text{m})$

$f = 138\ \text{Hz}$

8. A is correct.

$v = d / t$

$v = (0.6\ \text{m}) / (0.00014\ \text{s})$

$v = 4{,}286\ \text{m/s}$

$\lambda = v / f$

$\lambda = (4{,}286\ \text{m/s}) / (1.5 \times 10^6\ \text{Hz})$

$\lambda = 0.0029\ \text{m} = 2.9\ \text{mm}$

9. C is correct.

The wave velocity is increased by a factor of 1.3.

$v^2 = T / \rho_L$

$T = v^2 \times \rho_L$

Increasing v by a factor of 1.3:

$T = (1.3v)^2\rho_L$

$T = 1.69v^2\rho_L$

T increases by 69%

10. D is correct.

$\rho_L = \rho A$

$\rho_L = \rho(\pi r^2)$

If the diameter decreases by a factor of 2, the radius decreases by a factor of 2, and the area by a factor of 4.

The linear mass density decreases by a factor of 4.

11. B is correct.

The v and period (T) of wire C are equal to wire A, so the ρ_L must be equal.

$\rho_{LA} = \rho_{LC}$

$\rho_A A_A = \rho_C A_C$

$A_C = (\rho_A A_A) / \rho_C$

$(\pi / 4)\cdot(d_C)^2 = (7\ g/cm^3)(\pi / 4)\cdot(0.6\ mm)^2 / (3\ g/cm^3)$

$(d_C)^2 = (7\ g/cm^3)\cdot(0.6\ mm)^2 / (3\ g/cm^3)$

$d_C{}^2 = 0.84\ mm^2$

$d_C = \sqrt{(0.84\ mm^2)}$

$d_C = 0.92\ mm$

12. A is correct.

$A = \pi r^2$

If d increases by a factor of 4, r increases by a factor of 4.

A increases by a factor of 16.

13. B is correct.

Since the bird is moving toward the observer, the $f_{observed}$ must be higher than f_{source}.

Doppler shift for an approaching sound source:

$f_{observed} = (v_{sound} / v_{sound} - v_{source})f_{source}$

$f_{observed} = [340\ m/s / (340\ m/s - 10\ m/s)]f_{source}$

$f_{observed} = (340\ m/s / 330\ m/s)\cdot(60\ kHz)$

$f_{observed} = (1.03)\cdot(60\ kHz)$

$f_{observed} = 62\ kHz$

14. C is correct.

When an approaching sound source is heard, the observed frequency is higher than the frequency from the source due to the Doppler effect.

15. D is correct.

Sound requires a medium of solid, liquid, or gas substances to be propagated through. A vacuum is none of these.

16. A is correct.

According to the Doppler effect, frequency increases as the sound source moves towards the observer.

Higher frequency is perceived as a higher pitch.

Conversely, as the sound source moves *away* from the observer, the perceived pitch decreases.

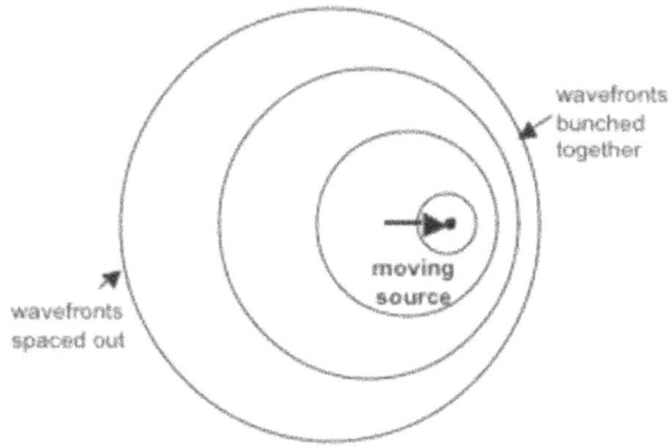

17. C is correct.

If waves are *out of phase*, the combination has its minimum amplitude of (0.6 – 0.4) Pa = 0.2 Pa.

If waves are *in phase*, the combination has its maximum amplitude of (0.6 + 0.4) Pa = 1.0 Pa.

When the phase difference has a value between in- and out-of-phase, the amplitude is between 0.2 and 1.0 Pa.

18. B is correct.

$I = P / A$

$I = P / \pi d^2$

Intensity at $2d$:

$I_2 = P / \pi(2d)^2$

$I_2 = P / 4\pi d^2$

$I_2 = \frac{1}{4}P / \pi d^2$

The new intensity is ¼ the original.

19. A is correct.

speed of sound = √[resistance to compression / density]

$v_{sound} = \sqrt{(E / \rho)}$

Low resistance to compression and high density results in low velocity because this minimizes the term under the radical and thus minimizes velocity.

20. B is correct.

A pipe open at each end has no constraint on displacement at the ends.

Furthermore, the pressure at the ends must equal the ambient pressure.

Thus, the pressure is maximum at the ends: an antinode.

21. C is correct.

For a pipe open at both ends, the resonance frequency:

$$f_n = nf_1$$

where n = 1, 2, 3, 4…

Therefore, only a multiple of 200 Hz can be a resonant frequency.

22. D is correct.

Unlike light, sound waves require a medium to travel through, and their speed depends on the medium.

Sound is fastest in solids, then liquids, and slowest in the air.

$$v_{solid} > v_{liquid} > v_{air}$$

23. B is correct.

Currents or moving charges induce magnetic fields.

24. C is correct.

$$\lambda = v / f$$

$$\lambda = (5{,}000 \text{ m/s}) / (620 \text{ Hz})$$

$$\lambda = 8.1 \text{ m}$$

25. A is correct.

Sound intensity radiating spherically:

$$I = P / 4\pi r^2$$

If r is doubled:

$$I = P / 4\pi(2r)^2$$

$$I = \tfrac{1}{4}P / 4\pi r^2$$

The intensity is reduced by a factor of ¼.

26. D is correct.

As the sound propagates through a medium, it spreads out in an approximately spherical pattern.

Thus, the power is radiated along the surface of the sphere, and the intensity can be given by:

$I = P / (4\pi r^2)$ ← for the surface area of a sphere

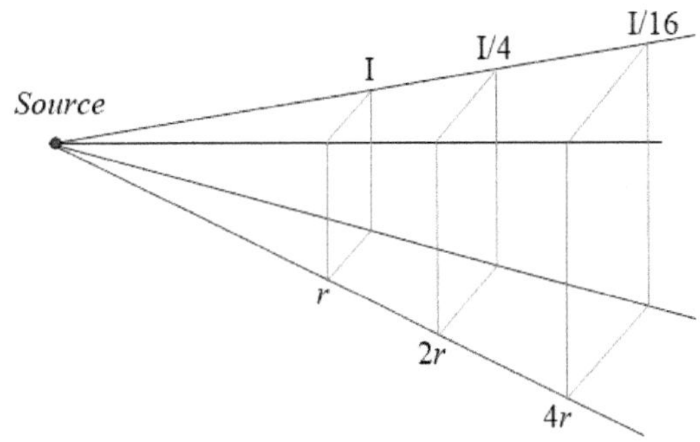

Because the surface area of a sphere contains a square component with regards to radial distance, the sound intensity is inversely proportional to the square of distance from the sound source.

27. B is correct.

The closed end is a node, and the open end is an antinode.

$\lambda = (4 / n)L$

where n = 1, 3, 5 …

For the fundamental n = 1:

$\lambda = (4 / 1)\cdot(1.5 \text{ m})$

$\lambda = 6 \text{ m}$

The 1.5 m tube (open at one end) is a quarter of a full wave, so the wavelength is 6 m.

28. A is correct.

The 1.5 m is ¼ a full wave, so the wavelength is 6 m, for the fundamental.

$f = v / \lambda$

$f = (960 \text{ m/s}) / 6 \text{ m}$

$f = 160 \text{ Hz}$

29. B is correct.

For a closed-ended pipe, the wavelength to the harmonic relationship is:

$\lambda = (4 / n)L$

where n = 1, 3, 5…

For the 5th harmonic n = 5

$\lambda = (4 / 5)\cdot(1.5 \text{ m})$

$\lambda_n = 1.2 \text{ m}$

Closed end tube

Harmonic # (n)	# of waves in a tube	# of nodes	# of antinodes	Wavelength to length
1	1/4	1	1	$\lambda = 4\ L$
3	3/4	2	2	$\lambda = 4/3\ L$
5	5/4	3	3	$\lambda = 4/5\ L$
7	7/4	4	4	$\lambda = 4/7\ L$

30. A is correct.

Find frequency:

$f = v / \lambda$

$f = (340 \text{ m/s}) / (6 \text{ m})$

$f = 57 \text{ Hz}$

31. C is correct.

Wavelength to harmonic number relationship in a standing wave on a string:

$\lambda = (2L / \text{n})$

where n = 1, 2, 3, 4, 5 …

For the 3rd harmonic:

$\lambda = (2)\cdot(0.34 \text{ m}) / 3$

$\lambda = 0.23 \text{ m}$

32. D is correct.

Beat frequency equation:

$f_{\text{beat}} = | f_2 - f_1 |$

If one tone increases in frequency, the beat frequency increases or decreases, but this cannot be determined unless the two tones are known.

33. A is correct.

For a closed-ended pipe, the wavelength to harmonic relationship is:

$\lambda = (4 / \text{n})L$

where n = 1, 3, 5, 7…

The lowest three tones are n = 1, 3, 5

$\lambda = (4 / 1)L$

$\lambda = (4 / 3)L$

$\lambda = (4 / 5)L$

34. D is correct.

The sound was barely perceptible, the intensity at Mary's ear is:

$$I_0 = 9.8 \times 10^{-12}\ \text{W/m}^2$$

Since the mosquito is 1 m away, imagine a sphere 1 m in a radius around the mosquito.

If 9.8×10^{-12} W emanates from each area 1 m^2, then the surface area is:

$$4\pi(1\ \text{m})^2$$

The power produced by one mosquito:

$$P = 4\pi r^2 I_0$$

$$P = 4\pi(1\ \text{m})^2 \times (9.8 \times 10^{-12}\ \text{W/m}^2)$$

$$P = 1.2 \times 10^{-10}\ \text{W}$$

$$\text{energy} = \text{power} \times \text{time}$$

$$E = Pt$$

Energy produced in 200 s:

$$Pt = (1.2 \times 10^{-10}\ \text{W})\cdot(200\ \text{s})$$

$$E = 2.5 \times 10^{-8}\ \text{J}$$

35. A is correct.

$$v = c / \text{n}$$

where c is the speed of light in a vacuum

$$v = \Delta x / \Delta t$$

$$\Delta x / \Delta t = c / \text{n}$$

$$\Delta t = \text{n}\Delta x / c$$

$$\Delta t = (1.33)\cdot(10^3\ \text{m}) / (3 \times 10^8\ \text{m/s})$$

$$\Delta t = 4.4 \times 10^{-6}\ \text{s}$$

36. A is correct.

When waves interfere constructively (i.e., in-phase), the sound level is amplified.

When they interfere destructively (i.e., out of phase), they cancel, and no sound is heard.

Acoustic engineers work to ensure that there are no "dead spots" and the sound waves add.

An engineer should minimize destructive interference, which can distort the sound.

37. B is correct.

The velocity of a wave on a string in tension can be calculated by:

$$v = \sqrt{(TL / m)}$$

Graph B gives a curve of a square root relationship which is how velocity and tension are related.

$$y = x^{1/2}$$

38. D is correct.

From the diagram, the wave is a 6th harmonic standing wave.

Find wavelength:

$$\lambda = (2L / n)$$

$$\lambda = (2)\cdot(4 \text{ m}) / (6)$$

$$\lambda = 1.3 \text{ m}$$

Find *frequency*:

$$f = v / \lambda$$

$$f = (20 \text{ m/s}) / (1.3 \text{ m})$$

$$f = 15.4 \text{ Hz}$$

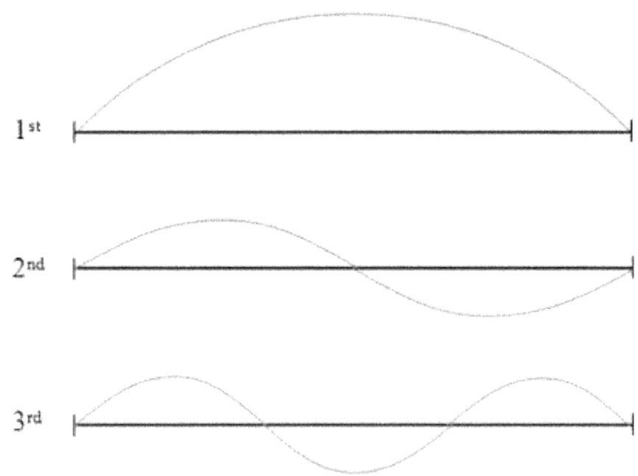

39. B is correct.

Soundwave velocity is independent of frequency and does not change.

40. C is correct.

Find the frequency of the string, then the length of the pipe excited to the second overtone using that frequency.

The speed of sound in the string is:

$$v_{string} = \sqrt{T/\mu}$$

where T is the tension in the string, and μ is the linear mass density

$$v_{string} = \sqrt{[(75 \text{ N}) / (0.00040 \text{ kg})]}$$

$$v_{string} = 433.01 \text{ m/s}$$

The wavelength of a string of length L_{string} vibrating in harmonic n_{string} is:

$$\lambda_{string} = 2L_{string} / n_{string}$$

Therefore, the vibration frequency of the string is:

$$f = v_{string} / \lambda_{string}$$

$$f = [(n_{string})(v_{string})] / 2L_{string}$$

continued…

$f = [(6)(433.01\ \text{m/s})] / (2 \times 0.50\ \text{m})$

$f = (2{,}598.06\ \text{m/s}) / 1\ \text{m}$

$f = 2{,}598.1\ \text{Hz}$

Now, consider the open pipe.

The relationship between length, wavelength and harmonic number for an open pipe is the same as for a string.

Therefore:

$L_{pipe} = n_{pipe}\ (\lambda_{pipe} / 2)$

However, since $\lambda_{pipe} = v_{air} / f$:

$L_{pipe} = n_{pipe}\ (v_{air} / 2f)$

Noting that the second overtone is the third harmonic ($n_{pipe} = 3$):

$L_{pipe} = (3 \times 345\ \text{m/s}) / (2 \times 2{,}598.1\ \text{Hz})$

$L_{pipe} = 0.20\ \text{m}$

Note that it is not necessary to calculate the frequency; its value cancels.

There is less chance for error if the two steps that use frequency are skipped.

In $L_{pipe} = n_{pipe}\ (v_{air} / 2f)$

substitute $f = [(n_{string})\cdot(v_{string})] / 2L_{string}$

In $L_{pipe} = (n_{string})\cdot(v_{string})] / 2L_{string}$

which gives:

$L_{pipe} = L_{string}\ (v_{air} / v_{string})\cdot(n_{pipe} / n_{string})$

It yields the same answer but with fewer calculations.

41. A is correct.

$v = \sqrt{(T / \mu)}$

$\mu = m / L$

$v = \sqrt{(TL / m)}$

$v_2 = \sqrt{(T(2L) / m)}$

$v_2 = \sqrt{2}\ \sqrt{(TL / m)}$

$v_2 = v\sqrt{2}$

42. C is correct.

For a standing wave, the resonance frequency:

$$f_n = nf_1$$

where n is the harmonic number, n = 1, 2, 3, 4 …

Therefore, only a multiple of 500 Hz can be a resonant frequency.

43. D is correct.

The angle of incidence always equals the angle of reflection.

A light beam entering a medium with a refractive index > than the incident medium refracts *toward* the normal.

Thus, the angle of refraction is less than the angles of incidence and reflection.

Snell's law:

$$n_1 \sin \theta_1 = n_2 \sin \theta_2$$

where

$$n_1 < n_2$$

then:

$$\theta_1 > \theta_2$$

44. A is correct.

Speed of sound in gas:

$$v_{sound} = \sqrt{(yRT / M)}$$

where y = adiabatic constant, R = gas constant, T = temperature and M = molecular mass

The speed of sound in a gas is only dependent upon temperature and not frequency or wavelength.

45. B is correct.

Waves only transport energy and not matter.

46. D is correct.

$$v = \lambda f$$

$$\lambda = v / f$$

$$\lambda = (344 \text{ m/s}) / (700 \text{ s}^{-1})$$

$$\lambda = 0.5 \text{ m}$$

The information about the string is unnecessary.

The contributor to the wavelength of the sound in air is the frequency and the speed.

47. C is correct.

$v = \lambda f$

$f = v / \lambda$

Distance from a sound source is not part of the equation for frequency.

48. A is correct.

Velocity of a wave in a rope:

$v = \sqrt{[T / (m / L)]}$

$t = d / v$

$d = L$

$t = d / \sqrt{[T / (m / L)]}$

$t = (8\text{ m}) / [40\text{ N} / (2.5\text{ kg} / 8\text{ m})]^{½}$

$t = 0.71\text{ s}$

49. C is correct.

Intensity to decibel relationship:

$I\text{ (dB)} = 10 \log_{10} (I_1 / I_0)$

where I_0 = threshold of hearing

$\text{dB} = 10\log_{10}[(10^{-5}\text{ W/m}^2) / (10^{-12}\text{ W/m}^2)]$

$I = 70$ decibels

50. B is correct.

The diagram represents the described scenario.

The wave is in the second harmonic with a wavelength of:

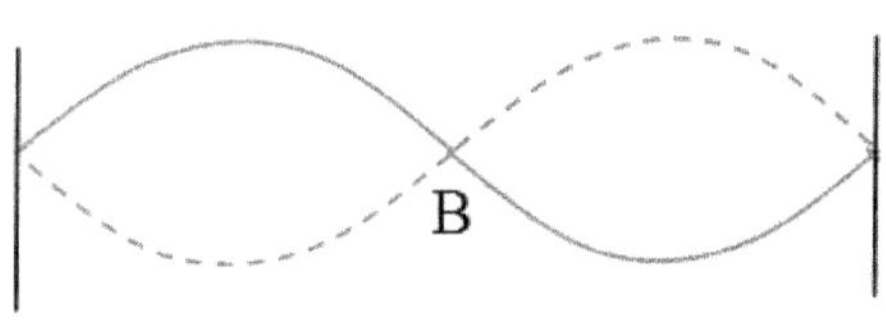

$\lambda = (2 / n)L$

$\lambda = (2 / 2)\cdot(1\text{ m})$

$\lambda = 1\text{ m}$

$f = v / \lambda$

$f = (3.8 \times 10^4\text{ m/s}) / (1\text{ m})$

$f = 3.8 \times 10^4\text{ Hz}$

The lowest frequency corresponds to the lowest possible harmonic number.

For this problem, n = 2.

51. D is correct.

The speed of light traveling in a vacuum is c.

$$c = \lambda v$$

$$c = \lambda f$$

$$f = c / \lambda$$

Frequency and wavelength are inversely proportional; increasing frequency decreases the wavelength.

52. B is correct.

Radio waves are electromagnetic waves, while all other choices are mechanical waves.

53. D is correct.

Since the microphone is equidistant from each speaker (i.e., equal path lengths), the sound waves take equal time to reach the microphone. The speakers emit sound waves in phase with each other (i.e., peaks are emitted simultaneously).

Since those peaks reach the microphone simultaneously (because of the equal path length), they combine constructively and add, forming a large peak or antinode.

54. C is correct.

Doppler equation for receding source of sound:

$$f_{observed} = [v_{sound} / (v_{sound} + v_{source})]f_{source}$$

$$f_{observed} = [(342 \text{ m/s}) / (342 \text{ m/s} + 30 \text{ m/s})]\cdot(1{,}200 \text{ Hz})$$

$$f_{observed} = 1{,}103 \text{ Hz}$$

The observed frequency is lower when the source is receding.

55. D is correct.

$$f_1 = 600 \text{ Hz}$$

$$f_2 = 300 \text{ Hz}$$

$$f_2 = \tfrac{1}{2}f_1$$

$$\lambda_1 = v / f_1$$

$$\lambda_2 = v / (\tfrac{1}{2}f_1)$$

$$\lambda_2 = 2\,(v / f_1)$$

The wavelength of 300 Hz frequency is twice as long as the 600 Hz frequency.

56. C is correct.

$f_2 = 2f_1$

$f = v / \lambda$

$v / \lambda_2 = (2)v / \lambda_1$

cancel v from both sides of the expression

$1 / \lambda_2 = 2 / \lambda_1$

$\lambda = (2 / n)L$, for open-ended pipes

$1 / (2 / n)L_2 = 2 / (2 / n)L_1$

$L_1 = 2L_2$

$L_1 / L_2 = 2$

57. A is correct.

Resonance occurs when energy gets transferred from one oscillator to another of similar f by a weak coupling.

Dispersion is the spreading of waves due to the dependence of wave speed on frequency.

Interference is the addition of two waves in the same medium, which happens when waves from both strings combine, but that is not the excitation of the C_4 string.

58. C is correct.

frequency = 1 / period

$f = 1 / T$

$f = 1 / 10\text{ s}$

$f = 0.1\text{ Hz}$

Find wavelength:

$\lambda = v / f$

$\lambda = (4.5\text{ m/s}) / (0.1\text{ Hz})$

$\lambda = 45\text{ m}$

59. D is correct.

$v = \sqrt{K / \rho}$

where K = bulk modulus (i.e., resistance to compression) and ρ = density

Since ρ for water is greater than for air, the greater v for water implies that water's bulk modulus (K) must be much greater than for air.

60. C is correct.

When visible light strikes glass, it causes the electrons in the glass to vibrate at their non-resonant frequency.

The vibration is passed from one atom to the next, transferring the energy of the light.

Finally, the energy is passed to the last atom before the light is re-mitted out of the glass at its original frequency.

If the light energy were converted into internal energy, the glass would heat up and not transfer the light.

Notes or active learning

Notes or active learning

Light and Geometrical Optics – Detailed Explanations

1. A is correct.

Soap film that reflects a given wavelength of light exhibits constructive interference.

The expression for constructive interference of a thin film:

$$2t = (m + ½)\lambda$$

where t = thickness, m = 0, 1, 2, 3… and λ = wavelength

To find the minimum thickness set m = 0:

$$2t = (0 + ½)\lambda$$

$$2t = ½\,\lambda$$

$$t = ¼\,\lambda$$

2. A is correct.

By the law of reflection, the angle of incidence = angle of reflection.

Thus, as the angle of incidence increases, the angle of reflection increases to be equal to the angle of incidence.

3. B is correct.

If image is twice her height and upright, then:

$$2h_o = h_i$$

$$m = h_i / h_o$$

Linear magnification of lens:

$$m = -d_i / d_o$$

$$m = 2h_o / h_o$$

$$m = 2$$

$$2 = -d_i / d_o$$

$$-2d_o = d_i$$

Use *lens equation* to solve:

$$1 / f = 1 / d_o + 1 / d_i$$

$$1 / 100\text{ cm} = 1 / d_o + (-1 / 2\,d_o)$$

$$1 / 100\text{ cm} = 1 / 2\,d_o$$

$$2d_o = 100\text{ cm}$$

$$d_o = 50\text{ cm}$$

4. D is correct.

If a person's eye is too long, the light entering the eye is focused in front of the retina, causing myopia. This condition is *nearsightedness* (i.e., distant objects are blurred).

Hyperopia is *farsightedness* (i.e., close objects are blurred).

5. C is correct.

Visible light:

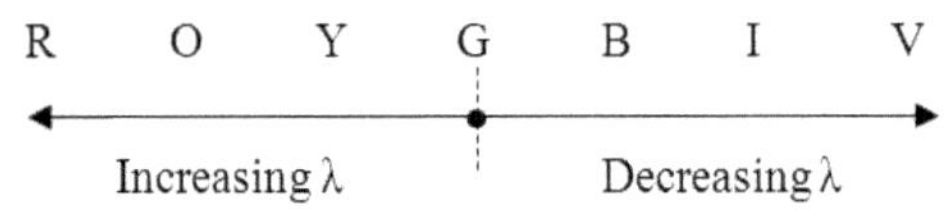

speed of light = wavelength × frequency

$c = \lambda f$

Wavelength to frequency:

$f = c / \lambda$

Frequency and wavelength are inversely proportional:

As λ increases, f decreases.

As λ decreases, f increases.

Thus, because $E = hf$:

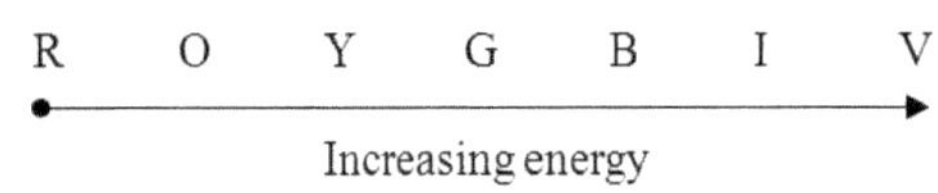

6. B is correct.

Lens equation:

$1 / f = 1 / d_o + 1 / d_i$

$1 / d_i = 1 / f - 1 / d_o$

$1 / d_i = -1 / 3 \text{ m} - 1 / 4 \text{ m}$

$1 / d_i = (-3 \text{ m} - 4 \text{ m}) / 12 \text{ m}$

$1 / d_i = -7 \text{ m} / 12 \text{ m}$

$d_i = -12 / 7 \text{ m}$

Magnification:

$m = -d_i / d_o$

$m = -(-12 / 7 \text{ m}) / 4 \text{ m}$

$m = 3 / 7$

continued…

Height of the candle image:

$h_i = mh_o$

$h_i = (3/7)\cdot(18\text{ cm})$

$h_i = 54 / 7\text{ cm}$

$h_i = 7.7\text{ cm}$

7. C is correct.

$\theta_{syrup} = \tan^{-1}(0.9\text{ m} / 0.66\text{ m})$

$\theta_s = \tan^{-1}(1.36)$

$\theta_s = 53.7°$

$\theta_{oil} = \tan^{-1}[(2\text{ m} - 0.9\text{ m}) / 1.58\text{ m}]$

$\theta_o = \tan^{-1}(0.7)$

$\theta_o = 34.8°$

$n_o \sin\theta_o = n_{air} \sin\theta_{air}$

$n_o \sin 34.8° = (1) \sin 90°$

$n_o = 1 / (\sin 34.8°)$

$n_o = 1.75$

8. D is correct.

$\theta_{syrup} = \tan^{-1}(0.9\text{ m} / 0.66\text{ m})$

$\theta_s = \tan^{-1}(1.36)$

$\theta_s = 53.7°$

$\theta_{oil} = \tan^{-1}[(2\text{ m} - 0.9\text{ m}) / 1.58\text{ m}]$

$\theta_o = \tan^{-1}(0.7)$

$\theta_o = 34.8°$

$n_o \sin\theta_o = n_{air} \sin\theta_{air}$

$n_o \sin 34.8° = (1) \sin 90°$

$n_o = 1 / (\sin 34.8°)$

$n_o = 1.75$

$n_s \sin\theta_s = n_o \sin\theta_o$

$n_s = n_o \sin\theta_o / \sin\theta_s$

$n_s = (1.75)\cdot(\sin 34.8°) / (\sin 53.7°) = 1.24$

9. A is correct.

The photoelectric effect (i.e., emission of electrons when light shines on a material) cannot be explained with the wave theory of light.

10. D is correct.

Geometrical optics, or ray optics, describes light propagation in terms of rays and fronts to approximate the path along which light propagates in certain circumstances.

11. D is correct.

Find the critical angle:

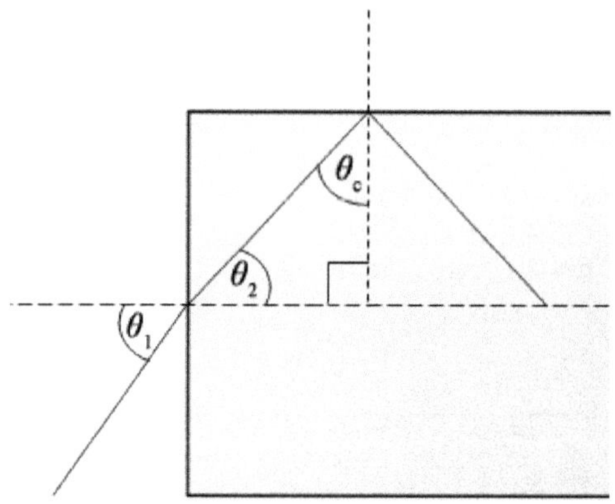

$n_{fiber} \sin \theta_c = n_{air} \sin \theta_{air}$

$(1.26) \sin \theta_c = (1) \sin 90°$

$\sin \theta_c = 1 / 1.26$

$\theta_c = \sin^{-1} (1 / 1.26)$

$\theta_c = 52.5°$

Find θ_2:

$\theta_2 + \theta_c + 90° = 180°$

$(\theta_2 + 52.5° + 90°) = 180°$

$\theta_2 = 37.5°$

Find θ_1:

$n_{air} \sin \theta_1 = n_{fiber} \sin \theta_2$

$(1) \sin \theta_1 = (1.26) \sin 37.5°$

$\sin \theta_1 = 0.77$

$\theta_1 = \sin^{-1} (0.77)$

$\theta_1 = 50°$

12. A is correct.

If the power of the lens is 10 diopters,

$1 / f = 10$ D

where f is the focal length in m

Thin Lens equation:

$1 / f = 1 / d_o + 1 / d_i$

$10 \text{ m}^{-1} = 1 / 0.5 \text{ m} + 1 / d_i$

$1 / d_i = 10 \text{ m}^{-1} - 1 / 0.5 \text{ m}$

$1 / d_i = 8 \text{ m}^{-1}$

$d_i = 1 / 8 \text{ m}$

$d_i = 0.13 \text{ m}$

13. D is correct.

Most objects observed by humans are virtual images or objects which reflect incoming light to project an image.

14. D is correct.

An image from a *convex* mirror has the following characteristics, regardless of object distance:

- located behind the convex mirror
- virtual
- upright
- reduced in size from the object (image < object)

15. A is correct.

The mirror has a positive focal length which indicates that the mirror is concave.

The object is at a distance greater than the focal length.

Therefore, it is inverted.

Use *lens equation* to solve image distance:

$1 / f = 1 / d_o + 1 / d_i$

$1 / 10 \text{ m} = 1 / 20 \text{ m} + 1 / d_i$

$d_i = 20$ cm

The image distance is positive so the image is real.

continued...

Use the *magnification equation* to determine if it is upright or inverted.

$m = -d_i / d_o$

$m = h_i / h_o$

$-(20\text{ m} / 20\text{ m}) = h_i / h_o$

$-1 = h_i / h_o$

The object height h_o is always positive, so the image height h_i must be negative to satisfy the equation.

A negative image height indicates an inverted image.

16. B is correct.

If an object is beyond $2f$ from the lens, the image is real, inverted, and reduced for a converging lens.

Use the lens equation to determine if the image is real (or virtual):

Assume $f = 1$ m and $d_o = 3f$ (because $d_o > 2f$)

$1 / f = 1 / d_o + 1/ d_i$

$1 / f = 1 / 3f + 1/ d_i$

$d_i = 1.5$

A positive d_i indicates a real image.

Use the *magnification equation* to determine if the image is inverted and reduced.

$m = -d_i / d_o$

$m = -(1.5\text{ m} / 3\text{ m})$

$m = -½$

$| m | = ½$

$| m | < 1$

A negative magnification factor with an absolute value less than 1 is a reduced and inverted image.

17. C is correct.

Radio waves range from 3 kHz to 300 GHz, lower than all forms of radiation listed.

Since the energy of radiation is proportional to frequency ($E = hf$), radio waves have the lowest energy.

18. B is correct.

A medium's index of refraction is the ratio of the speed of refracted light in a vacuum to its speed in the reference medium.

$n = c / v$

$n = 2.43$

$2.43 = c / v_{\text{diamond}}$

$c = 2.43(v_{\text{diamond}})$

19. D is correct.

Thin Lens equation:

$1 / f = 1 / d_o + 1 / d_i$

$1 / 20\text{ cm} = 1 / 15\text{ cm} + 1 / d_i$

$3 / 60\text{ cm} - 4 / 60\text{ cm} = 1 / d_i$

$-1 / 60\text{ cm} = 1 / d_i$

$d_i = -60\text{ cm}$

The negative sign indicates that the image is projected back the way it came.

20. B is correct.

Red paper absorbs all colors but reflects only red light giving it the appearance of being red.

Cyan is the complementary color to red, so when the cyan light shines upon the red paper, no light is reflected, and the paper appears black.

21. B is correct.

Thin Lens equation:

$1 / f = 1 / d_o + 1 / d_i$

If

$d_i = f$

$1 / d_o = 0$

Thus, d_o must be large.

22. A is correct.

Since the index of refraction depends on the frequency, and the focal length depends on the refraction of the beam in the lens, dispersion causes the focal length to depend on frequency.

23. C is correct.

Use the equation for *magnification*:

$$m = -d_i / d_o$$

$$d_i = d_o$$

$$m = 1$$

Thus, there is no magnification, so the image is the same size as the object.

24. D is correct.

When viewed straight down (90° to the surface), an incident light ray moving from water to air is refracted 0°.

25. C is correct.

The rotating of one polarized lens 90° with respect to the other lens results in complete darkness since no light would be transmitted.

26. C is correct.

First, find the angle that the ray makes with the normal of the glass:

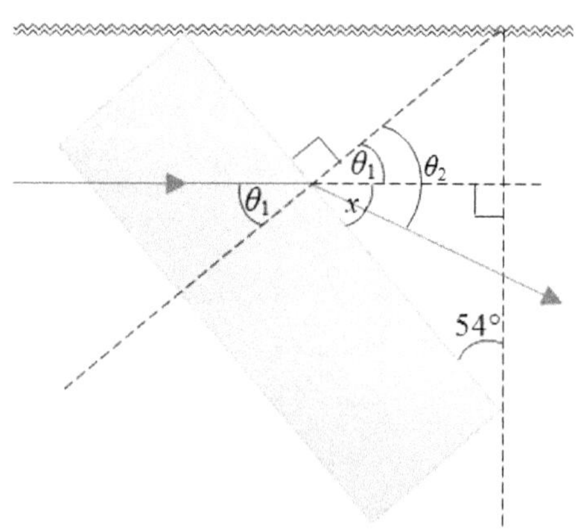

$$180° = x + 90° + 54°$$

$$x = 36°$$

Find θ_1:

$$\theta_1 = 90° - 36°$$

$$\theta_1 = 54°$$

Referring to the diagram, $\theta_1 = 54°$

Snell's Law:

$$n_1 \sin \theta_1 = n_2 \sin \theta_2$$

$$\sin^{-1} [(n_1 / n_2) \sin \theta_1] = \theta_2$$

$$\theta_2 = \sin^{-1}[(1.45 / 1.35) \sin 54°]$$

$$\theta_2 = 60°$$

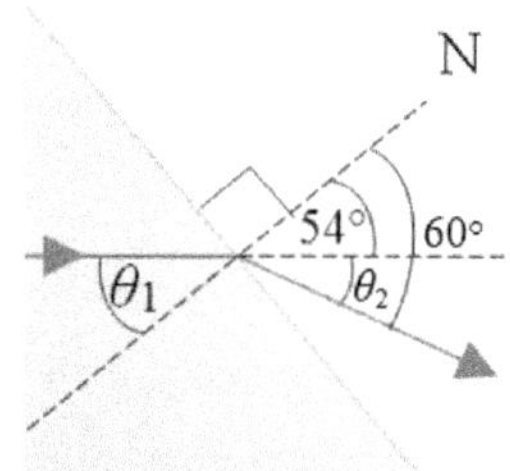

Solve for the angle with the horizontal:

$$\theta_H = 60° - 54°$$

$$\theta_H = 6°$$

27. A is correct.

The angle at which the ray is turned is the sum of the angles if reflected off each mirror once:

$\theta_{turned} = \theta_1 + \theta_2 + \theta_3 + \theta_4$

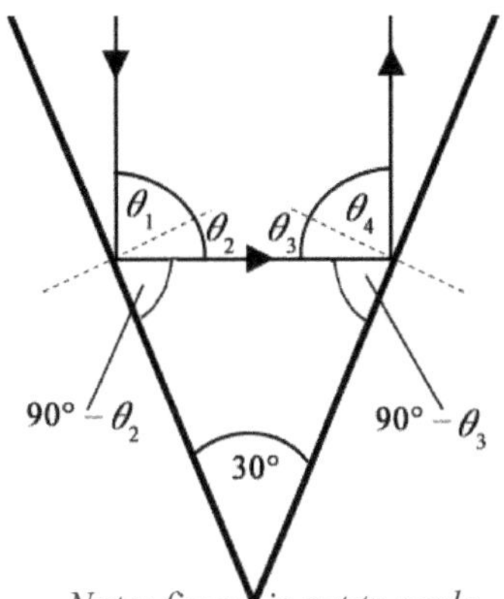

Note: figure is not to scale

By law of reflection:

$\theta_1 = \theta_2$

$\theta_3 = \theta_4$

Note the triangle formed (sum of interior angles is 180°):

$30° + (90° - \theta_2) + (90° - \theta_3) = 180°$

$\theta_2 + \theta_3 = 30°$

Given: $\theta_2 + \theta_3 = \theta_1 + \theta_4$

Thus: $\theta_{turned} = 30° + 30°$

$\theta_{turned} = 60°$

In general: for two plane mirrors that meet at an angle of $\theta \leq 90°$, the ray deflected off both mirrors is deflected through an angle of 2θ.

28. D is correct.

The following statements about light are true: a packet of light energy is known as a photon.

Color can be used to determine the approximate energy of visible light, and light travels through space at a speed of 3.0×10^8 m/s.

29. A is correct.

The angle of incidence is < the angle of refraction if the light travels into a *less dense medium.*

The angle of incidence is > the angle of refraction if the light travels into a *denser medium.*

The angle of incidence is = the angle of refraction if the densities of the mediums are *equal.*

30. B is correct.

Plane mirrors do not distort the size or shape of an object since light is reflected at the same angle as received.

Magnification equation:

$m = h_i / h_o$

For a plane mirror m = 1:

$1 = h_i / h_o$

$h_i = h_o$

Therefore, the image size is the same as object size, and the image is virtual since it is located behind the mirror.

31. C is correct.

A spherical concave mirror has a focal length of:

$f = R / 2$

32. B is correct.

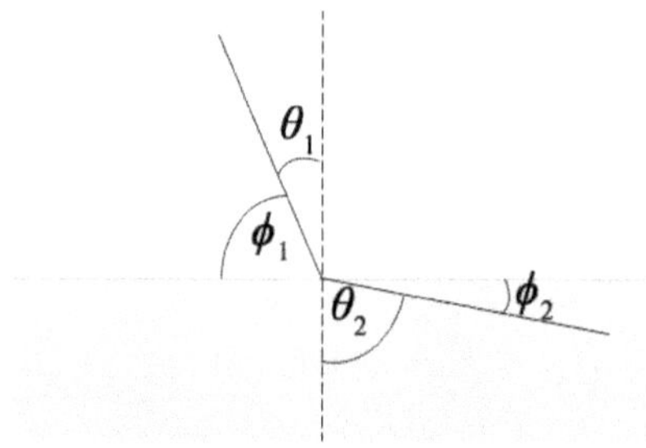

Refracted rays bend further from the normal than the original incident angle when the refracting medium is optically less dense than the incident medium.

Therefore,

$n_1 > n_2$

The index of refraction for a medium can never be less than 1.

33. B is correct.

If the eye is too short, light entering the eye is focused behind the retina, causing farsightedness (hyperopia).

34. A is correct.

Hot air is less dense than cold air.

Light traveling through both types of air experiences refractions, which appear as shimmering or "wavy" air.

35. C is correct.

Chromatic aberration occurs when a lens focuses on different wavelengths of color at various positions in the focal plane.

It occurs in the following pattern for converging lens:

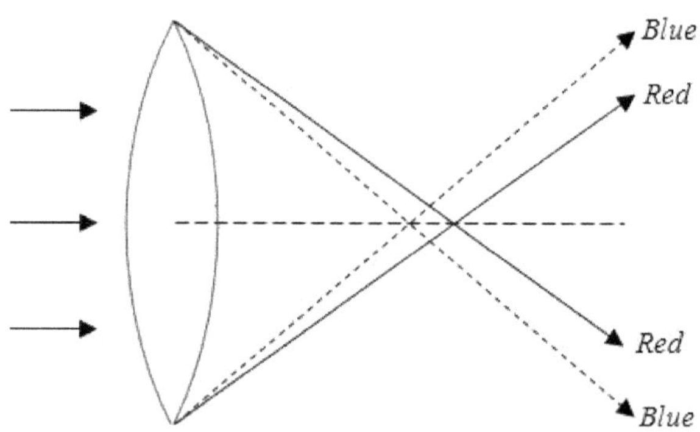

36. B is correct.

$1 / f_{total} = 1 / f_1 + 1 / f_2$

$1 / f_{total} = 1 / 2 \text{ m} + 1 / 4 \text{ m}$

$1 / f_{total} = 3 / 4 \text{ m}$

$f_{total} = 4 / 3 \text{ m}$

37. C is correct.

The angle in the water respective to the normal:

$\theta = \tan^{-1}(37.5 \text{ ft} / 50 \text{ ft})$

$\theta = \tan^{-1}(0.75)$

$\theta = 36.9°$

$n_{air} \sin(90 - \theta) = n_{water} \sin \theta$

$(1) \sin(90 - \theta) = (1.33) \sin 36.9°$

$\sin(90 - \theta) = 0.8$

$(90 - \theta) = \sin^{-1}(0.8)$

$(90 - \theta) = 52.9$

$\theta = 37.1° \approx 37°$

38. A is correct.

Violet light has the highest energy and frequency; the shortest wavelength.

39. D is correct.

Objects directly in front of plane mirrors are reflected in their likeness since plane mirrors are not curved and reflect light perpendicularly to their surface.

40. B is correct.

A virtual image is always upright and can be formed by a diverging and converging lens.

Diverging lens → reduced and virtual image

Converging lens → enlarged and virtual image

41. B is correct.

Neon light is emitted from neon atoms as their energized electrons cascade back down to ground level.

When this occurs, energy is released as light at particular wavelengths known as the *emission spectrum*.

When this light passes through a prism, a series of bright discontinuous spots or lines are seen due to the specific wavelengths of the emission spectrum of neon.

42. D is correct.

The law of reflection states that the angle of incidence is equal to the angle of reflection (with respect to the normal) and is valid for all mirrors.

$\theta_i = \theta_r$

43. C is correct.

A concave lens always forms an image that is virtual, upright, and reduced in size.

44. B is correct.

Virtual images are upright.

There is no correlation between the size and nature – virtual or real – of an image.

Images may be larger, smaller, or the same size as the object.

45. A is correct.

Red is the light with the lowest frequency (i.e., longest wavelength) detected by the eyes from the choices listed. (ROY G BIV)

46. D is correct.

Thin lens formula:

$$1 / f = 1 / d_o + 1 / d_i$$

$$1 / 6 \text{ m} = 1 / 3 \text{ m} + 1 / d_i$$

$$1 / d_i = 1 / 6 \text{ m} - 1 / 3 \text{ m}$$

$$1 / d_i = -1 / 6 \text{ m}$$

$$d_i = -6 \text{ m}$$

where the negative sign indicates the image is on the same side as the object

The image is upright and virtual since the rays must be extended to intersect.

47. A is correct.

A diverging lens (concave) produces a virtual, upright image and is reduced in size.

48. C is correct.

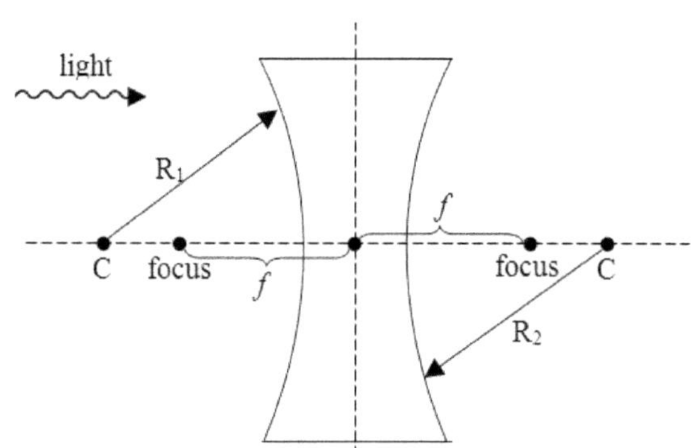

Thin lens formula:

$$1 / f = 1 / d_o + 1 / d_i$$

d_i is negative because the image is virtual

$$1 / f = 1 / 14 \text{ cm} + 1 / -5 \text{ cm}$$

$$f = -7.8 \text{ cm}$$

The focus is negative because the lens is diverging.

Lens maker formula:

$$1 / f = (n - 1) \cdot (1 / R_1 - 1 / R_2)$$

continued…

R_1 is negative by convention because the light ray passes its center of curvature before the curved surface.

$1 / (-7.8 \text{ cm}) = (n - 1)\cdot(1 / -15 \text{ cm} - 1 / 15 \text{ cm})$

$(1 / -7.8 \text{ cm})\cdot(15 \text{ cm} / -2) + 1 = n$

$n = 2$

49. B is correct.

The *magnification equation* relates the image and object *distance*:

$m = -d_i / d_o$

or

The *magnification equation* relates the image and *object height*:

$m = h_i / h_o$

50. D is correct.

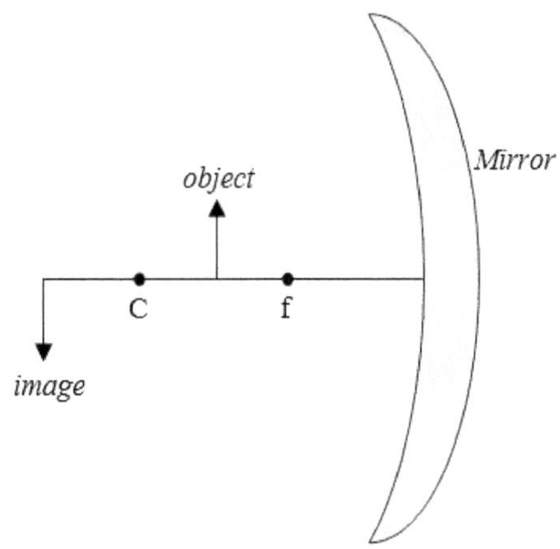

For a concave mirror, if an object is located between the focal point and center of curvature, the image is formed beyond the center of curvature.

In this problem, Mike (i.e., object) does not see his image because he is in front of where it forms.

51. D is correct.

For a concave spherical mirror, the produced image characteristics depend upon the placement of the object concerning the focal point and center of curvature.

The image can be smaller, larger, or the same size as the object.

52. A is correct.

Lens power is the reciprocal of the focal length in meters:

$P = 1 / f$

When the effective focal length of the lens combination is less than the focal length of the individual lens, the power of the combination must be greater than the power of either individual lens.

53. B is correct.

A medium's index of refraction is the ratio of the speed of refracted light in a vacuum to its speed in the reference medium.

$n = c / v$

54. D is correct.

As it is a plane mirror, the image is not distorted.

Only some light rays are reflected; the others create an image behind the mirror's surface.

For a plane mirror:

$$m = 1$$

$$m = -d_i / d_o$$

$$1 = -d_i / d_o$$

$$d_o = -d_i$$

The negative indicates the image is virtual and behind the mirror.

55. C is correct.

The radius length is the center of curvature:

$$r = 50 \text{ cm}$$

Find the focal length:

$$f = r / 2$$

$$f = 50 \text{ cm} / 2$$

$$f = 25 \text{ cm}$$

The resulting image is real and inverted for a concave mirror with an object between the center of curvature and the focal length.

56. B is correct.

Find index of refraction of glass:

Snell's Law:

$$n_1 \sin \theta_1 = n_2 \sin \theta_2$$

$$n_g \sin 48° = (1.33) \sin 68°$$

$$n_g = (1.33) \sin 68° / \sin 48°$$

$$n_g = 1.66$$

Find refracted angle of ray:

$$(1.66) \sin 29° = (1.33) \sin \theta$$

$$\sin \theta = (1.66) \sin 29° / (1.33)$$

$$\sin \theta = 0.605$$

$$\theta = \sin^{-1} (0.605) = 37°$$

57. A is correct.

In a compound microscope, the image of the objective serves as the object for the eyepiece.

58. D is correct.

The refractive index is given by:

$$n = c / v$$

Because $v \approx c$ in air:

$$n_{air} \approx n_{vacuum}$$

$$n_{air} = 1$$

$$n_{vacuum} = 1$$

All other transparent materials slow the speed of light.

Thus, n is greater than 1 because $v_{other\ materials} < c$.

59. D is correct.

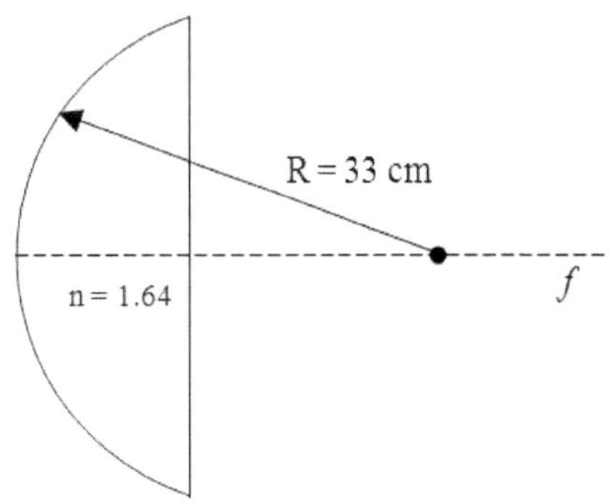

Lens maker formula:

$$1 / f = (n - 1)\cdot(1 / R_1 - 1 / R_2)$$

For a flat surface:

$$R_2 = \infty$$

$$1 / f = (1.64 - 1)\cdot[(1 / 33\text{ cm}) - (1 / \infty)]$$

$$1 / f = (0.64)\cdot(1 / 33\text{ cm})$$

$$f = 51.6\text{ cm} \approx 52\text{ cm}$$

60. C is correct.

Water does not absorb visible light very easily (λ = 400 to 700 nm) but absorbs infrared light (λ = 700 nm to 1 mm) from the vibrational motion of the molecule.

Water absorbs microwaves through rotational motion.

Notes for active learning

Thermodynamics – Detailed Explanations

1. B is correct.

Ideal gas law:

$PV = nRT$

$P_0 = nRT / V_0$

If isothermal expansion, then n, R and T are constant

$P = nRT / (1/3\ V_0)$

$P = 3(nRT / V_0)$

$P = 3P_0$

2. C is correct.

Area expansion equation:

$\Delta A = A_0(2\alpha\Delta T)$

$\Delta A = (\pi / 4)\cdot(1.2\ \text{cm})^2\cdot(2)\cdot(19 \times 10^{-6}\ \text{K}^{-1})\cdot(200\ °\text{C})$

$\Delta A = 8.6 \times 10^{-3}\ \text{cm}^2$

$\Delta A = A_f - A_0$

$8.6 \times 10^{-3}\ \text{cm}^2 = (\pi / 4)\cdot[d_f{}^2 - (1.2\ \text{cm})^2]$

$d_f = 1.2\ \text{cm}$

3. D is correct.

1 Watt = 1 J/s

Power × Time = Q

Heat needed to change the temperature of a mass:

$Q = mc\Delta T$

$P \times t = mc\Delta T$

$t = (mc\Delta T) / P$

$t = (90\ \text{g})\cdot(4.186\ \text{J/g}\cdot°\text{C})\cdot(30\ °\text{C} - 10\ °\text{C}) / (50\ \text{W})$

$t = 151\ \text{s}$

4. A is correct.

Convert 15 minutes to seconds:

$t = (15 \text{ min}/1)\cdot(60 \text{ s}/1 \text{ min})$

$t = 900 \text{ s}$

Find total energy generated:

$Q = P \times t$

$Q = (1{,}260 \text{ J/s})\cdot(900 \text{ s})$

$Q = 1{,}134 \text{ kJ}$

Find mass of water needed to carry away energy:

$Q = mL_v$

$m = Q / L_v$

$m = (1{,}134 \text{ kJ}) / (22.6 \times 10^2 \text{ kJ/kg})$

$m = 0.5 \text{ kg} = 500 \text{ g}$

5. C is correct.

Phase changes occur at a constant temperature.

Once the phase change is complete, the temperature of the substance then increases or decreases.

For example, water remains at 0 °C until it has completely changed phase to ice before the temperature decreases further.

6. B is correct.

The amount of energy needed to melt a sample of mass *m* is:

$Q = m L_f$

where L_f is the latent heat of fusion

$Q = (55 \text{ kg})\cdot(334 \text{ kJ/kg})$

$Q = 1.8 \times 10^4 \text{ kJ}$

7. D is correct.

Metals are good heat and electrical conductors because of their bonding structure.

In metallic bonding, the outer electrons are held loosely and can travel freely.

Electricity and heat require high electron mobility.

Thus, the looseness of the outer electrons in the materials allows them to be excellent conductors.

8. A is correct.

Find heat of phase change from *steam to liquid*:

$$Q_1 = mL_v$$

Find heat of phase change from *liquid to solid*:

$$Q_2 = mL_f$$

Find heat of temperature from 100 °C to 0 °C:

$$Q_3 = mc\Delta T$$

Total heat:

$$Q_{net} = Q_1 + Q_2 + Q_3$$

$$Q_{net} = mL_v + mL_f + mc\Delta T$$

To find mass:

$$Q_{net} = m(L_v + c\Delta T + L_f)$$

$$m = Q_{net} / (L_v + c\Delta T + L_f)$$

Solve:

$$Q_{net} = 200 \text{ kJ}$$

$$Q_{net} = 2 \times 10^5 \text{ J}$$

$$m = (2 \times 10^5 \text{ J}) / [(22.6 \times 10^5 \text{ J/kg}) + (4{,}186 \text{ J/kg·K})·(100 \text{ °C} - 0 \text{ °C}) + (33.5 \times 10^4 \text{ J/kg})]$$

$$m = 0.066 \text{ kg}$$

9. C is correct.

Fusion is the process whereby a substance changes from a solid to liquid (i.e., melting).

Condensation is the process whereby a substance changes from vapor to liquid.

Sublimation is when a substance changes directly from a solid to the gas phase without passing through the liquid phase.

10. D is correct.

Heat needed to change the temperature of a mass:

$$Q = mc\Delta T$$

$$Q = (0.2 \text{ kg})·(14.3 \text{ J/g·K})·(1{,}000 \text{ g/kg})·(280 \text{ K} - 250 \text{ K})$$

$$Q = 86{,}000 \text{ J}$$

$$Q = 86 \text{ kJ}$$

11. B is correct.

Heat needed to raise the temperature of aluminum:

$$Q_A = m_A c_A \Delta T$$

Heat needed to raise temperature of water:

$$Q_W = m_W c_W \Delta T$$

Total heat to raise temperature of system:

$$Q_{net} = Q_A + Q_W$$

$$Q_{net} = m_A c_A \Delta T + m_W c_W \Delta T$$

$$Q_{net} = \Delta T(m_A c_A + m_W c_W)$$

$$Q_{net} = (98\ °C - 18\ °C)\cdot[(0.5\ kg)\cdot(900\ J/kg\cdot K) + (1\ kg)\cdot(4{,}186\ J/kg\cdot K)]$$

$$Q_{net} = 370{,}880\ J$$

Time to produce Q_{net} with 500 W:

$$Q_{net} = (500\ W)t$$

$$t = Q_{net} / (500\ W)$$

$$t = (370{,}880\ J) / 500\ W$$

$$t = 741.8\ s$$

Convert to minutes:

$$t = (741.8\ s/1)\cdot(1\ min/60\ s)$$

$$t = 12.4\ min \approx 12\ min$$

12. A is correct.

When a substance goes through a phase change, the temperature does not change.

It can be assumed that the lower plateau is L_f and the upper plateau is L_v.

Count the columns:

$$L_f = 2,\ L_v = 7$$

Solve:

$$L_v / L_f = 7 / 2$$

$$L_v / L_f = 3.5$$

13. B is correct.

Specific heat is the amount of heat (i.e., energy) needed to raise the temperature of the unit mass of a substance by a given amount (usually one degree).

14. D is correct.

Find ½ of KE of the BB:

$KE = ½mv^2$

$½KE = ½(½mv^2)$

$½KE_{BB} = ½(½)·(0.0045 \text{ kg})·(46 \text{ m/s})^2$

$½KE_{BB} = 2.38 \text{ J}$

The $½KE_{BB}$ is equal to energy taken to change temperature:

$Q = ½KE_{BB}$
$Q = mc\Delta T$

$mc\Delta T = ½KE_{BB}$

$\Delta T = ½KE_{BB} / mc$

Calculate to find ΔT:

$\Delta T = (2.38 \text{ J}) / (0.0045 \text{ kg})·(128 \text{ J/kg·K})$

$\Delta T = 4.1 \text{ K}$

15. C is correct.

Vaporization is the process whereby a substance changes from a liquid to a gas. The process can be boiling or evaporation.

Sublimation is the process whereby a substance changes from a solid to a gas.

16. A is correct.

Carnot efficiency:

η = work done / total energy

$\eta = W / Q_H$

$\eta = 5 \text{ J} / 18 \text{ J}$

$\eta = 0.28$

The engine's efficiency:

$\eta = (T_H - T_C) / T_H$

$0.28 = (233 \text{ K} - T_C) / 233 \text{ K}$

$(0.28)·(233 \text{ K}) = (233 \text{ K} - T_C)$

$65.2 \text{ K} = 233 \text{ K} - T_C$

$T_C = 168 \text{ K}$

17. D is correct.

Heat needed to change the temperature of a mass:

$$Q = mc\Delta T$$

Calculate to find Q:

$$Q = (0.92\text{ kg})\cdot(113\text{ cal/kg}\cdot°\text{C})\cdot(96\text{ °C} - 18\text{ °C})$$

$$Q = 8{,}108.9\text{ cal}$$

Convert to joules:

$$Q = (8{,}108.9\text{ cal})\cdot(4.186\text{ J/cal})$$

$$Q = 33{,}940\text{ J}$$

18. B is correct.

During a state change, the addition of heat does not change the temperature (i.e., a measure of kinetic energy).

The heat energy added only adds to the potential energy of the substance until the substance completely changes state.

19. C is correct.

A pressure *vs.* volume graph of the work done for a cyclic process carried out by gas is equal to the area enclosed by the cyclic process.

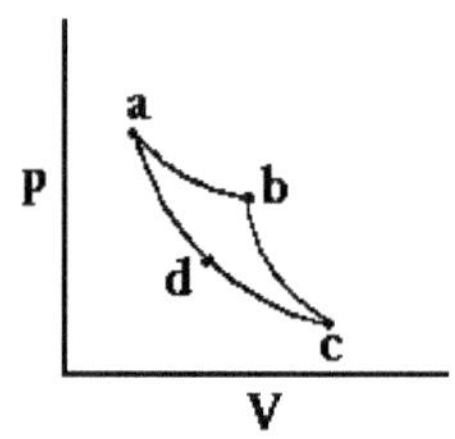

20. A is correct.

Specific heat of A is larger than B:

$$c_A > c_B$$

Energy to raise the temperature:

$$Q = mc\Delta T$$

If m and ΔT are equal for A and B:

$$Q_A = m_A c_A \Delta T_A$$

$$Q_B = m_B c_B \Delta T_B$$

$$Q_A > Q_B$$

This is valid because other factors are equal, and the magnitude of Q only depends on c.

21. D is correct.

Find kinetic energy of meteor:

$KE = ½mv^2$

$KE = ½(0.0065\ kg)·(300\ m/s)^2$

$KE = 292.5\ J$

Find temperature rise:

$Q = KE$

$Q = mcΔT$

$mcΔT = KE$

$ΔT = KE / mc$

Convert KE to calories:

$KE = (292.5\ J/1)·(1\ cal/4.186\ J)$

$KE = 69.9\ cal$

Calculate ΔT:

$ΔT = (69.9\ cal) / [(0.0065\ kg)·(120\ cal/kg·°C)]$

$ΔT = 89.6\ °C ≈ 90\ °C$

22. A is correct.

When a liquid freezes, it undergoes a phase change from liquid to solid.

For this to occur, heat energy must be dissipated (removed).

During a phase change, the temperature remains constant.

23. B is correct.

For an isothermal process:

$ΔU = 0$

$ΔU = Q – W$

$Q = W$

Work to expand an ideal gas in an isothermal process:

$W = nRT\ \ln(V_f / V_i)$

From the ideal gas law,

$nRT = P_fV_f$

continued…

Giving:

$W = P_f V_f \ln(V_f / V_i)$

$W = (130 \text{ kPa})\cdot(0.2 \text{ m}^3) \ln[(0.2 \text{ m}^3) / (0.05 \text{ m}^3)]$

$W = 36 \text{ kJ}$

$Q = W$

$Q = 36 \text{ kJ}$

Since the process is isothermal, there is no change in the internal energy.

Since the surroundings are doing negative work on the system:

$\Delta E = 0$

$0 = Q + W$

$Q + W = Q - 36 \text{ kJ}$

$\Delta E = Q - 36 \text{ kJ}$

Therefore:

$Q = 36 \text{ kJ}$

24. D is correct.

Find the potential energy of 1 kg of water:

$PE = mgh$

$PE = (1 \text{ kg})\cdot(9.8 \text{ m/s}^2)\cdot(30 \text{ m})$

$PE = 294 \text{ J}$

Assume all potential energy is converted to heat for maximum temperature increase:

$PE = Q$

$Q = mc\Delta T$

$mc\Delta T = PE$

$\Delta T = PE / mc$

$\Delta T = (294 \text{ J}) / [(1 \text{ kg})\cdot(4{,}186 \text{ J/kg/K})]$

$\Delta T = 0.07 \text{ °C}$

It is not necessary to convert to Kelvin for temperature differences because a temperature change in Kelvin is equal to a temperature change in Celsius.

25. A is correct.

Find heat from phase change:

$Q = mL_f$

$Q = (0.75\text{ kg})\cdot(33{,}400\text{ J/kg})$

$Q = 25{,}050\text{ J}$

Because the water is freezing, Q should be negative due to heat being released.

$Q = -25{,}050\text{ J}$

Find change in entropy:

$\Delta S = Q / \text{T}$

$\Delta S = -25{,}050\text{ J} / (0\text{ °C} + 273\text{ K})$

$\Delta S = -92\text{ J / K}$

A negative change in entropy indicates that the disorder of the isolated system has decreased.

When water freezes, the entropy is negative because water is more disordered than ice.

Thus, the disorder has decreased.

26. A is correct.

Copper has a larger coefficient of linear expansion than iron, so it expands more than iron during a given temperature change.

The bimetallic bar bends due to the difference in expansion between the copper and iron.

27. D is correct.

Calculate heat needed to raise temperature:

$Q = mc\Delta\text{T}$

$Q = (0.110\text{ kg})\cdot(4{,}186\text{ J/kg·K})\cdot(30\text{ °C} - 20\text{ °C})$

$Q = 4{,}605\text{ J}$

Calculate time needed to raise temperature with 60 W power source:

$Q = P \times t$

$Q = (60\text{ W})t$

$t = \text{Q} / (60\text{ W})$

$t = (4{,}605\text{ J}) / (60\text{ W})$

$t = 77\text{ s}$

28. B is correct.

During a state change, the addition of heat does not change the temperature (i.e., the kinetic energy).

The heat (i.e., energy) added only adds to the substance's potential energy (PE) until the substance completely changes state.

29. C is correct.

Convert to Kelvin:

$$T = -243 \text{ °C} + 273$$

$$T = 30 \text{ K}$$

Double temperature:

$$T_2 = (30 \text{ K})\cdot(2)$$

$$T_2 = 60 \text{ K}$$

Convert back to Celsius:

$$T_2 = 60 \text{ K} - 273$$

$$T_2 = -213 \text{ °C}$$

30. D is correct.

If a researcher attempts to determine how much the temperature of a particular piece of material would rise when a known amount of heat is added to it, knowing the specific heat would be most helpful.

31. B is correct.

Convert units:

$$1.7 \times 10^5 \text{ J/kg} = 170 \text{ kJ/kg}$$

Change in internal energy = heat added (Q)

$$Q = mL_v$$

$$Q = (1 \text{ kg})\cdot(170 \text{ kJ/kg})$$

$$Q = 170 \text{ kJ}$$

32. D is correct.

$Q = mc\Delta T$

If m and c are constant, the relationship is directly proportional.

To double Q, T must be doubled:

5 C + 273 = 278 K

278 K × 2 = 556 K

556 K − 273 = 283 C

33. B is correct.

The mass of each material is required to determine the time the system takes to reach thermal equilibrium.

34. B is correct.

Body heat gives energy to the water molecules in the sweat.

This energy is transferred via collisions until some molecules have enough energy to break the hydrogen bonds and escape the liquid (evaporation).

However, if a body stayed dry, the heat would not be given to the water, and the person would stay hot because the heat is not lost due to the evaporation of the water.

35. D is correct.

Calculate heat released when 0 °C water converts to 0 °C ice:

$Q_1 = mL_f$

$Q_1 = (2{,}200 \text{ kg})\cdot(334 \times 10^3 \text{ J/kg})$

$Q_1 = 734{,}800 \text{ kJ}$

Calculate heat released for temperature drop ΔT

$Q_2 = mc\Delta T$

$Q_2 = (2{,}200 \text{ kg})\cdot(2{,}050 \text{ J/kg K})\cdot[(0\ °\text{C} - (-30\ °\text{C})]$

ΔK = Δ°C, so units cancel:

$Q_2 = 135{,}300 \text{ kJ}$

Add heat released to get Q_{net}:

$Q_{net} = Q_1 + Q_2$

$Q_{net} = (734{,}800 \text{ kJ}) + (135{,}300 \text{ kJ})$

$Q_{net} = 870{,}100 \text{ kJ}$

36. A is correct.

Object 1 has three times the specific heat capacity and four times the mass of Object 2:

$c_1 = 3c_2$; $m_1 = 4m_2$

A single-phase substance obeys the *specific heat* equation:

$Q = mc\Delta T$

In this case, the same amount of heat is added to each substance.

Therefore:

$Q_1 = Q_2$

$m_1c_1\Delta T_1 = m_2c_2\Delta T_2$

$(4m_2)(3c_2)\Delta T_1 = m_2 c_2\Delta T_2$

$12m_2c_2\Delta T_1 = m_2c_2\Delta T_2$

$12 \Delta T_1 = \Delta T_2$

37. C is correct.

Conduction is a form of heat transfer in which the collisions of the molecules of the material transfer energy through the material.

The higher temperature of the material causes the molecules to collide with more energy which eventually is transferred throughout the material through subsequent collisions.

Radiation is a form of heat transfer in which electromagnetic waves carry energy from the emitting object and deposit the energy to the object that absorbs the radiation.

Convection is a form of heat transfer in which the mass motion of a fluid (i.e., liquids and gases) transfers energy from the source of heat.

38. A is correct.

From the ideal gas law:

$p_3V_3 = nRT_3$

$T_3 = p_3V_3 / nR$

$T_3 = 1.5p_1V_3 / nR$

$T_3 = 1.5V_3(p_1 / nR)$

Also from the ideal gas law:

$(p_1 / nR) = T_1 / V_1$

$(p_1 / nR) = (293.2 \text{ K}) / (100 \text{ cm}^3)$

$(p_1 / nR) = 2.932 \text{ K/cm}^3$

continued…

Calculate T_3:

$$T_3 = 1.5V_3\,(2.932\ \text{K/cm}^3)$$

$$T_3 = 1.5\,(50\ \text{cm}^3)\cdot(2.932\ \text{K/cm}^3)$$

$$T_3 = 219.9\ \text{K}$$

$$T_3 = -53.3\ °\text{C} \approx -53\ °\text{C}$$

Calculate T_4:

$$T_4 = 1.5V_4\,(2.932\ \text{K/cm}^3)$$

$$T_4 = 1.5\,(150\ \text{cm}^3)\cdot(2.932\ \text{K/cm}^3)$$

$$T_4 = 659.6\ \text{K}$$

$$T_4 = 386.5\ °\text{C} \approx 387\ °\text{C}$$

39. B is correct.

Steel is a very conductive material that can transfer thermal energy very well.

The steel feels colder than the plastic because its higher thermal conductivity allows it to remove more heat and thus makes touching it feel colder.

40. B is correct.

An *isobaric* process involves constant pressure.

An *isochoric* (also isometric) process involves a closed system at constant volume.

An *adiabatic* process occurs without the transfer of heat or matter between a system and its surroundings.

An *isothermal* process involves the change of a system in which the temperature remains constant.

41. D is correct. Carnot coefficient of performance of a refrigeration cycle:

$$C_P = T_C / (T_H - T_C)$$

$$C_P = Q_C / W$$

$$Q_C / W = T_C / (T_H - T_C)$$

$$W = (Q_C / T_C)\cdot(T_H - T_C)$$

$$W = (20 \times 10^3\ \text{J} / 293\ \text{K})\cdot(307\ \text{K} - 293\ \text{K})$$

$$W = 955.6\ \text{J} = 0.956\ \text{kJ}$$

Power = Work / time

$$P = W / t$$

$$P = 0.956\ \text{kJ} / 1\ \text{s}$$

$$P = 0.956\ \text{kW} \approx 0.96\ \text{kW}$$

42. C is correct.

Heat energy is measured in units of Joules and calories.

43. B is correct.

Convection is a form of heat transfer in which the mass motion of a fluid (i.e., liquids and gases) transfers energy from the source of heat.

44. A is correct.

Radiation is the transmission of energy as particles or waves through space or a material medium.

Examples include electromagnetic radiations (e.g., X-rays, alpha and beta particles, radio waves, visible light).

45. D is correct.

Convert P_3 to Pascals:

$$P_3 = (2 \text{ atm} / 1)\cdot(101{,}325 \text{ Pa} / 1 \text{ atm})$$

$$P_3 = 202{,}650 \text{ Pa}$$

Use the ideal gas law to find V_3:

$$PV = nRT$$

$$V = (nRT) / P$$

$$V_3 = [(0.008 \text{ mol})\cdot(8.314 \text{ J/mol}\cdot\text{K})\cdot(2{,}438 \text{ K})] / (202{,}650 \text{ Pa})$$

$$V_3 = 8 \times 10^{-4} \text{ m}^3$$

Convert to cm^3:

$$V_3 = (8 \times 10^{-4} \text{ m}^3/1)\cdot(100^3 \text{ cm}^3/1 \text{ m}^3)$$

$$V_3 = 800 \text{ cm}^3$$

46. C is correct.

An adiabatic process involves no heat added or removed from the system.

From the First Law of Thermodynamics:

$$\Delta U = Q + W$$

If $Q = 0$, then:

$$\Delta U = W$$

Because work is being done to expand the gas, it is considered negative, and then the change in internal energy is negative (decreases).

$$-\Delta U = -W$$

47. B is correct.

Standing in a breeze while wet feels colder than when dry because of water evaporation from the skin.

Water requires heat to evaporate, so this is taken from the body, making a person feel colder than if they were dry and the evaporation did not occur.

48. A is correct.

Conduction is a form of heat transfer in which the collisions of the molecules of the material transfer energy through the material.

The higher temperature of the material causes the molecules to collide with more energy which eventually is transferred throughout the material through subsequent collisions.

49. D is correct.

An *isobaric* process is a constant pressure process, so the resulting pressure is the same.

50. B is correct.

This question asks which surface has a higher emissivity and can radiate more energy over a set period.

A *blackbody* is an idealized radiator and has the highest emissivity.

A surface most similar to a blackbody (the black surface) is the best radiator of thermal energy.

A black surface is considered an ideal blackbody and therefore has an emissivity of 1 (i.e., perfect emissivity).

The black surface will be the best radiator compared to another surface that cannot be considered blackbodies and emissivity of < 1.

51. C is correct.

Calculate the gap between the rods.

The gap between the rods will be filled by both expanding, so total thermal expansion length is equal to 1.1 cm.

$$\Delta L = L_0\alpha\Delta T$$

$$\Delta L_{tot} = \Delta L_B + \Delta L_A$$

$$\Delta L_{tot} = (\alpha_B L_B + \alpha_A L_A)\,\Delta T$$

Rearrange the equation for ΔT:

$$\Delta T = \Delta L_{tot} \,/\, (\alpha_B L_B + \alpha_A L_A)$$

$$\Delta T = 1.1 \text{ cm} \,/\, [(2 \times 10^{-5}\text{ K}^{-1})\cdot(59.1 \text{ cm}) + (2.4 \times 10^{-5}\text{ K}^{-1})\cdot(39.3 \text{ cm})]$$

$$\Delta T = 517.6 \text{ K} \approx 518 \text{ K}$$

Measuring the difference in temperature in K is the same as in °C, so it is not required to convert:

$$\Delta T = 518\ °\text{C}$$

52. A is correct.

Find seconds in a day:

$t = (24 \text{ h} / 1 \text{ day})\cdot(60 \text{ min} / 1 \text{ h})\cdot(60 \text{ s} / 1 \text{ min})]$

$t = 86{,}400 \text{ s}$

Find energy lost in a day:

$E = \text{Power} \times \text{time}$

$E = (60 \text{ W})t$

$E = (60 \text{ W})\cdot(86{,}400 \text{ s})$

$E = 5{,}184{,}000 \text{ J}$

Convert to kcal:

$E = (5{,}184{,}000 \text{ J}/1)\cdot(1 \text{ cal}/4.186 \text{ J})\cdot(1 \text{ kcal}/10^3 \text{ cal})$

$E = 1{,}240 \text{ kcal}$

53. D is correct.

Conduction is a form of heat transfer in which the collisions of the molecules of the material transfer energy through the material.

The higher temperature of the material causes the molecules to collide with more energy which eventually is transferred throughout the material through subsequent collisions.

54. B is correct.

Heat given off by warmer water is equal to that absorbed by the frozen cube.

This heat is split into heat needed to melt the cube and bring the temperature to equilibrium.

$Q_{H2O,1} + Q_{alcohol,Temp1} + Q_{alcohol,Phase1} = 0$

$(mc\Delta T)_{H2O,1} + (mc\Delta T)_{alcohol,Temp1} + (mL_f)_{alcohol} = 0$

$Q_{H2O,2} + Q_{alcohol,Temp2} + Q_{alcohol,Phase2} = 0$

$(mc\Delta T)_{H2O,2} + (mc\Delta T)_{alcohol,Temp2} + (mL_f)_{alcohol} = 0$

Set equal to cancel heat from phase change (since they are equal):

$(mc\Delta T)_{H2O,1} + (mc\Delta T)_{alcohol,Temp1} = (mc\Delta T)_{H2O,2} + (mc\Delta T)_{alcohol,Temp2}$

$(m_{alcohol})(c_{alcohol})\cdot(\Delta T_{alcohol1} - \Delta T_{alcohol2}) = c_{H2O}(m\Delta T_{H2O,2} - m\Delta T_{H2O,1})$

$c_{alcohol} = (c_{H2O} / m_{alcohol})\cdot[(m\Delta T_{H2O,2} - m\Delta T_{H2O,1}) / (\Delta T_{alcohol1} - \Delta T_{alcohol2})]$

continued…

Solving for $c_{alcohol}$:

$$c_{alc} = [(4{,}190 \text{ J/kg·K}) / (0.22 \text{ kg})]·[(0.4 \text{ kg})·(10 - 30 \text{ °C}) - (0.35 \text{ kg})·(5 - 26 \text{ °C})] / [(5 \text{ °C} - (-10 \text{ °C}) - (10 \text{ °C} - (-10 \text{ °C})]$$

$$c_{alc} = (19{,}045 \text{ J/kg·K})·[(-0.65 \text{ °C}) / (-5 \text{ °C})]$$

$$c_{alc} = 2{,}475 \text{ J/kg·K}$$

55. D is correct.

Using the calculated value for c_{alc} in the problem above:

$$Q_{H2O,1} + Q_{alcohol,Temp1} + Q_{alcohol,Phase1} = 0$$

$$(mc\Delta T)_{H2O,1} + (mc\Delta T)_{alcohol} + (mL_f)_{alcohol} = 0$$

$$L_{f\,alcohol} = [-(mc\Delta T)_{H2O,1} - (mc\Delta T)_{alcohol}] / m_{alcohol}$$

Solve:

$$L_{f\,alcohol} = -[(0.35 \text{ kg})·(4{,}190 \text{ J/kg·K})·(5 \text{ °C} - 26 \text{ °C}) - (0.22 \text{ kg})·(2{,}475 \text{ J/kg·K})·(5 \text{ °C} - (-10 \text{ °C})] / (0.22 \text{ kg})$$

$$L_{f\,alcohol} = (30{,}796.5 \text{ J} - 8{,}167.5 \text{ J}) / (0.22 \text{ kg})$$

$$L_{f\,alcohol} = 103 \times 10^3 \text{ J/kg} = 10.3 \times 10^4 \text{ J/kg}$$

56. C is correct.

The silver coating reflects thermal radiation into the bottle to reduce heat loss by radiation.

Radiation is a form of heat transfer in which electromagnetic waves carry energy from an emitting object and deposit the energy in an object absorbing the radiation.

57. B is correct.

Convert calories to Joules:

$$E = (16 \text{ kcal/1})·(10^3 \text{ cal/1 kcal})·(4.186 \text{ J/1 cal})$$

$$E = 66{,}976 \text{ J}$$

Convert hours to seconds:

$$t = (5 \text{ h})·(60 \text{ min/h})·(60 \text{ s/min})$$

$$t = 18{,}000 \text{ s}$$

Find power expended:

$$P = E / t$$

$$P = (66{,}976 \text{ J}) / (18{,}000 \text{ s})$$

$$P = 3.7 \text{ W}$$

58. A is correct.

Specific heat capacity:

$\Delta Q = cm\Delta T$

$\Delta T = \Delta Q / cm$

$\Delta T = (50 \text{ kcal}) / [(1 \text{ kcal/kg·°C})·(5 \text{ kg})]$

$\Delta T = 10 \text{ °C}$

59. C is correct.

For melting, use L_f:

$Q = mL_f$

$Q = (30 \text{ kg})·(334 \text{ kJ/kg})$

$Q = 1 \times 10^4 \text{ kJ}$

60. B is correct.

Temperature change:

$Q = mc\Delta T$

$\Delta T = Q / mc$

$\Delta T = (160 \times 10^3 \text{ J}) / [(6 \text{ kg})·(910 \text{ J/kg·K})]$

$\Delta T = 29 \text{ °C}$

Final temperature:

$\Delta T = T_f - T_i$

$T_f = T_i + \Delta T$

$T_f = 12 \text{ °C} + 29 \text{ °C}$

$T_f = 41 \text{ °C}$

Notes for active learning

Notes for active learning

Atomic Nucleus and Electronic Structure – Detailed Explanations

1. A is correct.

Though *alpha particles* have low penetrating power and high ionizing power, they are not harmless.

All forms of radiation have risks and are not entirely harmless.

2. C is correct.

A *beta particle* (β) is a high-energy, high-speed electron (β^-) or positron (β^+) emitted in the radioactive decay of an atomic nucleus.

Electron emission (β^- decay) occurs in an unstable atomic nucleus with an excess of neutrons, whereby a neutron is converted into a proton, an electron, and an electron antineutrino.

Positron emission (β^+ decay) occurs in an unstable atomic nucleus with an excess of protons, whereby a proton is converted into a neutron, a positron, and an electron neutrino.

3. D is correct.

The *de Broglie wavelength* is given as:

$$\lambda = h / p$$

where h is Planck's constant and p is momentum

$$p = mv$$

$$\lambda_1 = h / mv$$

$$\lambda_2 = h / m(2v)$$

$$\lambda_2 = \tfrac{1}{2}h / mv$$

$$\lambda_2 = \tfrac{1}{2}\,\lambda_1$$

λ decreases by factor of 2

4. B is correct.

The *Bohr model* places electrons around the nucleus of the atom at discrete energy levels.

The *Balmer series* line spectra agreed with the Bohr model because the energy of the observed photons in each spectrum matched the transition energy of electrons within these discrete predicted states.

5. C is correct.

A *radioactive element* is an element that spontaneously emits radiation in the form of one or a combination of the following: alpha radiation, beta radiation, gamma radiation.

6. D is correct.

This is an example of an *electron capture nuclear reaction.*

When this happens, the atomic number decreases by one, but the mass number stays the same.

$$^{100}_{44}\text{Ru} + {}^{0}_{-1}\text{e}^{-} \rightarrow {}^{100}_{43}\text{Tc}$$

Ru: 100 = mass number (# protons + # neutrons)

Ru: 44 = atomic number (# protons)

From the periodic table, Tc is the element with 1 less proton than Ru.

7. B is correct.

A *nucleon* is a particle that makes up the nucleus of an atom.

The two known nucleons are protons and neutrons.

8. D is correct.

An *alpha particle* is composed of two neutrons and two protons and is identical to the nucleus of a ^{4}He atom.

Total mass of two alpha particles:

$2 \times (2 \text{ neutrons} + 2 \text{ protons}) = 8$

Mass of a ^{9}Be atom:

$5 \text{ neutrons} + 4 \text{ protons} = 9$

Mass of a ^{9}Be atom > total mass of two alpha particles

The mass of a ^{9}Be atom is greater than the mass of two alpha particles, so its mass is greater than twice the mass of a ^{4}He atom.

9. C is correct.

The *superscript* is the mass number (atomic weight), which is both neutrons and protons.

The *subscript* is the atomic number, which is the number of protons.

Therefore, the number of neutrons is equal to the superscript minus the subscript.

$181 - 86 = 95$, which is the greatest number of neutrons among the choices.

10. A is correct.

The number of neutrons and protons must be equal after the reaction.

Thus, the sum of the atomic number before and mass number before should be equal after the reaction.

Mass number (superscript):

$(1 + 235) - (131 + 3) = 102$

continued...

Atomic number (subscript):

$$92 - (53) = 39$$

${}^{102}_{39}$Y balances the reaction.

11. B is correct.

Balmer equation is given by:

$$\lambda = B[(n^2) / (n^2 - 2^2)]$$

$$\lambda = (3.6 \times 10^{-7})\cdot[(12^2) / (12^2 - 2^2)]$$

$$\lambda = 3.7 \times 10^{-7}\,\text{m}$$

where c is the speed of light

$$c = \lambda f$$

Convert wavelength to frequency:

$$f = c / \lambda$$

$$f = (3 \times 10^8\,\text{m/s}) / (3.7 \times 10^{-7}\,\text{m})$$

$$f = 8.1 \times 10^{14}\,\text{s}^{-1} = 8.1 \times 10^{14}\,\text{Hz}$$

12. D is correct.

Gamma rays are the most penetrating form of radiation because they are the highest energy and least ionizing.

A gamma-ray passes through a given amount of material without imparting as much of its energy into removing electrons from atoms and ionizing them as other forms of radiation do.

Gamma rays retain more of their energy passing through matter and can penetrate further.

13. A is correct. Use the Rydberg Formula:

$$E = hf$$

$$f = c / \lambda$$

$$E = (hc)\cdot(1 / \lambda)$$

$$1 / \lambda = R(1 / n_1^2 - 1 / n_2^2)$$

where $n_1 = 1$ and $n_2 = 2$

$$E = hcR[(1 / n_1^2) - (1 / n_2^2)]$$

$$E = (4.14 \times 10^{-15}\,\text{eV·s})\cdot(3 \times 10^8\,\text{m/s})\cdot(1.097 \times 10^7\,\text{m}^{-1})\cdot[(1 / 1^2) - (1 / 2^2)]$$

$$E = 13.6[1 - (1 / 4)]$$

$$E = 10.2\,\text{eV}$$

The positive energy indicates that a photon was absorbed and not emitted.

14. C is correct.

In β^- (beta minus) decay, the atomic number (subscript) increases by 1, but the atomic mass stays constant.

$$^{87}_{37}\text{Rb} \rightarrow {}^{87}_{38}\text{Sr} + {}^{0}_{-1}\text{e} + {}^{0}_{0}\nu$$

Sr is the element with 1 more proton (subscript) than Rb.

${}^{0}_{0}\nu$ represents an electron antineutrino.

15. C is correct.

The *nucleus* of an atom is bound by the strong nuclear force from the nucleons within it. The strong nuclear force must overcome the Coulomb repulsion of the protons (due to their like charges).

Neutrons help stabilize and bind the nucleus by contributing to the strong nuclear force so that it is greater than the Coulomb repulsion experienced by the protons.

16. A is correct.

Geiger-Muller counters operate using a Geiger-Muller tube, consisting of a high voltage shell and small rod in the center, filled with a low-pressure inert gas (e.g., argon).

When exposed to radiation (specifically particle radiation), the particles ionize the argon atoms, allowing for a brief charge between the high voltage rod and outer shell.

The electric pulse is then displayed visually or via audio to indicate radioactivity.

17. B is correct.

The *half-life* calculation:

$$A = A_0(½)^{t\,/\,h}$$

where A_0 = original amount, t = time elapse and h = half-life

If three half-life pass,

$$t = 3h$$

$$A = (1)\cdot(½)^{3h\,/\,h}$$

$$A = (1)\cdot(½)^3$$

$$A = 0.125 = 12.5\%$$

18. D is correct.

In the *Lyman series*, electron transitions go from $n \geq 2$ to $n = 1$.

19. B is correct.

Most of the volume of an atom is occupied by space.

20. A is correct.

Alpha particles are positively charged due to their protons, while beta minus particles are negatively charged (i.e., electrons). When exposed to a magnetic field, alpha particles and electrons deflect in opposite directions due to their opposite charges and thus experience opposite forces due to the magnetic field.

21. D is correct.

Chemical reactions store and release energy in their chemical bonds but do not convert mass into energy.

Nuclear reactions convert some mass into energy, which can be measured and calculated via the equation:

$$E = \Delta mc^2$$

22. C is correct.

The *Curie* is a non-SI unit of radioactivity equivalent to 3.7×10^{10} decays (disintegrations) per second; named after the early radioactivity researchers Marie and Pierre Curie.

23. B is correct.

The atomic numbers:

$$^{235}_{92}\text{U} \rightarrow {}^{141}_{56}\text{Ba} + {}^{92}_{36}\text{Kr}$$

The subscripts on each side sum to 92, so adding a proton ($^{1}_{1}\text{H}$) to the right side would not balance.

The superscripts sum to 235 on the left and sum to 233 on the right.

Add two neutrons ($^{1}_{0}\text{n} + {}^{1}_{0}\text{n}$) to the right side to balance each side of the equation.

24. D is correct.

The atomic number is the subscript and represents the number of protons. The superscript is the mass number and is the sum of protons and neutrons.

The number of neutrons can be found by taking the difference between the mass number and atomic number.

In this example, there are 16 protons and 18 neutrons.

25. A is correct.

The *half-life* calculation:

$$A = A_0(½)^{t\,/\,h}$$

where A_0 = original amount, t = time elapse and h = half-life

Consider: (2 days)·(24 hours / 1 day) = 48 hours

$$A = (1)·(½)^{(48\,/\,12)}$$

$$A = 0.5^4$$

$$A = 0.0625 = 1/16$$

26. C is correct.

An *alpha particle* consists of two protons and two neutrons and is identical to a helium nucleus so that it can be written as ${}^{4}_{2}He$

For a nuclear reaction to be written correctly, it must be balanced, and the sum of superscripts and subscripts must be equal on both sides of the reaction. The superscripts add to 238, and the subscripts add to 92. Therefore, it is the only balanced answer.

27. D is correct.

The question is asking for the λ of the emitted photon so use the *Rydberg Formula*:

$$1 / \lambda = R(1 / n_1^2 - 1 / n_2^2)$$

$$\lambda = 1 / [R(1 / n_1^2 - 1 / n_2^2)]$$

Use $n_1 = 5$ and $n_2 = 20$ because we are solving for λ of an emitted (not absorbed) photon.

$$\lambda = 1 / [(1.097 \times 10^7 \text{ m}^{-1})\cdot(1 / 5^2 - 1 / 20^2)]$$

$$\lambda = 2.43 \text{ μm}$$

28. D is correct.

The *superscript* represents the mass number when writing a nuclear reaction.

The *subscript* represents the atomic number.

A correct nuclear reaction is balanced when the sum of superscripts (mass number) and subscripts (atomic number) is equal on each side of the reaction.

29. C is correct.

A *blackbody* is an ideal system that absorbs 100% of light incident upon it and reflects none.

A blackbody emits 100% of the radiation it generates; therefore, it has perfect absorption and emissivity.

30. D is correct.

Carbon dating relies upon a steady creation of ^{14}C and knowledge of the creation rate at various time points to determine the approximate age of objects.

If a future archeologist is unaware of nuclear bomb testing and the higher levels of ^{14}C created, then the dates they calculate for an object would be too young.

This is because a higher amount of ^{14}C would be present in samples and make them seem like they had not had time to decay and thus appear younger.

31. A is correct.

Gamma radiation is an electromagnetic wave and is not a particle. Thus, when gamma radiation is emitted, the atomic number and mass number remain the same.

32. C is correct.

In β^- decay, a neutron is converted to a proton, and an electron and electron antineutrino are emitted.

In β^+ decay, a proton is converted to a neutron, and a positron and an electron neutrino are emitted.

$$^{14}_{6}C \rightarrow ^{14}_{7}N + e^+ + \nu_e$$

33. D is correct.

In a *balanced nuclear equation*, the number of nucleons must be conserved.

The sum of mass numbers and atomic numbers must be equal on both sides of the equation.

The product (i.e., daughter nuclei) should be on the right side of the equation.

34. B is correct.

The *atomic number* is the number of protons within the nucleus that characterizes the element's nuclear and chemical properties.

35. C is correct.

The *Balmer series* is the emission spectrum of H when electrons transition from a higher state to the n = 2 state.

Four visible spectral lines within the Balmer series with colors range from red to violet (i.e., ROY G BIV).

36. D is correct.

Electrons were discovered through early experiments with electricity, specifically in high voltage vacuum tubes (cathode ray tubes).

Beams of electrons were observed traveling through these tubes when high voltage was applied between the anode and cathode, and the electrons struck fluorescent material at the back of the tube.

37. B is correct.

Beams a and c deflect when an electric field is applied, indicating they have a net charge and must be particles.

Beam b is undisturbed by the applied electric field, indicating it has no net charge and must be a high-energy electromagnetic wave since other forms of radioactivity (alpha and beta radiation) are charged particles.

38. C is correct.

Beam a is composed of negatively charged particles, while beam c is composed of positively charged particles; therefore, both beams are deflected by the electric field.

Beam b is composed of particles; however, they are neutral because the electric field does not deflect them. An example of this kind of radiation would be a gamma-ray, which consists of neutral photons.

39. C is correct.

A *helium nucleus* is positively charged, deflected from the top plate, and attracted toward the negative plate.

40. B is correct.

5.37 eV is the energy required to excite the electron from the ground state to the zero-energy state.

Calculate the wavelength of a photon with this energy:

$E = hf$

$f = c / \lambda$

$E = hc / \lambda$

$\lambda = hc / E$

$\lambda = (4.14 \times 10^{-15}\ \text{eV·s})·(3 \times 10^{8}\ \text{m/s}) / (5.37\ \text{eV})$

$\lambda = 2.3 \times 10^{-7}\ \text{m}$

41. D is correct.

Elements with atomic numbers 84 and higher are radioactive because the strong nuclear force binding the nucleus cannot overcome the Coulomb repulsion from the high number of protons within the atom.

Thus, these nuclei are unstable and emit alpha radiation to decrease the number of protons within the nucleus.

42. D is correct.

Beta particles, like all forms of ionizing radiation, cannot be considered harmless.

43. D is correct.

The *half-life* calculation:

$A = A_0(½)^{t / h}$

where A_o = original amount, t = time elapse and h = half-life

$0.03 = (1)·(½)^{(t / 20{,}000\ \text{years})}$

$\ln (0.03) = (t / 20{,}000\ \text{years})·\ln (½)$

$t = 101{,}179$ years

44. B is correct.

Planck's constant quantizes the amount of energy that can be absorbed or emitted.

Therefore, it sets a discrete lowest amount of energy for energy transfer.

45. D is correct.

The *decay rate* of a radioactive isotope is constant and independent of temperature, pressure, or surface area.

46. B is correct.

Larger nuclei (atomic number above 83) tend to decay because the attractive force of the nucleons (strong nuclear force) has a limited range, and the nucleus is larger than this range.

Therefore, these nuclei tend to emit alpha particles to decrease the size of the nucleus.

Smaller nuclei are not large enough to encounter this problem, but some isotopes have an irregular ratio of neutrons to protons and become unstable.

^{14}Carbon has 8 neutrons and 6 protons, and its neutron to proton ratio is too large; it is unstable and radioactive.

47. C is correct.

Positron emission occurs during β^+ decay.

In β^+ decay, a proton converts to a neutron and emits a positron and electron neutrino.

The decay can be expressed as:

$$^{44}_{21}\text{Sc} \rightarrow \, ^{44}_{20}\text{Ca} + e^+ + \nu_e$$

48. C is correct.

A scintillation counter operates by detecting light flashes from the scintillator material.

When radiation strikes the scintillator crystal (often NaI), a light flash is emitted and detected by a photomultiplier tube, which then passes an electronic signal to audio or visual identification equipment.

49. B is correct.

When a Geiger counter clicks, it indicates that it has detected the radiation from one nucleus decaying.

The click could be from an alpha, beta, or gamma-ray source but cannot be determined without other information.

50. A is correct.

The *Pauli Exclusion Principle* states that no two electrons have the same quantum numbers in an atom.

Thus, every electron in an atom has a unique set of quantum numbers, and a particular set belongs to only one electron.

51. D is correct.

The atomic number indicates the number of protons within the nucleus of an atom.

52. C is correct.

Gamma rays are high-energy electromagnetic radiation rays and have no charge or mass.

Due to their high energy and speed, they have a high penetrating power.

53. B is correct.

Calculate mass defect:

m_1 = (2 protons)·(1.0072764669 amu) + (2 neutrons)·(1.0086649156 amu)

m_1 = 4.031882765 amu

Δm = 4.031882765 amu – 4.002602 amu

Δm = 0.029280765 amu

Convert to kg:

Δm = (0.029280765 amu / 1)·(1.6606 × 10^{-27} kg / 1 amu)

Δm = 4.86236 × 10^{-29} kg

Find binding energy:

$E = \Delta mc^2$

E = (4.86236 × 10^{-29} kg)·(3 × 10^8 m/s)^2

E = 4.38 × 10^{-12} J ≈ 4.4 × 10^{-12} J

54. D is correct.

Boron has an atomic number of five and thus has five protons and five electrons.

Two electrons fill the 1s orbital; then the $2s$ orbital is filled with two electrons.

Only one electron remains, and it is in the $2p$ orbital, leaving it partially filled.

The electron configuration of boron: $1s^22s^22p$

55. C is correct.

Einstein's theory of special relativity that expresses the equivalence of mass and energy.

$E = mc^2$

$m = E / c^2$

m = (3.85 × 10^{26} J) / (3 × 10^8 m/s)^2

m = 4.3 × 10^9 kg

56. D is correct.

When ionizing radiation interacts with the body, the high-energy particles or electromagnetic waves cause atoms within the tissue to ionize.

This process creates unstable ions and free radicals that are damaging and pose serious health risks.

57. B is correct.

When a gamma-ray is emitted, the atom must lose energy due to the conservation of energy.

Thus, the atom will have less energy than before.

58. A is correct.

The lower-case symbol "n" signifies a neutron.

An *atomic number* of zero indicates that there are no protons.

A *mass number* of one indicates that it contains a neutron.

59. C is correct.

The uncertainty principle states that the position and momentum of a particle cannot be simultaneously measured with precision.

Additionally, energy and time cannot be simultaneously known.

Mathematically this is stated as:

$$\Delta x \Delta p > \hbar / 2$$

$$\Delta E \Delta t > \hbar / 2$$

where x = position, p = momentum, $\hbar$ = reduced Planck's constant, E = energy, t = time

Notes for active learning

Particle Physics – Detailed Explanations

1. C is correct.

Leptons do not interact via the strong force, so they cannot comprise the hadrons.

Quarks do interact via the strong force and are the building blocks of hadrons.

Electrons are leptons, while protons and neutrons are hadrons (since they are made of quarks).

2. D is correct.

One defining characteristic of hadrons is that they interact by the strong nuclear force.

Electrons are leptons, not hadrons (because electrons do not interact by the strong force since they have no quarks), although neutrons and protons are hadrons.

Quarks, which are neither leptons nor hadrons, are the building blocks of hadrons.

3. B is correct.

Leptons interact by all forces except the strong nuclear force.

4. B is correct.

The positron (antielectron) is a lepton and is not made of quarks.

The neutron is a hadron.

The alpha particle is composed of two protons and two neutrons, which are hadrons made of quarks.

5. B is correct.

Hadrons have charge +1e, 0e, or −1e. They are combinations of two (mesons) or three (baryons) quarks.

The only possible way of combining two or three fractionally charged objects so that the net charge is +1e, 0e or −1e is if the charges of the objects are taken from sets (1/3e, −2/3e) or (−1/3e, +2/3e).

The former set represents the quarks, and the latter the antiquarks.

Note that it is not possible to ever see a quark by itself (i.e., not bound to other quarks/antiquarks).

Thus, any actual particle that can be observed will have an integer charge; the fractional charge for any particle that can exist by itself is never measured.

6. D is correct.

The *deuteron* is composed of two baryons, one proton and one neutron.

The *baryons* each have three quarks, so the total number of quarks in the deuteron is six.

7. D is correct.

The tritium nucleus, the triton, comprises three baryons, one proton, and two neutrons.

The baryons each have three quarks, so the total number of quarks in the triton is nine.

8. B is correct.

The more massive the particle, the more decay channels are available, resulting in a shorter lifetime and reducing the range of the exchange mediated force.

One can also think of this using the uncertainty principle.

The exchange particle is a "virtual particle," meaning it appears temporarily and disappears once the exchange is complete.

The uncertainty principle limits how long such a particle can exist: $\Delta E\ \Delta t > \hbar/2$.

Since the exchange particle's energy increases with greater mass, a heavier particle can only survive for a shorter time Δt.

A shorter lifetime means it cannot cover as much distance as a lighter particle, so the range for a heavier particle will be less.

9. A is correct.

The total mass-energy of the system is the rest mass of the electron and positron combined.

Since two photons are emitted, each photon will have energy equal to the mass-energy of one electron.

The energy of each photon is, therefore:

$$E = mc^2$$

where m is the mass of an electron.

The relationship between the energy of a photon and frequency and wavelength is:

$$E = hf = hc/\lambda$$

Combining these formulas:

$$\lambda = (hc) / (mc^2) = h / mc$$

Entering the values given:

$$\lambda = 2.42 \times 10^{-12}\ \text{m}$$

$$\lambda = 2.42\ \text{pm}$$

10. A is correct.

The proton has a charge of $+1e$.

Answer A is the only choice in which the sum of the quark charges is $+1e$.

A proton is made of two up quarks and one down quark.

11. C is correct.

The positron is quickly annihilated by an electron in the sample, producing two gamma rays, each having energy equal to the mass-energy of an electron or positron (the mass of a positron is the same as the mass of an electron).

The mass-energy of an electron (and thus a positron) is 0.511 MeV.

Note that *two* photons must be emitted because a final state with only one photon would violate the conservation of linear momentum.

To see this, imagine a reference frame where the center-of-mass of the electron-positron pair is stationary.

Thus, before the annihilation, the total momentum is zero (since the center-of-mass is not moving in this reference frame).

The final state of just one photon cannot have zero momentum.

12. C is correct.

To produce a proton-antiproton pair, an amount of energy is necessarily equal, at least to the combined mass of the proton and antiproton.

The proton and antiproton both have mass-energy of 938 MeV, so the minimum energy to produce a pair is twice this, 1876 MeV.

13. C is correct.

Each gamma-ray will have energy equal to half the mass of η.

The energy of each gamma is, therefore:

$$E = mc^2 / 2$$

where m is the mass of an η meson.

The relationship between the energy of a photon and frequency and wavelength is:

$$E = hf = hc/\lambda$$

Combining these:

$$\lambda = (2hc) / (mc^2)$$

Entering the values given for hc and rest energy mc^2 and converting units:

$$\lambda = (2 \times 1240 \text{ eV nm} \times 10^{-9} \text{ m/nm}) / (548 \times 10^6 \text{ eV})$$

$$\lambda = 4.5 \times 10^{-15} \text{ m}$$

14. D is correct.

The kinetic energy of a relativistic particle is the total energy less the rest energy.

$$K = E - mc^2 = \gamma mc^2 - mc^2 = (\gamma - 1)mc^2$$

where,

$$\gamma = 1 / \sqrt{(1 - \beta^2)} \text{ and } \beta = v / c$$

Rearranging:

$$\gamma - 1 = K / mc^2$$

$$\gamma = 1 + K / mc^2$$

$$1 / \sqrt{(1 - \beta^2)} = 1 + K / mc^2$$

Square both sides and invert:

$$1 - \beta^2 = 1 / (1 + K / mc^2)^2$$

$$\beta^2 = 1 - [1 / (1 + K / mc^2)^2]$$

$$\beta = \sqrt{\{1 - [1 / (1 + K / mc^2)^2]\}}$$

The mass of a proton is $mc^2 = 0.938$ GeV.

Using this and the given value of kinetic energy, $K = 2$ GeV:

$$\beta = 0.948$$

so,

$$v = 0.948c \approx 0.95c$$

15. D is correct.

The energy of a relativistic particle is:

$$E = \gamma mc^2 = (mc^2) / \sqrt{(1 - \beta^2)}$$

where $\beta = v / c$. Divide by mc^2 and square:

$$(E / mc^2)^2 = 1 / (1 - \beta^2)$$

Solve for β:

$$\beta = \sqrt{[1 - (mc^2 / E)^2]}$$

For an electron, $mc^2 = 0.511$ MeV.

Using this and the given value $E = 2$ MeV, gives:

$$\beta = 0.967$$

or

$$v = 0.97\ c$$

16. B is correct.

Using the definition of Boltzmann's constant, the relationship between temperature and energy is:

$kT = E$

$T = E / k$

where k is Boltzmann's constant, $k = 8.617 \times 10^{-5}$ eV/K.

With an energy of 50 GeV = 50×10^9 eV:

$T = (50 \times 10^9 \text{ eV}) / (8.617 \times 10^{-5} \text{ eV/K})$

$T = 5.8 \times 10^{14}$ K

The commonly quoted conversion factor is 1eV = 11600K.

17. D is correct.

Hubble's law relates the speed of a receding galaxy to its distance from Earth.

$d = v / H$

$d = 0.5\ (3.0 \times 10^8 \text{ m/s}) / (0.022 \text{ m/s/lightyear})$

$d = 6.8 \times 10^9$ lightyears

18. A is correct.

Hubble's law relates the speed of a receding galaxy to its distance from Earth.

$v = dH$

$v = (2.7 \times 10^9 \text{ lightyears})\cdot(0.022 \text{ m/s/lightyear})$

$v = 5.94 \times 10^7$ m/s

In terms of the speed of light $c = 3.00 \times 10^8$ m/s:

$v = 0.2\ c$

19. B is correct.

The minimum energy state of an electron and a positron is one in which neither has any kinetic energy.

The total energy of the system is then the rest mass-energy of the electron and positron.

Each has a rest energy of 0.511 MeV, so the total minimum energy is 1.022 MeV.

20. C is correct.

The creation of the meson would violate the principle of conservation of energy unless its lifetime were short enough that its mass-energy fell within the limit set by Heisenberg's uncertainty principle.

$$\Delta E \Delta t \geq h / 2\pi$$

Take the minimum uncertainty to be the rest mass of the meson, as the uncertainty has to be at least sufficient to create the meson.

$$\Delta t \geq h / 2\pi mc^2$$

The meson must decay before this amount of time has elapsed.

Since no particle can travel faster than the speed of light, the maximum range the meson can have is:

$$r = c\Delta t = hc / 2\pi mc^2$$

$$r = (1.24 \text{ eV } \mu\text{m}) / [2\pi \, (140 \text{ MeV})]$$

$$r = 1.4 \times 10^{-15} \text{ m}$$

21. C is correct.

The creation of the boson would violate the principle of conservation of energy unless its lifetime was short enough that its mass-energy fell within the limit set by Heisenberg's uncertainty principle.

$$\Delta E \Delta t \geq h / 2\pi$$

Take the minimum uncertainty to be the rest mass of the boson, as the uncertainty must be at least sufficient to create the boson.

$$\Delta t \geq h / 2\pi mc^2$$

For conservation of energy not to be violated, the boson must decay before this time interval has elapsed.

So, the interaction time is:

$$t = h / 2\pi mc^2$$

The rest energy of the meson in SI units is:

$$mc^2 = 91 \times 10^9 \text{ eV } (1.6 \times 10^{-19} \text{ J/eV})$$

$$mc^2 = 1.456 \times 10^{-8} \text{ J}$$

Therefore:

$$t = (6.63 \times 10^{-34} \text{J s}) / [2\pi \, (1.456 \times 10^{-8} \text{J})]$$

$$t = 7.2 \times 10^{-27} \text{s}$$

22. A is correct.

Since the muons are at rest (no kinetic energy), the total energy of the system is the sum of the rest mass energies of the muons, that is, twice the mass of a single muon.

The annihilation produces two photons of equal energy, so the energy of each equals the mass-energy of one muon.

The wavelength of the photon of energy E is:

$$\lambda = hc / E$$

$$\lambda = (1.24 \text{ eV } \mu\text{m}) / (106 \text{ MeV})$$

$$\lambda = 1.17 \times 10^{-14} \text{ m}$$

23. B is correct.

The creation of the boson would violate the principle of conservation of energy unless its lifetime were short enough that its mass-energy fell within the limit set by Heisenberg's uncertainty principle.

$$\Delta E \Delta t \geq h / 2\pi$$

Take the minimum uncertainty to be the rest mass of the boson, as the uncertainty must be at least sufficient to create the boson.

$$\Delta t \geq h / 2\pi mc^2$$

The meson must decay before this amount of time has elapsed.

The maximum range the boson can have is:

$$r = c\Delta t = hc / 2\pi mc^2$$

$$r = (1.24 \text{ eV } \mu\text{m}) / [2\pi \cdot (80 \text{ GeV})]$$

$$r = 2.5 \times 10^{-18} \text{ m}$$

Notes for active learning

Quantum Mechanics – Detailed Explanations

1. D is correct.

The energy of each incident photon is transferred to an electron, which must then overcome the material's work function to be ejected.

Therefore:

hc / λ = Work function

$\lambda = (6.626 \times 10^{-34}\ \text{J·s} \cdot 3.00 \times 10^{8}\ \text{m/s}) / (1.90\ \text{eV} \cdot 1.60 \times 10^{-19}\ \text{J/eV})$

$\lambda = 6.53 \times 10^{-7}\ \text{m}$

$\lambda = 653\ \text{nm}$

2. B is correct.

The power varies with the number of photons per second and the energy of the photons; therefore, I is incorrect.

The energy of the photons varies with the frequency of the light; therefore, III is incorrect.

The intensity of a laser beam depends on the number of photons and the energy of each photon.

The energy of the photons varies linearly with the frequency.

If the photon frequency doubles, the energy doubles.

Since the number of photons is unchanged, the intensity doubles.

3. C is correct.

The energy of each incident photon is transferred to an electron, which must then overcome the material's work function to be ejected.

Therefore, the maximum kinetic energy remaining in any electron is:

E = Energy in 240 nm photon – work function

2.5 eV = hc / λ – work function

Work function = $(6.626 \times 10^{-34}\ \text{J·s} \cdot 3.00 \times 10^{8}\ \text{m/s}) / (240 \times 10^{-9}\ \text{m}) - 2.58\ \text{eV}$

Work function = $(8.28 \times 10^{-19}\ \text{J}) / (1.60 \times 10^{-19}\ \text{J/eV}) - 2.58\ \text{eV}$

Work function = 2.6 eV

4. C is correct.

The minimum energy of the electron-positron pair is its rest mass ($E = 2m_{electron}c^2$).

The photon that creates the pair must have more than the minimum energy.

The energy of the photon (hv) must therefore be:

$$hv > 2m_{electron}c^2$$

$$v > 2m_{electron}c^2 / h$$

$$v > 2\cdot(9.11 \times 10^{-31}\ \text{kg})\cdot(3.00 \times 10^8\ \text{m/s})^2 / (6.626 \times 10^{-34}\ \text{J}\cdot\text{s})$$

$$v > 2.47 \times 10^{20}\ \text{Hz}$$

5. D is correct.

The energy of each incident photon is transferred to an electron, which must then overcome the material's work function to be ejected.

Therefore:

$$hc / \lambda > \text{Work function}$$

$$\lambda < (6.626 \times 10^{-34}\ \text{J}\cdot\text{s} \cdot 3.00 \times 10^8\ \text{m/s}) / (2.20\ \text{eV} \cdot 1.60 \times 10^{-19}\ \text{J/eV})$$

$$\lambda < 564\ \text{nm}$$

6. C is correct.

Anderson used a cloud chamber to observe the effects of cosmic rays.

A cloud chamber is a device filled with saturated water or alcohol vapor and held in a magnetic field.

As the cosmic rays interact with the molecules in the vapor, they form high-energy charged particles.

These high-energy charged particles move out from their point of creation in curved paths because of the magnetic field.

Their path is visualized as they condense droplets from the saturated vapor.

Anderson observed an event in which a pair of charged particles was created at the same point, moving out in opposite directions and with equal but opposite curvature.

This corresponded to the earlier prediction by Dirac of a positively charged electron (a positron) based upon an extra solution to the equations of a relativistic invariant form of Schrodinger's Equation.

The electron and positron pair (pair production) was produced in a collision of a cosmic ray photon with a heavy nucleus in the vapor.

7. C is correct.

$E = hv$

$E = (6.626 \times 10^{-34}\ J{\cdot}s){\cdot}(6.43 \times 10^{14}\ Hz) / (1.60 \times 10^{-19}\ J/eV)$

$E = 2.66\ eV$

8. B is correct.

The kinetic energy of the emitted electrons (KE) is the energy of the incident photon minus, at least, the work function of the photocathode surface.

Therefore:

$KE = 3.4\ eV - 2.4\ eV$

$KE = (1.0\ eV){\cdot}(1.60 \times 10^{-19}\ J / eV)$

$KE = 1.60 \times 10^{-19}\ J$

9. D is correct.

The energy of each incident photon is transferred to an electron, which must then overcome the material's work function to be ejected.

Therefore, to eject an electron:

$hv = hc / \lambda >$ Work function

$\lambda < hc /$ (Work function)

$\lambda < (6.626 \times 10^{-34}\ J{\cdot}s){\cdot}(3.00 \times 10^{8}\ m/s) / (2.9\ eV \cdot 1.60 \times 10^{-19}\ J/eV)$

$\lambda < 428 \times 10^{-9}\ m$

The illumination range of 400 nm–700 nm that does not satisfy this requirement is:

$\lambda > 428\ nm$

10. A is correct.

Photons move at the speed of light because they are light.

Since the momentum of photon A is twice as great as the momentum of photon B, its energy is twice as great, and therefore its wavelength is half as great.

11. A is correct.

The Balmer formula for Hydrogen is:

$$1 / \lambda = (1 / 91.2 \text{ nm})\cdot(1 / m^2 - 1 / n^2)$$

The energy difference of the n = 20 and n = 7 state corresponds to a photon of wavelength:

$$1 / \lambda = (1 / 91.2 \text{ nm})\cdot(1 / 7^2 - 1 / 20^2)$$

$$\lambda = (91.2 \text{ nm}) / (1 / 7^2 - 1 / 20^2)$$

$$\lambda = 5092 \text{ nm}$$

The energy of a photon with wavelength λ is:

$$E = hc / \lambda$$

$$E = (6.626 \times 10^{-34} \text{ J·s})\cdot(3.00 \times 10^8 \text{ m/s}) / (5092 \times 10^{-9} \text{ m})$$

$$E = (3.93 \times 10^{-20} \text{ J}) / (1.60 \times 10^{-19} \text{ J/eV})$$

$$E = 0.244 \text{ eV}$$

12. D is correct.

The *de Broglie wavelength* is given by:

$$\lambda = h / p$$

When the energy ($E = p^2 / 2m$ for a non-relativistic proton) is doubled, the momentum is increased by √2.

Therefore, its de Broglie wavelength decreases by √2.

13. D is correct.

Trivially, it is known that the incoming photon must lose energy to the electron in the scattering, and therefore its wavelength must increase.

There is only one answer with a longer wavelength.

Using the Compton equation at 120°:

$$\Delta\lambda = h / m_0c \, (1 - \cos \theta)$$

$$\Delta\lambda = \lambda_{Compton} \, (1 - \cos \theta)$$

where $\lambda_{Compton} = 2.43 \times 10^{-12}$ m

$$\Delta\lambda = 0.00243 \text{ nm } (1.5)$$

$$\Delta\lambda = 0.00365 \text{ nm}$$

$$\lambda = 0.591 \text{ nm} + 0.00365 \text{ nm}$$

$$\lambda = 0.595 \text{ nm}$$

14. C is correct.

The *brightness* of a beam of light is linearly proportional to the energy of the photons in the beam and the number of photons in the beam.

If the *color* of the light beam, which is dependent on its frequency distribution or energy distribution, is unchanged, then the frequency and energy distribution of the light beam must be unchanged.

15. B is correct.

The Balmer formula for Hydrogen is:

$$1 / \lambda = (1 / 91.2 \text{ nm})(1 / m^2 - 1 / n^2)$$

$$1 / \lambda = (1 / 91.2 \text{ nm})(1 / 4^2 - 1 / 9^2)$$

$$1 / \lambda = (1 / 91.2 \text{ nm})(1 / m^2 - 1 / n^2)$$

$$1 / \lambda = (1 / 1818 \text{ nm})$$

The frequency of light is given by:

$$v = c / \lambda$$

$$v = (3.00 \times 10^8 \text{ m/s}) / (1818 \times 10^{-9} \text{ m})$$

$$v = 1.65 \times 10^{14} /\text{s} = 1.65 \times 10^{14} \text{ Hz}$$

16. D is correct.

The *Balmer formula for Hydrogen* for the emission of radiation from the nth level down to the mth (i.e., $n > m$):

$$1 / \lambda = (1 / 91.2 \text{ nm}) \cdot (1 / m^2 - 1 / n^2)$$

Therefore:

$$(1 / m^2 - 1 / n^2) = 91.2 \text{ nm} / 377 \text{ nm}$$

$$(1 / m^2 - 1 / n^2) = 0.2419$$

$$1 / n^2 = 1 / m^2 - 0.2419$$

For n to be real, it must be m = 1 or m = 2.

If m = 1:

$$1 / n^2 = 1 - 0.2419 = 0.7581$$

n is not an integer (i.e., the photon has less energy than the emission photon expected in a transition from n = 2 to m = 1).

Therefore, m must be 2 (i.e., the emission is from the n level down to the m = 2 level) and:

$$1 / n^2 = 1 / 4 - 0.2419 = 0.0081$$

$$n = 11$$

17. C is correct.

Wein's displacement law describes the wavelength of maximum emission of radiation of a black body at temperature T:

$$\lambda_{max} \cdot T = \text{constant} = 0.00290\ \text{m·K}$$

At T = 5000K:

$$\lambda_{max} = 0.00290\ \text{m·K} / 5000\ \text{K}$$

$$\lambda_{max} = 580\ \text{nm}$$

18. D is correct.

Dirac factored the Schrödinger equation into a simpler equation with two solutions; one solution described the electron, and the other described an identical particle with an opposite electric charge.

Four years later, such a "positively charged" electron was found in cloud chamber pictures of cosmic rays.

19. B is correct.

The electron absorbs the total energy of the photon and loses the work function energy as it escapes from the material.

Therefore, it has a maximum kinetic energy of:

$$\text{Max KE} = \text{E} - \text{Work function}$$

$$\text{Max KE} = 3.4\ \text{eV} - 2.4\ \text{eV} = 1.0\ \text{eV} \cdot (1.60 \times 10^{-19}\ \text{J/eV})$$

$$\text{Max KE} = 1.60 \times 10^{-19}\ \text{J}$$

20. D is correct.

Consider the problem from the center of the mass frame, a frame moving in the same direction as the incident photon. In this frame, the electron starts moving with the speed of the center of the mass frame and in the opposite direction of the center of the mass frame.

In the center of the mass frame, the electron is scattered with the same speed into one direction, and the photon is scattered in the opposite direction, the scattering angle.

Transforming back into the laboratory frame, the electron has a final velocity equal to the vector sum of its scattered velocity in the center of the mass frame and the velocity of the center of mass.

This sum is greatest when the scattering angle of the electron is in the same direction as the velocity of the center of mass frame, the direction of the incident photon. The photon is then scattered in the opposite direction of the center of the mass frame at a scattering angle of 180°.

If the velocity of the electron is maximal at that scattering angle, then the energy of the photon is minimal at that scattering angle.

Since wavelength is inversely proportional to energy, the maximal change in wavelength occurs when the photon is scattered at 180°.

21. C is correct.

This is a trick question that has nothing to do with the photocathode or the work function.

If the radiation has energy 3.5 eV, then its wavelength is given by:

$E = hv = hc / \lambda = 3.5$ eV

$\lambda = hc / (3.5\ \text{eV})$

$\lambda = (6.626 \times 10^{-34}\ \text{J·s})·(3.00 \times 10^{8}\ \text{m/s}) / (3.5\ \text{eV})$

$\lambda = (6.626 \times 10^{-34}\ \text{J·s})·(3.00 \times 10^{8}\ \text{m/s}) / (3.5\ \text{eV})$

$\lambda = (5.679 \times 10^{-26}\ \text{J·m /eV}) / (1.6 \times 10^{-19}\ \text{J/eV})$

$\lambda = 355$ nm

22. A is correct.

The energy of a photon is given by:

$E = hv$

$E = hc / \lambda$

Therefore, if the wavelength is doubled, the energy is halved.

23. D is correct.

The energy of each incident photon is transferred to an electron, which must then overcome the material's work function to be ejected.

Therefore:

$hv >$ Work function

$v > (2.8\ \text{eV} \cdot 1.60 \times 10^{-19}\ \text{J/eV}) / (6.626 \times 10^{-34}\ \text{J·s})$

$v > 6.8 \times 10^{-14}$ /s

24. B is correct.

The de Broglie wavelength of a matter wave is:

$\lambda = h / p$

$\lambda = h / mv$

$v = h / (m\lambda)$

$v = (6.626 \times 10^{-34}\ \text{J·s}) / [(9.11 \times 10^{-31}\ \text{kg})·(380 \times 10^{-9}\ \text{m})]$

$v = 1.91 \times 10^{3}$ m/s

25. B is correct.

Using the Compton Equation:

$$\Delta\lambda = h / m_0c\ (1 - \cos\theta)$$

$$\Delta\lambda = \lambda_{Compton}\ (1 - \cos\theta)$$

where $\lambda_{Compton} = 2.43 \times 10^{-12}$ m

For $\theta = 90°$:

$$\Delta\lambda = 2.43 \times 10^{-12}\ \text{m}$$

and

$$\lambda_{scattered} = \lambda_{incident} + \Delta\lambda$$

$$\lambda_{scattered} = (1.50 \times 10^{-10}\ \text{m}) + (2.43 \times 10^{-12}\ \text{m})$$

$$\lambda_{scattered} = 1.5243 \times 10^{-10}\ \text{m}$$

26. B is correct.

As the *intensity of light* increases, the photon flux increases but not the energy of the photons.

The *kinetic energy* of the ejected electrons depends solely on the energy of the incident photons and the work function of the metal.

Since the energy of the incident photons is unchanged, the kinetic energy of the ejected electrons does not change.

However, the *probability of an electron being ejected*, and therefore the number of electrons ejected per second, depends on the flux of incident photons. It increases as the intensity of the light increases.

The electron is ejected at the same instance as the light is absorbed, independent of the intensity of the light.

Therefore, the time lag does not change.

Note: the time lag between the illumination of the surface (not the absorption of light) and the ejection of the first electron depends on the probability of ejection, which depends on the intensity of the incident light.

27. D is correct.

The *de Broglie wavelength* of a matter wave is:

$$\lambda = h / p$$

$$\lambda = h / (mv)$$

The *energy of a photon* with this wavelength is:

$$E = hv$$

$$E = hc / \lambda$$

continued…

Substituting in for λ:

$$E = hc / [h / (mv)]$$

$$E = mcv$$

$$E = (1.67 \times 10^{-27}\ \text{kg})\cdot(3.00 \times 10^{8}\ \text{m/s})\cdot(7.2 \times 10^{4}\ \text{m/s})$$

$$E = 36.1 \times 10^{-15}\ \text{J}$$

$$E = (36.1 \times 10^{-15}\ \text{J}) / (1.60 \times 10^{-19}\ \text{J/eV})$$

$$E = 225 \times 10^{3}\ \text{eV}$$

28. B is correct.

The Compton effect measures the change in energy as an x-ray scatters off an electron.

The total momentum and total energy of the x-ray and the electron, initially at rest, must be conserved.

As the scattering angle of the x-ray increases monotonically, its change in momentum increases, and therefore the momentum imparted to the electron must increase.

If the momentum of the electron increases, then its energy must increase, and the energy of the X-ray decreases.

If the energy of the X-ray decreases, then the frequency, which is proportional to its energy, must decrease.

29. A is correct.

The Rydberg formula for Hydrogen for the emission of radiation from the nth level down to m^{th} (i.e., $n > m$) is:

$$1 / \lambda = (1 / 91.2\ \text{nm})\cdot(1 / m^2 - 1 / n^2)$$

If $n = 16$ (since the first spectral line is from $n = 2$) and $m = 1$, then:

$$1 / \lambda = (1 / 91.2\ \text{nm})\cdot(1 / 1^2 - 1 / 16^2)$$

$$\lambda = 91.2\ \text{nm} / (1 - 0.004)$$

$$\lambda = 91.6\ \text{nm}$$

30. D is correct.

If the frequency of light in a laser beam is doubled, the energy of each photon in that laser beam doubles since each photon has energy $E = hv$.

The wavelength of each photon is divided by 2, since $\lambda = c / v$.

The intensity of the laser beam is the power per unit area.

The power is the energy delivered per unit time.

The energy delivered is proportional to the energy of each photon times the number of photons. If the energy of each photon is doubled and the number of photons remains unchanged (assuming the area of the laser beam is unchanged), then the intensity doubles.

Therefore, I and III are correct.

31. B is correct.

Order diffraction occurs when the incident and scattered beams hit the crystal at the same angle (i.e., in this case, the crystal planes are oriented at 58° / 2 to the normal).

The neutron matter waves add coherently (i.e., constructive or destructive interference) if the extra distance that the neutrons must travel as they scatter off the next plane of the crystal is equal to an integer number of de Broglie wavelengths.

At 58°, the extra distance is:

$$2 \cdot 159.0 \text{ pm} \cdot \cos(58° / 2) = 278 \text{ pm}$$

The de Broglie wavelength of a matter wave is:

$$\lambda = h / p$$

Therefore:

$$p = h / \lambda$$

and

$$E = p^2 / 2m = (6.626 \times 10^{-34} \text{ J·s} / 278 \times 10^{-12} \text{ m})^2 / (2 \cdot 1.67 \times 10^{-27} \text{ kg})$$

$$E = (1.70 \times 10^{-21} \text{ J}) / (1.6 \times 10^{-19} \text{ J/eV})$$

$$E = 0.0106 \text{ eV}$$

32. A is correct.

The energy of each incident photon is transferred to an electron, and the electron's energy goes into overcoming the photocathode's 2.5 eV work function.

The remaining energy is then stopped by the stopping potential.

The greatest amount of energy an electron can have comes from a photon with a wavelength of 360 nm.

The remaining energy is then:

$$\text{remaining energy} = hv - 2.5 \text{ eV}$$

$$\text{remaining energy} = hc / \lambda - 2.5 \text{ eV}$$

$$\text{remaining energy} = (6.626 \times 10^{-34} \text{ J·s})·(3.00 \times 10^8 \text{ m/s}) / (360 \times 10^{-9} \text{ m}) - 2.5\text{eV}$$

$$\text{remaining energy} = [(5.52 \times 10^{-19} \text{ J}) / (1.6 \times 10^{-19} \text{ J/eV})] - 2.5 \text{ eV}$$

$$\text{remaining energy} = 3.45 \text{ eV} - 2.5 \text{ eV}$$

$$\text{remaining energy} = 0.95 \text{ eV}$$

The electron has a charge of –1 e.

Therefore, this energy can be stopped by a voltage of 0.95 V.

33. D is correct.

The de Broglie wavelength is given by:

$\lambda = h / p$

$\lambda = (6.626 \times 10^{-34}\ \text{J·s}) / (1.95 \times 10^{-27}\ \text{kg·m/s})$

$\lambda = 340\ \text{nm}$

34. C is correct.

The Balmer formula for Hydrogen for the emission of radiation from the nth level down to the mth (i.e., $n > m$):

$1 / \lambda = (1 / 91.2\ \text{nm})·(1 / m^2 - 1 / n^2)$

If $n = 9$ and $m = 6$, then:

$1 / \lambda = (1 / 91.2\ \text{nm})·(1 / 6^2 - 1 / 9^2)$

$1 / \lambda = (1 / 91.2\ \text{nm})·(0.01543)$

Thus:

$v = c / \lambda$

$v = (3.00 \times 10^8\ \text{m/s})·(1 / 91.2\ \text{nm})·(0.01543)$

$v = 5.08 \times 10^{13}\ /\text{s}$

$v = 5.08 \times 10^{13}\ \text{Hz}$

35. A is correct.

$E = hv$

$E = (6.626 \times 10^{-34}\ \text{J·s})·(110\ \text{GHz})$

$E = (6.626 \times 10^{-34}\ \text{J·s})·(110 \times 10^9 / \text{s})$

$E = 7.29 \times 10^{-23}\ \text{J}$

36. C is correct.

The visible spectrum ranges from 400 nm to 700 nm.

The Balmer formula for Hydrogen for the emission of radiation from the nth level down to the mth (i.e., $n > m$):

$1 / \lambda = (1 / 91.2\ \text{nm})·(1 / m^2 - 1 / n^2)$

Consider emission from any level down to the $m = 1$ level.

The lowest energy is from the $n = 2$ level, and its wavelength is:

$91.2\ \text{nm} · (4 / 3) = 121\ \text{nm}$, which is not visible.

Therefore, no visible lines are radiating down to the $m = 1$ level.

continued…

Consider emission from any level down to the m = 3 level.

The highest energy comes from $n = \infty$, and its wavelength is:

$$91.2 \text{ nm} \cdot 9 = 820 \text{ nm}$$, which is not visible.

Therefore, no visible lines are radiating down to the m = 3 level, nor are any visible lines radiating down to any m level higher than 3.

Consider radiation from various levels down to the m = 2 level.

From n = 3, the wavelength is:

$$91.2 \text{ nm} / (1 / 4 - 1 / 9) = 91.2 \text{ nm} \cdot 7.2 = 656.6 \text{ nm}$$

From n = 4, the wavelength is:

$$91.2 \text{ nm} / (1 / 4 - 1 / 16) = 91.2 \text{ nm} \cdot 5.333 = 486.4 \text{ nm}$$

From n = 5, the wavelength is:

$$91.2 \text{ nm} / (1 / 4 - 1 / 25) = 91.2 \text{ nm} \cdot 4.762 = 434.3 \text{ nm}$$

From n = 6, the wavelength is:

$$91.2 \text{ nm} / (1 / 4 - 1 / 36) = 91.2 \text{ nm} \cdot 4.5 = 410.4 \text{ nm}$$

From n = 7, the wavelength is:

$$91.2 \text{ nm} / (1 / 4 - 1 / 49) = 91.2 \text{ nm} \cdot 4.355 = 397.2 \text{ nm}$$, which is not visible.

Therefore, there are 4 visible lines:

m = 2 to n = 3   656.6 nm

m = 2 to n = 4   486.4 nm

m = 2 to n = 5   434.3 nm

m = 2 to n = 6   410.4 nm

37. B is correct.

$$E = hv$$

$$E = hc / \lambda$$

$$\lambda = hc / E$$

$$\lambda = (6.626 \times 10^{-34} \text{ J·s})\cdot(3.00 \times 10^{8} \text{ m/s}) / (4.20 \text{ eV})$$

$$\lambda = (4.73 \times 10^{-26} \text{ J·m/eV}) / (1.60 \times 10^{-19} \text{ J/eV})$$

$$\lambda = 2.96 \times 10^{-7} \text{ m}$$

$$\lambda = 296 \text{ nm}$$

38. C is correct.

The energy of each incident photon is transferred to an electron, which must then overcome the material's work function to be ejected.

Therefore, if a wavelength of light is just able to eject an electron, then:

$hv = hc / \lambda$ = Work function

$(6.626 \times 10^{-34}\ \text{J·s})\cdot(3.00 \times 10^{8}\ \text{m/s}) / (500 \times 10^{-9}\ \text{m})$ = Work function

$(3.98 \times 10^{-19}\ \text{J}) / (1.6 \times 10^{-19}\ \text{J/eV})$ = Work function

2.48 eV = Work function

39. C is correct.

In the Bohr theory, there is a fixed number of de Broglie wavelengths of the electron in orbit.

This number is the principal quantum number:

$n = 2\pi r / \lambda$

The de Broglie wavelength is given by:

$\lambda = h / p$

Therefore:

$n = (2\pi / h)rp$

or:

$n^2 = (2\pi / h)^2 r^2 p^2$

The attractive force of the proton keeps the electron in a circular orbit.

The attractive force is proportional to $1 / r^2$, and if that force keeps the electron in orbit, $1 / r^2$ must be proportional to v^2 / r.

Therefore,

v^2 (or p^2) is proportional to $1 / r$

Since r^2p^2 is proportional to n^2, and p^2 is proportional to $1 / r$, then r (i.e., $r^2 \cdot 1 / r$) is proportional to n^2.

40. B is correct.

Heisenberg's Uncertainty Principle states that:

$\Delta p \Delta x \geq h / 2\pi$

Therefore:

$m \Delta v \Delta x \geq h / 2\pi$

$\Delta v \geq (6.626 \times 10^{-34}\ \text{J·s}) / (2\pi \cdot 0.053\ \text{nm} \times 1.67 \times 10^{-27}\ \text{kg})$

$\Delta v \geq 1.19 \times 10^{3}\ \text{m/s}$

41. C is correct.

The Balmer formula for Hydrogen for the emission of radiation from the nth level down to the mth (i.e., n > m):

$1 / \lambda = (1 / 91.2\ \text{nm})\cdot(1 / m^2 - 1 / n^2)$

$1 / \lambda = (1 / 91.2\ \text{nm})\cdot(1 / 9^2 - 1 / 11^2)$

$\lambda = (91.2\ \text{nm}) / (0.00408)$

$\lambda = 22{,}300\ \text{nm}$

42. D is correct.

Heisenberg's Uncertainty Principle states that:

$\Delta E \Delta t \geq h/2\pi$

Therefore:

$\Delta t \geq (h / 2\pi) / \Delta E$

$\Delta t \geq (6.626 \times 10^{-34}\ \text{J}\cdot\text{s} / 2\pi) / (10^{-18}\ \text{J})$

$\Delta t \geq 1.05 \times 10^{-16}\ \text{s}$

43. C is correct.

The energy of each incident photon is transferred to an electron, which must then overcome the material's work function to be ejected.

Therefore, if a wavelength of light can eject an electron with energy 2.58 eV, then:

$hv = hc / \lambda = \text{Work function} + 2.58\ \text{eV}$

$(6.626 \times 10^{-34}\ \text{J}\cdot\text{s})\cdot(3.00 \times 10^8\ \text{m/s}) / (240 \times 10^{-9}\ \text{m}) = \text{Work function} + 2.58\ \text{eV}$

$\text{Work function} = [(8.28 \times 10^{-19}\ \text{J}) / (1.6 \times 10^{-19}\ \text{J/eV})] - 2.58\text{eV}$

$\text{Work function} = 2.60\ \text{eV}$

44. C is correct.

The de Broglie wavelength of a matter wave is:

$\lambda = h / p$

$\lambda = h / (mv)$

$\lambda = (6.626 \times 10^{-34}\ \text{J}\cdot\text{s}) / [(1.30\ \text{kg})\cdot(28.10\ \text{m/s})]$

$\lambda = (6.626 \times 10^{-34}\ \text{J}\cdot\text{s}) / [(1.30\ \text{kg})\cdot(28.10\ \text{m/s})]$

$\lambda = 1.81 \times 10^{-35}\ \text{m}$

45. D is correct.

By doubling the frequency of the light, the energy of each photon doubles, but not the number of photons (hence B is wrong).

Although doubling the energy of each photon would cause more electrons to be ejected, it does not necessarily, double that number since the number is not linearly proportional to the incident energy (hence A is not always true).

The kinetic energy of the ejected electrons would more than double since the initial kinetic energy is less than the energy of the incident photons - the work function of the surface reduced it (hence C is wrong).

The kinetic energy would increase by at least 2 but not necessarily by 4 (hence D is wrong).

46. D is correct.

The Balmer formula for Hydrogen is:

$$1 / \lambda = (1 / 91.2 \text{ nm})\cdot(1 / m^2 - 1 / n^2)$$

From the n = 3 level, the transition to the n = 2 level has the lowest energy and, therefore the longest wavelength.

The Balmer formula for n = 3, m = 2 is:

$$1 / \lambda = (1 / 91.2 \text{ nm})\cdot(1 / 2^2 - 1 / 3^2)$$

$$1 / \lambda = (1 / 91.2 \text{ nm})\cdot(5 / 36)$$

$$\lambda = 656 \text{ nm}$$

47. C is correct.

Photons are light, and the speed of light is invariant. Therefore, I is wrong.

The energy of the photons in a beam of light determines the color of the light, and if the color is unchanged, then the average energy of the photons is unchanged, making II wrong.

Changing the brightness of a beam of light increases the number of photons in the beam of light.

48. B is correct.

Using the Compton Equation:

$$\Delta\lambda = h / m_0c\,(1 - \cos\theta)$$

$$\Delta\lambda = \lambda_{Compton}\,(1 - \cos\theta)$$

where $\lambda_{Compton} = 2.43 \times 10^{-12}$ m

At $\theta = 180°$, $\Delta\lambda = \lambda$.

Thus: $\lambda = 2\lambda_{Compton}$

$$\lambda = 4.86 \times 10^{-12} \text{ m}$$

49. D is correct.

The energy of the photons in the beam depends linearly on the frequency of the beam.

Since the frequency of beam B is twice the frequency of beam A, the photons in beam B have twice the energy as the photons in beam A.

The intensity depends on the energy of the photons and the number of photons per second carried by the beam.

Nothing is known about the intensity or the number of photons per second.

50. A is correct.

The Balmer formula for Hydrogen is:

$$1 / \lambda = (1 / 91.2 \text{ nm})\cdot(1 / m^2 - 1 / n^2)$$

From the n = 3 level, the transition to the n = 1 level has the greatest energy; therefore, the shortest wavelength.

The Balmer formula for n = 3, m = 1 is:

$$1 / \lambda = (1 / 91.2 \text{ nm})\cdot(1 / 1^2 - 1 / 3^2)$$

$$1 / \lambda = (1 / 91.2 \text{ nm})\cdot(8 / 9)$$

$$\lambda = 102.6 \text{ nm}$$

51. D is correct.

The uncertainty principle states that the uncertainty in position and momentum must be at least:

$$\Delta x \Delta p \approx h / 2\pi$$

$$\Delta p \approx h / (2\pi \Delta x)$$

If: $\Delta x = 5.0 \times 10^{-15}$ m

Then:

$$\Delta p \approx 6.626 \times 10^{-34} \text{ J·s} / (2\pi \cdot 5.0 \times 10^{-15} \text{ m})$$

$$\Delta p \approx 2.11 \times 10^{-20} \text{ kg·m/s}$$

The uncertainty in the (non-relativistic) kinetic energy of the proton is then:

$$\Delta KE \approx (\Delta p)^2 / (2m)$$

$$\Delta KE \approx [h / (2\pi \Delta x)]^2 / (2m)$$

$$\Delta KE \approx [(6.626 \times 10^{-34} \text{ J·s}) / (2\pi \cdot 5.0 \times 10^{-15} \text{ m})]^2 / (2 \cdot 1.67 \times 10^{-27} \text{ kg})$$

$$\Delta KE \approx (1.33 \times 10^{-13} \text{ J})$$

Converting to MeV:

$$\Delta KE \approx (1.33 \times 10^{-13} \text{ J}) / (1.6 \times 10^{-13} \text{ J/MeV})$$

$$\Delta KE \approx (0.83 \text{ MeV})$$

52. B is correct.

The electron absorbs the full energy of the photon, losing 2.4 eV as it escapes from the photocathode and an additional 1.1 eV of kinetic energy after it escapes from the photocathode.

This allows it to be stopped only with a potential exceeding 1.1 volts.

It therefore absorbs 2.4 eV + 1.1 eV = 3.5 eV from the incident photon.

A photon with energy 3.5 eV has a wavelength given by:

$$hv = hc / \lambda = E$$

$$\lambda = hc / E = (6.626 \times 10^{-34}\ J{\cdot}s){\cdot}(3.00 \times 10^{8}\ m/s) / (3.5\ eV \cdot 1.60 \times 10^{-19}\ J/eV)$$

$$\lambda = 355\ nm$$

53. B is correct.

The *Compton effect* measures the reduction in wavelength of an x-ray beam as it scatters off a sample.

The reduction in wavelength can be explained if the incident x-ray beam is considered as being made up of individual particles, each with momentum and energy, and electrons scatter those particles in the sample.

Hence, the Compton effect demonstrates the particle nature of electromagnetic radiation.

While the energy content and momenta of the individual x-rays are included in the scattering calculation, they are not directly demonstrated in the Compton effect.

54. B is correct.

The de Broglie wavelength of a matter wave is:

$$\lambda = h / p$$

$$\lambda = h / (m\gamma v)$$

(Since $v / c < 0.1$, this will equal about 1.005 and can be ignored)

$$\lambda = (6.626 \times 10^{-34}\ J{\cdot}s) / [(9.11 \times 10^{-31}\ kg){\cdot}(2.5 \times 10^{7}\ m/s)]$$

$$\lambda = 29 \times 10^{-12}\ m$$

$$\lambda = 29\ pm$$

55. D is correct.

The energy uncertainty is given by the uncertainty principle:

$$\Delta E \Delta t \geq h / 2\pi$$

$$\Delta E \geq (6.626 \times 10^{-34}\ J{\cdot}s / 2\pi) / (30 \times 10^{-12}\ s)$$

$$\Delta E \geq (3.5 \times 10^{-24}\ J)$$

Converting to eV (1.6×10^{-19} J/eV):

$$\Delta E \geq 2.2 \times 10^{-5}\ eV$$

Notes for active learning

Special Relativity – Detailed Explanations

1. A is correct.

Alpha sees Beta moving away at +0.7*c*, and Beta sees Gamma moving away at +0.7*c*.

To find the speed that Alpha sees Gamma moving at, apply the relativistic velocity addition expression:

$$\beta = (\beta_1 + \beta_2) / (1 + \beta_1\beta_2)$$

$$\beta = (0.7 + 0.7) / [1 + (0.7)\cdot(0.70)]$$

$$\beta = 0.94$$

$$v = 0.94c$$

2. D is correct.

The train is not moving in the vertical direction.

Therefore, there is no Lorentz Contraction of lengths in that direction.

3. D is correct.

As measured in the frame of Earth, the total distance is:

$$d = 2\cdot(4.367 \text{ years})\cdot c$$

The time is therefore:

$$t = d / v$$

$$t = (2 \cdot 4.367 \text{ years} \cdot c) / (0.5c)$$

$$t = 17.468 \text{ years}$$

4. D is correct.

From the point of view of an observer in the lab, the muon's time appears to be slowed by relativistic time dilation.

The lifetime for the muon in a frame in which it is at rest ($\Delta x = 0$) is $\Delta t = 2.2\mu s$.

The Lorentz transformation to the frame in which the muon is moving is:

$$\Delta t' = \gamma (\Delta t - v\,\Delta x / c^2)$$

where $\gamma = 1 / \sqrt{(1 - \beta^2)}$ and $\beta = v / c$

$$\Delta t' = \gamma\,\Delta x$$

$$\gamma = 1 / \sqrt{(1 - 0.99^2)}$$

$$\gamma = 7.09$$

$$\Delta t' = 15.6\ \mu s$$

5. D is correct.

The speed of the scout, as measured by the asteroid, is the relativistic addition of the speed of the spaceship as measured from the asteroid, plus the speed of the scout as measured from the spaceship.

$$\beta = (\beta_1 + \beta_2) / (1 + \beta_1\beta_2)$$

$$\beta = (0.6 + 0.4) / [1 + (0.6)\cdot(0.4)]$$

$$\beta = 0.81$$

$$v = 0.81c$$

6. D is correct.

The angle depends on the ratio of the elevation of the bed to the length of the bed along the floor.

The elevation of the bed does not appear different because the motion of the spaceship is zero in that direction.

The length along the floor appears shorter due to Lorentz Contraction when viewed from the observer's frame.

Therefore, the ratio appears to be larger, and the angle appears to have increased.

7. A is correct.

In the frame of the spaceship, the separation of the two events, Δx, is zero, and the time is dilated according to the Lorentz transformation:

$$\Delta t' = \gamma\ \Delta t$$

$$\gamma = 1 / \sqrt{(1 - \beta^2)}$$

$$\beta = v / c$$

If $v = 0.9640c$:

$$\gamma = 1 / \sqrt{(1 - 0.964^2)}$$

$$\gamma = 3.76$$

$$\Delta t = (10.5 \text{ years}) / (3.76)$$

$$\Delta t = 2.79 \text{ years}$$

8. C is correct.

Conservation of energy requires:

$$E_p + E_n = E_{\text{final}}$$

In the center of the mass frame, the proton and neutron start with insignificant kinetic energy, so their initial energies are approximately their rest energies.

The final energy is the rest energy of the deuterium plus the released energy:

$$m_pc^2 + m_nc^2 = m_dc^2 + E_{\text{released}}$$

continued…

So:

$$m_d c^2 = m_p c^2 + m_n c^2 - E_{\text{released}}$$

The proton rest mass energy is $m_p c^2 = 938.28$ MeV, and the neutron rest mass energy is:

$$m_n c^2 = 939.57 \text{ MeV}$$

The rest energy of the deuterium is then:

$$m_d c^2 = [(938.28 \text{ MeV}) + (939.57 \text{ MeV}) - (3.6 \times 10^{-13} \text{ J})] / [(1.6 \times 10^{-19} \text{ J/eV})\cdot(10^6 \text{ eV/MeV})]$$

$$m_d c^2 = (1877.85 \text{ MeV}) - (2.25 \text{ MeV})$$

$$m_d c^2 = 1875.6 \text{ MeV}$$

Converting to SI units:

$$m_d = (1875.6 \text{ MeV})\cdot(1.6 \times 10^{-19} \text{ J/eV}) / c^2$$

$$m_d = [(1875.6 \text{ MeV})\cdot(1.6 \times 10^{-19} \text{ J/eV})\cdot(10^6 \text{ eV/MeV})] / (3.0 \times 10^8 \text{ m/s})^2$$

$$m_d = 3.3 \times 10^{-27} \text{ kg.}$$

9. B is correct.

The time interval between events increases in the moving frame from the value measured in the rest frame because of time dilation.

Let primed variables represent quantities measured in the moving frame, and unprimed variables represent quantities measured in the rest frame.

Then, the expression for time dilation is:

$$\Delta t' = \gamma \Delta t$$

where $\gamma = 1 / \sqrt{(1 - \beta^2)}$ and $\beta = v^2 / c^2$

If some factor increases time intervals, rates are decreased by the same factor.

Let r be the heart rate measured in the rest frame, and r' be the heart rate measured in the moving frame.

Because of time dilation, the rate in the moving frame is reduced according to:

$$r' = r / \gamma$$

If $v = 0.99c$, then $\gamma = 7.09$.

Thus:

$$r' = 12 \text{ beats/min}$$

10. C is correct.

As measured by the first spaceship, the speed of Earth is $0.31c$, and the speed of the second spaceship as measured by the Earth is also $0.31c$.

The speed of the second spaceship, as measured by the first, is the relativistic sum of the two speeds.

$$\beta = (\beta_1 + \beta_2) / (1 + \beta_1\beta_2)$$

$$\beta = (0.31 + 0.31) / [1 + (0.31)\cdot(0.31)]$$

$$\beta = 0.566$$

$$v = 0.57\ c$$

11. D is correct.

The kinetic energy, total energy, and linear momentum have no limit.

They increase monotonically with increasing γ.

Although the speed of a particle has an upper limit of c, the value of γ goes to infinity as v approaches c.

12. B is correct.

This problem can be solved using Lorentz transformation, time dilation, or length contraction.

Using length contraction, the distance between the beacons as measured by the spaceship is contracted:

$$l' = l / \gamma$$

where $\beta = v / c$ and $\gamma = 1 / \sqrt{(1 - \beta^2)}$

$$\gamma = 1 / \sqrt{[1 - (0.5)^2]} = 1.155$$

$$l' = (49 \times 10^6\ \text{m}) / 1.155$$

$$l' = 42.42 \times 10^6\ \text{m}$$

The time interval for passage across this distance as measured in the spaceship frame is:

$$t' = l' / v = l' / (0.5\ c)$$

$$t' = (42.42 \times 10^6\ \text{m}) / [(0.5)\cdot(3 \times 10^8\ \text{m/s})]$$

$$t' = 283\ \text{ms}$$

13. A is correct.

The lifetime of the pion is greater as measured by clocks on Earth due to time dilation.

The lifetime measured on Earth is:

$$\Delta t' = \gamma\Delta t$$

where $\gamma = 1 / \sqrt{(1 - \beta^2)}$ and $\beta = v^2 / c^2$

continued…

If $\beta = 0.23$, then $\gamma = 1.028$, so the lifetime as measured on Earth is:

$$\Delta t' = 2.62 \times 10^{-8}\ \text{s}$$

The distance traveled in that time interval is:

$$d' = v\Delta t'$$

$$d' = 0.23(3 \times 10^{8}\ \text{m/s})\cdot(2.62 \times 10^{-8}\ \text{s})$$

$$d' = 1.81\ \text{m}$$

14. B is correct.

The problem can be solved in the rest frame of the planets or the rest frame of the astronaut.

In the rest frame of the planets, the light from the nuclear explosion from planet Y arrives at the astronaut when:

$$c\Delta t = (1\ \text{hour})\ c + v\Delta t$$

$$\Delta t = (1\ \text{hour})\ c / (c - v)$$

If $v = 0.6c$, then:

$$\Delta t = 2.5\ \text{hours}$$

From planet Z, the light arrives to the astronaut at:

$$c\Delta t = (1\ \text{hour})\ c - v\Delta t$$

$$\Delta t = (1\ \text{hour})\ c / (c + v)$$

$$\Delta t = 0.625\ \text{hours}$$

This is 1.875 hours sooner as measured in the frame of the planets.

In the planets' frame, the astronaut's clock appears to be moving slower because of time dilation.

Therefore, the time interval measured by the astronaut is:

$$\Delta t' = (1.875\ \text{hours}) / \gamma$$

$$\gamma = 1 / \sqrt{(1 - \beta^2)}$$

$$\gamma = 1.25$$

$$\Delta t' = 1.5\ \text{hours} = 90\ \text{minutes}$$

Alternatively

Consider the problem in the rest frame of the astronaut.

In that frame, the two explosions do not occur simultaneously but at a time interval given by:

$$\Delta t = \gamma\ (\Delta t' + v\ \Delta x' / c^2)$$

where $\Delta t'$, the time interval as measured in the frame of the planets, is 0 and $\Delta x'$ is 2 light-hours

continued…

Therefore:

$\Delta t = 1.25 \cdot (0.6c) \cdot (2 \text{ hours} \cdot c) / c^2$

$\Delta t = 1.5$ hours = 90 minutes

15. D is correct.

The speed of the Klingon ship, as measured by the Enterprise, is the relativistic addition of the speed of the Klingon ship as measured from Earth, plus the speed of Earth as measured from the Enterprise.

The formula for the relativistic addition of speeds is:

$\beta = (\beta_1 + \beta_2) / (1 + \beta_1\beta_2)$

$\beta = (0.9 + 0.8) / [1 + (0.9) \cdot (0.8)]$

$\beta = 0.988$

$v = 0.988\ c$

16. D is correct.

In Sofia's inertial frames of reference, one before the spaceship increased its speed and one after, the spaceship is at rest.

Therefore, the measurements of time and length must give the same result.

Otherwise, the laws of physics would be different in these two inertial frames of reference.

During the acceleration of the spaceship, there are some unusual effects caused by the acceleration, and those effects are the subject of General Relativity.

17. C is correct.

In the frame of the beacons:

$t = d / v$

$t = (40 \times 10^6 \text{ m}) / [(0.30) \cdot (3 \times 10^8 \text{ m/s})]$

$t = 44.4 \times 10^{-2}$ s

$t = 444$ ms

18. B is correct.

Kinetic energy is total energy minus rest energy:

$K = E - mc^2$

$K = \gamma mc^2 - mc^2$

$K = (\gamma - 1)mc^2$

where $\gamma = 1 / \sqrt{(1 - \beta^2)}$ and $\beta = v / c$

continued…

$\gamma = 1 / (\surd[1 - (0.737)^2]$

$\gamma = 1.480$

So:

$K = (1 - 1.480)\cdot(511 \text{ keV})$

$K = 245 \text{ keV}$

19. D is correct.

The space time interval, Δs, is given by:

$(\Delta s)^2 = (c\Delta t)^2 - (\Delta x)^2$

For these events,

$\Delta s = 90c$ years and $\Delta t = 100$ years

Solving for Δx:

$(\Delta x)^2 = (100c \text{ years})^2 - (90c \text{ years})^2$

$(\Delta x)^2 = 1900c^2 \text{ years}^2$

$\Delta x = 44$ light-years

20. A is correct.

Lorentz contraction and time dilation are proportional to:

$\gamma = 1 / \surd(1 - \beta^2) \qquad \beta = v / c$

Therefore, if $\gamma = 2$:

$\beta^2 = 0.75$

$\beta = 0.866$

$v = 0.866c$

21. A is correct.

As measured from the inertial frame of Earth, time on the spaceship is dilated, and a clock on the moving spaceship moves slower than a stationary clock on the Earth.

This is true when the spaceship is traveling away from Earth and traveling towards the Earth.

Therefore, the elapsed time on the spaceship and the consequential advance of the age of the twin on the spaceship is less than the elapsed time and age of the twin on Earth.

What is the difference between the two inertial frames? As viewed from the frame of Earth, nothing unusual happens while the spaceship is decelerating and re-accelerating as it changes directions and heads back to Earth.

As viewed from the spaceship frame, time on Earth advances quickly while the spaceship changes directions.

continued…

It is during this brief period that the spaceship-bound twin observes the Earth-bound twin aging rapidly.

If the twin calculates the apparent rate of time change on Earth on the decelerating and then accelerating spaceship, the result reflects the General Relativistic effect of a differing gravitational potential on time.

During the deceleration and acceleration, the twin on the spaceship is in a frame equivalent to a stationary frame with a uniform gravitational field.

In that frame, the Earth, situated at a very different gravitational potential than the spaceship, is experiencing time moving very quickly due to the perceived gravitational potential difference, according to the weak equivalence principle.

22. D is correct.

Using subscript 1 and 2 for the two initial masses and subscript 3 for the final combined mass, conservation of momentum gives:

$$\gamma_1 m v_1 - \gamma_2 m v_2 = \gamma_3 m_3 v_3$$

Conservation of energy gives:

$$\gamma_1 m c^2 + \gamma_2 m c^2 = \gamma_3 m_3 c^2$$

Using these two equations to solve for v_3:

$$v_3 = (\gamma_1 v_1 - \gamma_2 v_2) / (\gamma_1 + \gamma_2)$$

$$\gamma_1 = 1 / \sqrt{(1 - 0.650^2)}$$

$$\gamma_1 = 1.316$$

$$\gamma_2 = 1 / \sqrt{(1 - 0.850^2)}$$

$$\gamma_2 = 1.898$$

$$v_3 = [(1.316 \cdot 0.650c) - (1.898 \cdot 0.850c)] / (1.316 + 1.898)$$

$$v_3 = -0.234c$$

$$\gamma_3 = 1 / \sqrt{(1 - 0.277^2)}$$

$$\gamma_3 = 1.029$$

Using the conservation of energy to solve for m_3:

$$m_3 = m(\gamma_1 + \gamma_2) / \gamma_3$$

$$m_3 = (100 \text{ kg}) \cdot (1.316 + 1.898) / 1.029$$

$$m_3 = 312 \text{ kg}$$

23. D is correct.

The relativistic expression for momentum is:

$$p = \gamma mv = \gamma m\beta c$$

where $\beta = v / c$ and $\gamma = 1/ \sqrt{(1 - \beta^2)}$

Therefore:

$$\beta = 0.60$$

$$\gamma = 1 / \sqrt{(1 - 0.60^2)}$$

$$\gamma = 1.25$$

$$p = (1.25)\cdot(1.67 \times 10^{-27}\ \text{kg})\cdot(0.60)\cdot(3 \times 10^8)$$

$$p = 3.8 \times 10^{-19}\ \text{kg}\cdot\text{m/s}$$

24. B is correct.

The total energy of the rock, $\gamma\, mc^2$, is the sum of its rest mass plus the added energy:

$$E_{\text{Total}} = \gamma\, mc^2$$

$$E_{\text{Total}} = \text{mc}^2 + E_{\text{Added}}$$

$$E_{\text{Added}} = (\gamma - 1)\, mc^2$$

$$\gamma = 1 / \sqrt{(1 - \beta^2)}$$

$$\beta = v / c$$

Given that $v = 0.866c$

$$\beta = 0.866$$

$$\gamma = 2.00$$

$$E_{\text{Added}} = (2.00 - 1)\cdot(1.0\ \text{kg})\cdot(3.0 \times 10^8\ \text{m/s})^2$$

$$E_{\text{Added}} = 9.00 \times 10^{16}\ \text{kg}\cdot\text{m/s}$$

$$E_{\text{Added}} = 9.00 \times 10^{16}\ \text{J}$$

25. C is correct.

The speed of the secondary rocket, as measured in the frame of Earth, is the relativistic addition of the speed of the secondary rocket as measured from the spaceship, plus the speed of the spaceship as measured from Earth. The formula for the *relativistic addition of speeds* is:

$$\beta = (\beta_1 + \beta_2) / (1 + \beta_1\beta_2)$$

$$\beta = (0.5 + 0.5) / [1 + (0.5)\cdot(0.5)]$$

$$\beta = 0.80$$

$$v = 0.80\ c$$

26. A is correct.

The observer moving with the rocket is making the time measurements at the same location:

$$\Delta x_B = 0$$

The *Lorentz transformation* is, therefore, simply:

$$T_A = \gamma T_B$$

Since γ is greater than one,

$$T_A > T_B$$

This is *time dilation*— the shortest time interval for two events occurs in the frame in which the two events occur at the same location, a frame stationary with respect to the two events.

If those events are clicking a clock, the clock appears to move the fastest in its rest frame.

27. B is correct.

As measured by an observer on Earth, the spaceship will have traveled for:

$$t = d / v$$

$$t = (4.367\ c\ \text{years}) / 0.50c$$

$$t = 8.734\ \text{years}$$

To an observer on the spaceship, $\Delta x' = 0$, and the time $\Delta t'$ is dilated by the Lorentz transformation:

$$\Delta t = \gamma\ \Delta t'$$

$$\gamma = 1 / \sqrt{(1 - \beta^2)}$$

$$\beta = v / c$$

$$\gamma = 1 / \sqrt{(1 - 0.50^2)}$$

$$\gamma = 1.15$$

$$\Delta t' = (8.734\ \text{years}) / 1.15$$

$$\Delta t' = 7.6\ \text{years}$$

28. D is correct.

The decay time Δt in the rest frame of the particle (where $\Delta x = 0$) is dilated to a shorter elapsed time.

Since,

$$\Delta x = 0$$

The Lorentz transformation for the two events is:

$$\Delta t' = \gamma \Delta t$$

continued...

$\gamma = 1 / \sqrt{(1 - \beta^2)}$

$\beta = v / c = 0.997$

$\gamma = 12.9$

$\Delta t = 37.0\ \mu s / 12.9$

$\Delta t = 2.9\ \mu s$

29. C is correct.

The total energy of a particle is:

$E_{Total} = \gamma\ mc^2$

When the kinetic energy is correctly defined as the difference between the total energy and the rest mass energy, then:

$E_{Total} = mc^2 + E_{Kinetic}$

$E_{Kinetic} = (\gamma - 1)\ mc^2$

$\gamma = 1 / \sqrt{(1 - \beta^2)}$

$\beta = v / c$

Given that $v = 0.5c$

$\beta = 0.5$

$\gamma = 1.1547$

$E_{Kinetic} = 0.1547\ mc^2$

$E_{Kinetic} = (mv^2 / 2)\cdot(0.1547\cdot 2)\cdot(c / v)^2$

$E_{Kinetic} = (mv^2 / 2)\cdot(1.24)$

Therefore,

by using $mv^2 / 2$ as the kinetic energy, the calculation is off by:

$mv^2 / 2 = 0.81 \times E_{Kinetic}$, which is a 19% error.

30. B is correct.

The separation $\Delta x'$ of the two stars in a frame moving with respect to the two stars is affected by Lorentz Contraction.

In the spaceship's frame, the measurement of $\Delta x'$ was made with $\Delta t' = 0$.

The Lorentz transformation is therefore:

$$\Delta x = \gamma \; \Delta x'$$

$$\Delta x' = \Delta x \,/\, \gamma$$

$$\gamma = 1 \,/\, \sqrt{(1 - \beta^2)} \qquad \beta = v \,/\, c$$

Since $\Delta x = 90$ light-years and $\Delta x' = 68.7$ light-years:

$$\gamma = 90 \,/\, 68.7$$

Solving for β:

$$\beta = \sqrt{[1 - (1 / \gamma^{2})]}$$

$$\beta = 0.646$$

Therefore:

$$v = 0.646c$$

31. D is correct.

One of the postulates of Special Relativity is that the laws of physics are the same in inertial frames.

If the laws of physics are the same in all inertial frames, there is no way to differentiate frames.

32. D is correct.

Using subscript 1 and 2 for the two initial masses and subscript 3 for the final combined mass, conservation of momentum gives:

$$\gamma_1 m v_1 - \gamma_2 m v_2 = \gamma_3 m_3 v_3$$

Conservation of energy gives:

$$\gamma_1 mc^2 + \gamma_2 mc^2 = \gamma_3 m_3 c^2$$

Using these two equations to solve for v_3:

$$v_3 = (\gamma_1 v_1 - \gamma_2 v_2) \,/\, (\gamma_1 + \gamma_2)$$

$$\gamma_1 = 1 \,/\, \sqrt{(1 - 0.550^2)} = 1.197$$

$$\gamma_2 = 1 \,/\, \sqrt{(1 - 0.750^2)} = 1.512$$

$$v_3 = [(1.197 \cdot 0.550c) - (1.512 \cdot 0.750c)] \,/\, (1.197 + 1.512)$$

$$v_3 = -0.1756c$$

(The direction is in the opposite direction of the first satellite.)

33. A is correct.

In the moving frame, the time interval between events is increased from the value measured in the rest frame because of time dilation.

Consider two heartbeats as two events on the spaceship.

They have a value of $\Delta x = 0$ and $\Delta t = (1 / 70)$ minutes.

In the frame of Earth, because Δx on the spaceship is 0, the Lorentz transformation for $\Delta t'$ is simply:

$$\Delta t' = \gamma\, \Delta t$$

where:

$$\gamma = 1 / \sqrt{(1 - \beta^2)}$$

$$\beta = v / c$$

$$\gamma = 1 / \sqrt{(1 - 0.50^2)}$$

$$\gamma = 1.15$$

$$\Delta t' = 1.15 \cdot (1 / 70) \text{ minutes}$$

$$\Delta t' = 0.0164 \text{ minutes}$$

The heart rate is, therefore, the inverse of this interval, 61 beats per minute.

34. B is correct.

The accelerator's length is contracted as measured in the frame of the particle:

$$l' = l / \gamma$$

where $\beta = v / c$ and $\gamma = 1 / \sqrt{(1 - \beta^2)}$

$$l' = (453 \text{ m}) [\sqrt{(1 - (0.875)^2}]$$

$$l' = 219 \text{ m}$$

35. C is correct.

In the spaceship's frame, the two events have a value of $\Delta x = 0$.

The time difference in the spaceship's frame appears to be dilated so that the time difference in the frame of Earth is given by:

$$\Delta t' = \gamma\, \Delta t$$

$$\gamma = 1 / \sqrt{(1 - \beta^2)} \qquad \beta = v / c$$

Since $\Delta t = 5.78$ years and $\Delta t' = 10$ years:

$$\gamma = 10 / 5.78$$

continued…

Solving for β:

$$\beta = \sqrt{(1 - 1/\gamma^2)}$$

$$\beta = 0.816$$

Therefore:

$$v = 0.816c$$

36. C is correct.

The length of a moving object appears to be shortened by Lorentz contraction.

In the stationary observer's frame on the ground, the length measurement involves two events, with $\Delta t' = 0$ and $\Delta x'$ being the length of the moving train as measured by the stationary observer.

In the frame of the moving train, those two events have an Δx value equal to the length of the train as measured in the rest frame of the train. (Note that the value of Δt is not zero.)

Since $\Delta t'$ is zero, the Lorentz transformation for Δx is simply:

$$\Delta x = \gamma\, \Delta x'$$

Since γ is greater than 1, $\Delta x'$, the length measured by the stationary observer on the ground is less than Δx, the length measured in the rest frame of the train.

37. C is correct.

At first glance, this is a complicated problem since the spaceship's velocity is unknown and cannot be readily determined.

However, the space-time interval, Δs^2, of the two events (i.e., the spaceship crossing the orbit of Jupiter and Mars) can be determined.

The space-time interval is the same in all frames.

$$\Delta s^2 = (c\Delta t)^2 + (\Delta x)^2$$

$$\Delta s'^2 = (c\Delta t')^2 + (\Delta x')^2$$

In the spaceship's frame, the distance between the two measurements is $\Delta x = 0$, and the time between the two measurements is Δt.

In the Earth's frame:

$$\Delta x' = (778 \times 10^9 \text{ m}) - (228 \times 10^9 \text{ m})$$

$$\Delta x' = 550 \times 10^9 \text{ m}$$

It is given that:

$$\Delta t' = 50.0 \text{ minutes}$$

continued…

Solving for Δt by equating the space-time interval in the two frames:

$(c\Delta t)^2 = (c\Delta t')^2 + (\Delta x')^2$

$\Delta t^2 = \Delta t'^2 + (\Delta x'/c)^2$

$\Delta t^2 = (50 \text{ minutes} \cdot 60 \text{ s/minute})^2 + (550 \times 10^9 \text{ m}) / (3.0 \times 10^8 \text{ m/s})$

$\Delta t^2 = (3000 \text{ s})^2 + (1833 \text{ s})^2$

$\Delta t^2 = (3515 \text{ s})^2$

$\Delta t = 58.6$ minutes

38. D is correct.

In the frame of the two beacons, in a time Δt, a radar signal goes from one beacon to the other beacon and back to the moving spaceship, while the moving spaceship travels from the first beacon towards the second beacon.

The total distance traveled by the spaceship plus the radar signal in Δt is twice the distance between the beacons:

$2 \cdot (65 \times 10^6 \text{ m}) = (0.90c + c)\Delta t$

$\Delta t = (68.4 \times 10^6 \text{ m}) / (3 \times 10^8 \text{ m})$

$\Delta t = 228$ ms

$\Delta t = 228$ ms

As measured in the frame of the moving spaceship, the time interval is transformed into the spaceship's frame for which a stationary clock has $\Delta x' = 0$ (note that Δx is not zero).

Therefore,

The Lorentz transformation is:

$\Delta t = \gamma \, \Delta t'$

$\Delta t' = \Delta t / \gamma$

$\gamma = 1 / \sqrt{(1 - \beta^2)}$

$\beta = v/c = 0.9$

$\gamma = 2.29$

$\Delta t' = 228 \text{ ms} / 2.29$

$\Delta t' = 100$ ms

39. B is correct.

The total energy of the electron, $\gamma\, mc^2$, is the sum of its rest mass plus the potential difference through which it is accelerated, V, times its charge, e:

$$\gamma\, mc^2 = mc^2 + eX$$

$$V = mc\,(\gamma - 1)$$

where $\gamma = 1 / \sqrt{(1 - \beta^2)}$ and $\beta = v / c$

$$\gamma = 1 / \sqrt{[1 - (0.648)^2]}$$

$$\gamma = 1.313$$

$$V = 511 \text{ keV} \cdot 0.313$$

$$V = 160 \text{ keV}$$

40. C is correct.

In the particle's frame, the two events (creation and annihilation) have a value of $\Delta x = 0$.

The time difference in the laboratory's frame appears to be dilated by:

$$\Delta t' = \gamma\, \Delta t$$

$$\gamma = 1 / \sqrt{(1 - \beta^2)} \qquad \beta = v / c$$

and:

$$\Delta x' = v \Delta t'$$

Since $\Delta t = 1.52 \times 10^{-6}$ s and $\Delta x' = 342$m:

$$342 \text{ m} = v\gamma\, (1.52 \times 10^{-6} \text{ s})$$

Dividing both sides by c:

$$(342 \text{ m}) / (3.00 \times 10^8 \text{ m/s}) = (v/c)\, \gamma\, (1.52 \times 10^{-6} \text{ s})$$

$$0.75 = \beta\, \gamma$$

$$0.75 = \sqrt{[\beta^2 / (1 - \beta^2)]}$$

$$0.75 = \sqrt{[1 / \beta^2 - 1)]}$$

Solving for B:

$$\beta = 0.6$$

$$v = 0.6c$$

41. C is correct.

The relativistic kinetic energy formula is derived from the conservation of energy:

$E_{\text{total}} = \gamma\, mc^2$

This is true for all values of γ (or velocity).

The kinetic energy formula acts as a definition of kinetic energy by dividing the total energy into a rest mass component and a kinetic component:

$E_{\text{total}} = mc^2 + E_{\text{kinetic}}$

$E_{\text{kinetic}} = (\gamma - 1)mc^2$

For $v < c$, a Taylor expansion for $(\gamma - 1)$ gives the Newtonian formula for kinetic energy:

$E_{\text{kinetic}} = mv^2 / 2$, which is true only for non-relativistic velocities.

42. B is correct.

The separation $\Delta x'$ of Earth and the star in a frame moving with respect to the spaceship is affected by Lorentz contraction.

Since the separation $\Delta x'$ is measured when $\Delta t'$ is zero, the Lorentz transformation is simply:

$\Delta x = \gamma \Delta x'$

$\gamma = 1 / \sqrt{(1 - \beta^2)}$ $\beta = v / c$

With $v = 0.800c$ and $\Delta x = 4.30$ light-years:

$\gamma = 1.67$

$\Delta x' = 2.58$ light-years

43. B is correct.

As measured in the frame of the asteroid, the speed of the missile is the relativistic addition of the speed of the spaceship as measured from the asteroid, plus the speed of the missile as measured from the spaceship.

The two velocities are in the opposite direction.

The formula for the relativistic addition of two speeds in the same directions is:

$\beta = (\beta_1 + \beta_2) / (1 + \beta_1\beta_2)$

$\beta = (0.8 + 0.5) / [1 + (0.8)\cdot(0.5)]$

$\beta = 0.93$

$v = 0.93c$

44. A is correct.

The total energy of a particle of mass m moving with velocity v is given by:

$E = \gamma\ mc^2$

$\gamma = 1 / \sqrt{(1 - \beta^2)}$

$\gamma = v / c$

$\gamma = 0.950$

$\gamma = 1 / \sqrt{(1 - 0.950^2)}$

$\gamma = 3.20$

$E = (3.20)\cdot(9.11 \times 10^{-31}\ \text{kg})\cdot(3.0 \times 10^8\ \text{m/s})^2$

$E = 2.6 \times 10^{-13}\ \text{J}$

45. D is correct.

In the spaceship's frame, the two events (i.e., two time measurements on the same clock) have a value of $\Delta x = 0$.

The time difference in the spaceship's frame appears to be dilated so that the time difference in the frame of Earth is given by:

$\Delta t' = \gamma\ \Delta t$

$\gamma = 1 / \sqrt{(1 - \beta^2)}$ $\beta = v / c$

Since $\Delta t = 1.0$ s and $v = 0.960c$:

$\gamma = 3.6$

$\Delta t = 3.6\ \text{s}$

46. A is correct.

In the frame of the first particle, the speed of the second particle is the relativistic sum of the speed of Earth (as measured from the frame of the first particle - note that Earth appears to be moving to the right), plus the speed of the second particle (as measured from the frame of Earth – note that it appears to be moving to the right).

The formula for the relativistic addition of two speeds in the same directions is:

$\beta = (\beta_1 + \beta_2) / (1 + \beta_1\beta_2)$

$\beta = (0.741 + 0.543) / [1 + (0.741)\cdot(0.543)]$

$\beta = 0.916$

$v = 0.916c$

47. A is correct.

Relativistic momentum is given by:

$$p = \gamma mv = \gamma m\beta c$$

where $\beta = v / c$ and $\gamma = 1 / \sqrt{(1 - \beta^2)}$

Newtonian momentum is given by:

$$p = mv$$

The *percentage difference* is:

$$(\gamma mv - mv) / \gamma mv = [(1 - (1 / \gamma)]$$

For $v = 0.86c$:

$$\gamma = 1 / \sqrt{(1 - 0.86^2)}$$

$$\gamma = 1.96$$

$$1 - (1 / \gamma) = 0.49 = 49\%$$

48. B is correct.

The speed of the secondary rocket as measured in the frame of the space platform is the relativistic addition of the speed of the secondary rocket as measured in the frame of the spaceship, plus the speed of the spaceship as measured in the frame of the space platform.

The two velocities are in the opposite direction.

The formula for the relativistic addition of two speeds in opposite directions is:

$$\beta = (\beta_1 - \beta_2) / (1 - \beta_1\beta_2)$$

$$\beta = (0.1 - 0.56) / [1 - (0.1)\cdot(0.56)]$$

$$\beta = 0.487$$

$$v = 0.487c$$

49. D is correct.

The relativistic expression for momentum is:

$$p = \gamma mv = \gamma m\beta c$$

where $\beta = v / c$ and $\gamma = 1 / \sqrt{(1 - \beta^2)}$

Therefore:

$$\Delta p = \gamma_1 v_1 m - \gamma_2 v_2 m$$

where subscript 1 refers to before the slowing and subscript 2 refers to after the slowing

continued…

$\Delta p = [(\gamma_1 v_1 / c) - (\gamma_2 v_2 / c)]mc^2/c$

$v_1 = 0.998c$

$\beta_1 = 0.998$

$\gamma_1 = 1 / \sqrt{(1 - 0.998^2)}$

$\gamma_1 = 15.82$

$v_2 = 0.500c$

$\beta_2 = 0.500$

$\gamma_2 = 1 / \sqrt{(1 - 0.500^2)}$

$\gamma_2 = 1.155$

$\Delta p = [(15.82 \cdot 0.998) - (1.155 \cdot 0.500)] \cdot (511 \text{ keV}) / c$

$\Delta p = (7772 / c \text{ keV})$

$\Delta p = 7.77 \text{ MeV/c}$

50. D is correct.

The speed of light has the same value for any observer, regardless of the state of motion of the source or the observer. This is a fundamental principle of Special Relativity.

Notes for active learning

Notes for active learning

Diagnostic Test 1 – Answer Key and Detailed Explanations

1	B	Kinematics & dynamics	31	B	Equilibrium & momentum
2	C	Force, motion, gravitation	32	C	Work & energy
3	E	Heat & thermodynamics	33	E	Waves & periodic motion
4	A	Work & energy	34	A	Heat & thermodynamics
5	B	Waves & periodic motion	35	A	Light & optics
6	C	Sound	36	B	Electric circuits
7	A	Waves & periodic motion	37	D	Electrostatics
8	B	Electrostatics	38	E	Fluids
9	E	Electric circuits	39	C	Sound
10	D	Light & optics	40	E	Waves & periodic motion
11	B	Rotational motion	41	D	Work & energy
12	D	Equilibrium & momentum	42	C	Equilibrium & momentum
13	C	Work & energy	43	B	Force, motion, gravitation
14	A	Waves & periodic motion	44	A	Kinematics & dynamics
15	C	Sound	45	C	Heat & thermodynamics
16	E	Fluids	46	E	Work & energy
17	C	Electrostatics	47	B	Sound
18	B	Electric circuits	48	B	Electrostatics
19	C	Light & optics	49	D	Light & optics
20	E	Rotational motion	50	E	Kinematics & dynamics
21	D	Heat & thermodynamics	51	A	Equilibrium & momentum
22	A	Force, motion, gravitation	52	B	Waves & periodic motion
23	B	Sound	53	E	Fluids
24	A	Fluids	54	D	Electric circuits
25	E	Electrostatics	55	B	Rotational motion
26	B	Electric circuits	56	D	Heat & thermodynamics
27	C	Light & optics	57	C	Kinematics & dynamics
28	E	Waves & periodic motion	58	C	Force, motion, gravitation
29	A	Kinematics & dynamics	59	B	Equilibrium & momentum
30	D	Force, motion, gravitation	60	E	Light & optics

1. B is correct.

Equation for displacement given an initial velocity and constant acceleration:

$d = v_it + ½at^2$

$a = (v_f - v_i) / t$

$a = (1.8 \text{ m/s} - 0.4 \text{ m/s}) / 4 \text{ s}$

$a = 0.35 \text{ m/s}^2$

Substituting $a = 0.35 \text{ m/s}^2$ into the equation for displacement,

$d = v_it + ½at^2$

$d = (0.4 \text{ m/s})\cdot(4 \text{ s}) + ½(0.35 \text{ m/s}^2)\cdot(4 \text{ s})^2$

$d = 4.4 \text{ m}$

2. C is correct.

The object experiencing the force increases in speed due to the acceleration from the force:

$F = ma$

However, because the force is decreasing, *the rate of increase of its speed* (not the speed itself) will decrease.

3. E is correct.

$Q = mc\Delta\text{T}$

$c = Q / m\Delta\text{T}$

$c = (150 \text{ kcal}) / [(3 \text{ kg})\cdot(200 \text{ °C})]$

$c = 0.25 \text{ kcal/kg}\cdot\text{°C}$

4. A is correct.

$\text{W} = F \times d \cos \theta$

$F = \text{W} / d$

$F = 40 \text{ J} / 4 \text{ m}$

$F = 10 \text{ N}$

5. B is correct.

Equation for the period T of a simple pendulum:

$$T = 2\pi\sqrt{(L / g)}$$

Rearranging the equation for g,

$$g = (2\pi / T)^2L$$

$$g = (4\pi^2)L / T^2$$

$$g = 4\pi^2(0.58 \text{ m}) / (2.5 \text{ s})^2$$

$$g = 3.7 \text{ m/s}^2$$

6. C is correct.

Electromagnetic waves are created by accelerating electric charges (change in velocity or speed).

7. A is correct.

Speed of the wave = wavelength × frequency

$$v = \lambda f$$

$$v = (0.5 \text{ m})\cdot(800 \text{ Hz})$$

$$v = 400 \text{ m/s}$$

Distance = velocity × time, the distance covered in one second is:

$$d = vt$$

$$d = (400 \text{ m/s})\cdot(1 \text{ s})$$

$$d = 400 \text{ m}$$

8. B is correct.

$$V = IR$$

$$R = V / I$$

$$R = 12 \text{ V} / 2 \text{ A}$$

$$R = 6 \text{ ohms}$$

9. E is correct.

Connecting two identical batteries in parallel yields the same voltage but twice the total available charge.

10. D is correct.

The power of the combination of the lenses is the sum of the powers.

$$5 \text{ D} + 3 \text{ D} = 8 \text{ D}$$

11. B is correct.

Consider this formula for rotational kinematics:

$$\omega_f = \omega_i + \alpha\Delta t$$

Solve for Δt:

$$\Delta t = (\omega_f - \omega_i) / \alpha$$

$$\Delta t = (33.3 \text{ rad/s} - 15.0 \text{ rad/s}) / 3.45 \text{ rad/s}^2$$

$$\Delta t = 5.30 \text{ s}$$

12. D is correct.

$$m_1v_1 = m_2v_2$$

$$(1{,}450 \text{ kg})v_1 = (90 \text{ kg})\cdot(30 \text{ m/s})$$

$$v_1 = (90 \text{ kg})\cdot(30 \text{ m/s}) / (1{,}450 \text{ kg})$$

$$v_1 = 1.9 \text{ m/s}$$

13. C is correct.

work = force × distance

$$W = Fd$$

Machines reduce the amount of force required, but the distance through which the force acts increases. They do not reduce the amount of work required.

If 900 J of energy is expended to lift the block to a certain height, then, even if a perfectly efficient pulley were used, the engineer expends a total of 900 J of work to lift the block to the same height using the pulley.

14. A is correct.

amplitude = ½(total displacement)

amplitude = ½(0.4 m)

amplitude = 0.2 m

The amplitude is the displacement from the position of equilibrium.

The amplitude is 0.2 m because the mass travels 0.2 m to the left of equilibrium and 0.2 m to the right in one oscillation.

15. C is correct.

One wave is completed at the 3rd node.

From the figure, the given wave has 1.5 λ:

$L = 1.5\ \lambda$

$\lambda = 2/3\ L$

$\lambda = (2/3)\cdot(\text{total length})$

$\lambda = (2/3)\cdot(0.6\ \text{m})$

$\lambda = 0.4\ \text{m}$

$v = f\lambda$

$1\ \text{Hz} = 1\ \text{s}^{-1}$

$v = (900\ \text{s}^{-1})\cdot(0.4\ \text{m})$

$v = 360\ \text{m/s}$

16. E is correct.

density = mass / volume

$\rho = m\ /\ \text{V}$

$\rho = \text{kg/m}^3$

17. C is correct.

The charge q is equally attracted to both Qs because q is equidistant.

$q_{\text{net}} = 0$

$F_{1\rightarrow 2} = kq_1q_2\ /\ r^2$

$F_{2\rightarrow 1} = -kq_1q_2\ /\ r^2$

One force is attractive to the first Q, but the other is attractive to the second Q and therefore points in the opposite direction.

The sign of the second force is reversed.

$F_{\text{net}} = 0$

18. B is correct.

A simple LC circuit with 0 resistance acts as an oscillator with charge flowing back and forth across the capacitor and inductor.

19. C is correct.

Light passing through a prism will be dispersed according to wavelength and display a continuous spectrum of colors (wavelengths).

20. E is correct.

Use the conservation of energy.

Note that the ball is at rest in the initial state, so the initial kinetic energy is zero. Take the zero of gravitational potential energy to be the height of the ball at its final position.

The final kinetic energy has a translational part and a rotational part:

$$E_f = E_i$$

$$KE_f = PE_i$$

$$KE_{\text{translation}} + KE_{\text{rotation}} = PE_i$$

$$½mv^2 + ½I\omega^2 = mgh$$

The ball is rolling without slipping, so the relationship between translational velocity and rotational velocity is:

$$v = r\omega$$

Then, the above expression becomes:

$$½mr^2\omega^2 + ½(2/5)mr^2\omega^2 = mgh$$

$$(7/10)mr^2\omega^2 = mgh$$

Solving for the angular speed:

$$\omega = \sqrt{(10gh / 7r^2)}$$

$$\omega = \sqrt{\{[10\cdot(9.8 \text{ m/s}^2)\cdot(5.3 \text{ m})] / [7\cdot(1.7 \text{ m})^2]\}}$$

$$\omega = 5.1 \text{ rad/s}$$

21. D is correct.

Because the path *ca* is adiabatic, no heat is added or lost by the system during this process.

Adiabatic process:

$$Q = 0 \text{ kJ}$$

22. A is correct.

$\Sigma F_x = 0$

$0 = F_{Ax} + F_{Bx} + F_{Cx} + F_x$

$0 = (30\text{ N})\cdot(\sin 35°) - (40\text{ N})\cdot(\cos 25°) + (50\text{ N})\cdot(\cos 40°) + F_x$

$0 = (30\text{ N})\cdot(0.57) - (40\text{ N})\cdot(0.91) + (50\text{ N})\cdot(0.77) + F_x$

$0 = (17\text{ N} - 36\text{ N} + 39\text{ N}) + F_x$

$F_x = -20\text{ N}$

23. B is correct.

The 4th harmonic has four nodes (i.e., three more than the fundamental), and the ends are antinodes.

The harmonic to wavelength relationship for an organ pipe is calculated by:

$\lambda_n = 2L / n$

where L is the length of the pipe and n is the harmonic number (1, 2, 3,…)

For n = 4:

$\lambda_4 = (2)\cdot(0.2\text{ m}) / 4$

$\lambda_4 = 0.1\text{ m}$

24. A is correct.

Volumetric flow rate = velocity × area

$V = vA$

Water is incompressible, so volumetric flow rate is constant.

$(vA)_{in} = (vA)_{out}$

$(0.02\text{ m/s})(\pi/4)(0.15\text{ m})^2 = v_{out}(\pi/4)(0.003\text{ m})^2$

$v_{out} = [(0.02\text{ m/s})(0.15\text{ m})^2] / (0.003\text{ m})^2$

$v_{out} = 50\text{ m/s}$

25. E is correct.

Microwaves and blue light are electromagnetic radiation.

The difference between them is with respect to their frequency, wavelength, and corresponding energy.

$c = \lambda f$

$E = hf$

26. B is correct.

The internal resistance is added to the external resistance of the circuit.

If the external resistance is large, the current is small, and the internal resistance can be ignored.

27. C is correct.

A plane mirror does not magnify the image and always produces an erect, virtual image.

$$m = 1$$

$$1 = d_i / d_o$$

$$1 = h_i / h_o$$

28. E is correct.

Constructive interference occurs when two or more waves of equal frequency and phase produce a single amplitude wave that is the sum of amplitudes of the individual waves.

If there is any phase difference, the interference will not be the total amplitude of each wave.

If the phase difference is 180°, there will be total destructive interference.

29. A is correct.

Average acceleration is a change in velocity over change in time:

$$\text{average acceleration} = \Delta v / \Delta t$$

30. D is correct.

Equilibrium occurs when the sum of the forces equals zero.

$$0 = F_1 + F_2 + F_3$$

Set down as the negative direction:

$$0 = 80\text{ N} - 24\text{ N} + F_3$$

$$F_3 = -56\text{ N}$$

A negative value corresponds to the downward direction, so 56 N points down.

31. B is correct.

Impulse:

$J = F\Delta t$

Increase in Δt = decrease in F

Decreased force means that it is less likely to break.

Impulse and change in momentum are directly related.

$F\Delta t = m\Delta v$

Because the wine glass always goes from terminal velocity to zero in both cases, and the mass is constant, the impulse is constant and equal.

Thus, the only way to decrease force is to increase Δt, the stopping time.

32. C is correct.

$W = Fd \cos \theta$

θ is perpendicular to the direction of the velocity

$\cos 90° = 0$

$Fd \cos \theta = 0$ J, no work is done against the force of gravity.

Since the velocity is constant, there is no net horizontal force, so no net work is done by horizontal forces.

33. E is correct.

velocity = wavelength × frequency

$v = \lambda f$

As frequency decreases, the wavelength must increase for velocity to remain constant.

34. A is correct.

$\Delta L / L_i = \alpha_L \Delta T$

$\alpha_L = \Delta L / L_i \Delta T$

$\alpha_L = (2.0005 \text{ m} - 2.0000 \text{ m}) / (2.0000 \text{ m}) \cdot (40 \text{ °C} - 20 \text{ °C})$

$\alpha_L = 0.0005 \text{ m} / 40 \text{ m} \cdot \text{°C}$

$\alpha_L = 1.25 \times 10^{-5} \text{ K}^{-1}$

35. A is correct.

$1 / f = 1 / d_i + 1 / d_o$

$1 / 12 \text{ m} = 1 / d_i + 1 / 6 \text{ m}$

$1 / 12 \text{ m} - 2 / 12 \text{ m} = 1 / d_i$

$-1 / 12 \text{ m} = 1 / d_i$

$d_i = -12 \text{ m}$

A negative sign indicates that the image is behind the mirror.

36. B is correct.

The total resistance for a circuit with parallel resistors decreases as more resistors are added.

A decrease in the total resistance increases the total current in the circuit.

37. D is correct.

$E = kQ / r^2$

E is directly proportional to Q and inversely proportional to r^2:

$2Q = 2E$

The only solution that doubles the magnitude of E is to double Q.

38. E is correct.

If an object's entire volume is suspended in a liquid, the object's density is equal to that of the liquid.

If 20% of the floating buoy is above the surface of the liquid, 80% of its volume is suspended in the liquid.

Since 80% of the buoy's volume is in the liquid, it is only 80% as dense as the liquid.

39. C is correct.

The wave of the fundamental frequency for an open pipe is shown.

Therefore:

$\lambda = 2L$

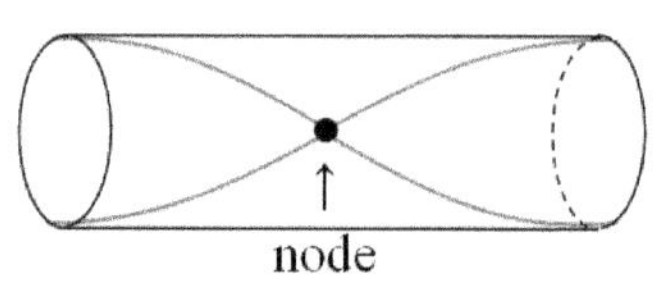

The two endpoints are antinodes, and the midpoint is a node.

40. E is correct.

Wave interference occurs for all types of waves, including mechanical and electromagnetic.

41. D is correct.

$F = -kx$

$-k = F / x$

$-k = 30\ \text{N} / -0.75\ \text{m}$

$k = 40\ \text{N/m}$

42. C is correct.

Since the two objects come to rest immediately, they have equal but opposite momenta before the collision:

$p_A = p_B$

$p = mv$

$m_A v_A = m_B v_B$

$v_A / v_B = m_B / m_A$

v_A is 5 times greater than v_B

$v_A / v_B = 5$

$5 = m_B / m_A$

take the reciprocal of each side

$m_A / m_B = 1 / 5$

The ratio of the mass of A to the mass of B is 1: 5

43. B is correct.

If mass and force are constant:

$F = ma$

$a = F / m$

$a = \text{constant}$

Therefore, *acceleration is constant.*

44. A is correct.

$v_{\text{Parallel}} = v \sin \theta$

The component of velocity in the direction of the slope is:

$v_P = v \sin \theta$

where $\theta = 20°$

continued…

The velocity parallel to the slope:

$v_P = v \sin \theta$

$v_P = v \sin 20°$

$v_P = (16 \text{ m/s})\cdot(0.34)$

$v_P = 5.47 \approx 5.5 \text{ m/s}$

45. C is correct.

Heat released to the surroundings *equals the work* needed to compress the gas in isothermal compression.

$Q_{released} = W$

$W = nRT \ln (V_2 / V_1)$

46. E is correct.

All three answer choices are correct because they describe PE:

$PE = mgh$

PE equals the work done by the pile driver:

$PE = W$

$PE = Fd$

$PE / d = F$

The force is directly proportional to PE, so all three choices are correct as they relate to the PE of the piledriver.

47. B is correct.

An organ pipe is either open at both ends or closed at one end.

Open pipe:

$f_n = nv / 2L$

where n = 1, 2, 3…

Closed pipe:

$f_n = nv / 4L$

where n = 1, 3, 5…

Observing the equation for frequencies produced by an open or closed pipe, there is an infinite number of frequencies possible depending on the value of n.

48. B is correct.

Ohms is the unit used to measure electrical resistance.

49. D is correct.

A diopter is a unit of measurement of the optical power of a lens or a curved mirror.

The power equals the reciprocal of the focal length (i.e., $1 / f$). Therefore, the diopter is a unit of reciprocal length.

For example, a 3 diopter lens brings parallel rays of light to focus at 1/3 m.

$$1 / d_o + 1 / d_i = 1 / f$$

Setting object at infinity (∞):

$$1 / \infty + 1 / d_i = 1 / f$$

$$0 + 1 / d_i = 1 / f$$

$$1 / d_i = 1 / f$$

Power of the lens:

$$P = 1 / f$$

$$3 = 1 / d_i$$

$$d_i = 1 / 3 \text{ m} = 0.33 \text{ m}$$

50. E is correct.

Accelerating uniformly for 8 s:

$$\Delta v = a\Delta t$$

$$v_2 - v_1 = a\Delta t$$

$$v_2 = v_1 + a\Delta t$$

$$v_2 = 0 \text{ m/s} + (2.5 \text{ m/s}^2)\cdot(8 \text{ s})$$

$$v_2 = 20 \text{ m/s}$$

51. A is correct.

Set North as the positive direction.

Conservation of momentum:

$$(m_1 + m_2)v = m_1v_1 + m_2v_2$$

$$(m_1 + m_2)v = (2{,}500 \text{ kg})\cdot(7 \text{ m/s}) + (2{,}000 \text{ kg})\cdot(-14 \text{ m/s})$$

$$(m_1 + m_2)v = 17{,}500 \text{ kg}\cdot\text{m/s} + (-28{,}000 \text{ kg}\cdot\text{m/s})$$

$$(m_1 + m_2)v = -10{,}500 \text{ kg}\cdot\text{m/s}$$

$$-10{,}500 \text{ kg}\cdot\text{m/s} = (4{,}500 \text{ kg})v$$

Velocity immediately after the inelastic collision:

$$v = (-10{,}500 \text{ kg}\cdot\text{m/s}) / (4{,}500 \text{ kg})$$

$v = -2.3$ m/s, the minus sign indicates v points South.

52. B is correct.

$$T = 2\pi\sqrt{(L / g)}$$

If the clock is too slow, the period (T) is too large.

Decrease T by raising the weight, which effectively decreases L.

53. E is correct.

When the air above the paper is blown, the air moves faster, and the decrease in pressure causes the air below the paper to attempt to fill that void, pushing the paper upward in the process.

54. D is correct.

voltage = current × resistance

$$V = IR$$

$$4\text{ V} = I(20\ \Omega)$$

$$I = 4\text{ V} / 20\ \Omega$$

$$I = 0.2\text{ A}$$

55. B is correct.

For three rotating bodies, the total angular momentum is the sum of their angular momenta.

$$L_{total} = L_1 + L_2 + L_3$$

Angular momentum of a rotating object:

$$L = I\omega$$

The flywheels are identical, so they each have the same rotational inertia I, so:

$$L = I(\omega_1 + \omega_2 + \omega_3)$$

Moment of inertia of the flywheel:

$$I = \tfrac{1}{2}mr^2$$

$$I = 0.5(65.0\text{ kg})(1.47\text{ m})^2$$

$$I = 70.23\text{ kg m}^2$$

Thus:

$$L = (70.23\text{ kg m}^2)\cdot(3.83\text{ rad/s} + 3.83\text{ rad/s} - 3.42\text{ rad/s})$$

$$L = 298\text{ kg m}^2\text{/s}$$

56. D is correct.

As the temperature rises the mercury and glass expand by:

$$\Delta V = \alpha V_i \Delta T$$

To reach 450 ml, the expansion of the glass and the mercury must be considered:

$$V_G + \Delta V_G = V_M + \Delta V_M$$

$$450\text{ ml} + \Delta V_G = 442\text{ ml} + \Delta V_M$$

$$8\text{ ml} = \Delta V_M - \Delta V_G$$

$$8\text{ ml} = \alpha_M V_{iM} \Delta T - \alpha_G V_{iG} \Delta T$$

$$8\text{ ml} = \Delta T(\alpha_M V_{iM} - \alpha V_{iG})$$

$$8\text{ ml} = \Delta T[(18 \times 10^{-5}\text{ K}^{-1})\cdot(442\text{ ml}) - (2 \times 10^{-5}\text{ K}^{-1})\cdot(450\text{ ml})]$$

$$8\text{ ml} = \Delta T(0.0706\text{ K}^{-1}\text{ ml})$$

$$\Delta T = 113.4\text{ °C}$$

$$T_f = T_i + \Delta T$$

$$T_f = 22\text{ °C} + 113.4\text{ °C}$$

$$T_f = 135\text{ °C}$$

57. C is correct.

A constant positive acceleration increases the velocity linearly upwards over time.

The velocity *vs.* time graph with a positive slope represents a constant positive acceleration.

58. C is correct.

First convert the final velocity from km/h to m/s:

$$v_f = (100\text{ km/1 h})\cdot(1{,}000\text{ m/1 km})\cdot(1\text{ h/3,600 s})$$

$$v_f = 100{,}000\text{ m / 3,600 s}$$

$$v_f = 27.8\text{ m/s}$$

$$a = \Delta v / \Delta t$$

$$a = (27.8\text{ m/s}) / (4.8\text{ s})$$

$$a = 5.8\text{ m/s}^2$$

$$F = ma$$

$$F = (68\text{ kg})\cdot(5.8\text{ m/s}^2)$$

$$F = 394\text{ N}$$

59. B is correct.

Torque can be written as:

$$\tau = I\alpha$$

where τ = torque, I = moment of inertia and α = angular acceleration

Thus, if torque is constant, then the angular acceleration must be constant as well because the moment of inertia does not change

60. E is correct.

Calculate the focal length:

$$1 / f = 1 / d_i + 1 / d_o$$

where f is focal length d_o is distance to the object and d_i is distance to the image

$$1 / f = 1 / 2\text{ m} + 1 / 4\text{ m}$$

$$1 / f = 2 / 4\text{ m} + 1 / 4\text{ m}$$

$$1 / f = 3 / 4\text{ m}$$

$$f = 4 / 3\text{ m}$$

Notes for active learning

Notes for active learning

Diagnostic Test 2 – Answer Key and Detailed Explanations

#	Answer	Topic
1	D	Kinematics & dynamics
2	C	Force, motion, gravitation
3	D	Heat & thermodynamics
4	A	Work & energy
5	C	Waves & periodic motion
6	B	Sound
7	A	Fluids
8	C	Electrostatics
9	E	Electric circuits
10	C	Light & optics
11	D	Rotational motion
12	D	Equilibrium & momentum
13	B	Work & energy
14	A	Waves & periodic motion
15	B	Sound
16	D	Fluids
17	A	Electrostatics
18	C	Electric circuits
19	E	Light & optics
20	D	Rotational motion
21	A	Heat & thermodynamics
22	E	Force, motion, gravitation
23	C	Sound
24	A	Fluids
25	B	Electrostatics
26	D	Electric circuits
27	B	Light & optics
28	C	Waves & periodic motion
29	B	Kinematics & dynamics
30	E	Force, motion, gravitation
31	D	Equilibrium & momentum
32	B	Work & energy
33	A	Waves & periodic motion
34	E	Heat & thermodynamics
35	C	Light & optics
36	C	Electric circuits
37	C	Electrostatics
38	C	Fluids
39	E	Sound
40	C	Waves & periodic motion
41	A	Work & energy
42	E	Equilibrium & momentum
43	E	Force, motion, gravitation
44	D	Kinematics & dynamics
45	C	Heat & thermodynamics
46	E	Work & energy
47	D	Sound
48	D	Electrostatics
49	A	Light & optics
50	D	Kinematics & dynamics
51	C	Equilibrium & momentum
52	B	Waves & periodic motion
53	D	Fluids
54	E	Electric circuits
55	B	Rotational motion
56	C	Heat & thermodynamics
57	A	Kinematics & dynamics
58	E	Force, motion, gravitation
59	A	Equilibrium & momentum
60	C	Light & optics

1. D is correct.

Let $y = 0$ be the ground, which is the final position of the package. Then:

$$y_f = y_i + v_i\Delta t - ½g(\Delta t)^2$$

$$0 = d + v_i\Delta t - ½g(\Delta t)^2$$

$$d = v_i\Delta t + ½g\Delta t^2$$

$$d = (-15\text{ m/s})\cdot(16\text{ s}) + ½(10\text{ m/s}^2)\cdot(16\text{ s})^2$$

$$d = (-240\text{ m} + 1{,}280\text{ m})$$

$$d = 1{,}040\text{ m}$$

2. C is correct.

$$F = ma$$

$$m = F / a$$

$$m = 3{,}000\text{ N} / (2\text{ m/s}^2)$$

$$m = 1{,}500\text{ kg}$$

3. D is correct.

If metal A has a larger coefficient of thermal expansion, it will elongate more than metal B when the strip is heated. Because the two strips are bound together, the bimetallic strip will curve downward when heated as the configuration gives strip A more room for expansion.

4. A is correct.

$$F = -kx$$

The negative sign is by convention; it indicates that the spring resting force is opposite the stretch direction.

$$x = F / k$$

$$x = mg / k$$

$$x = [(30\text{ kg})\cdot(10\text{ m/s}^2)] / 4{,}600\text{ N/m}$$

$$x = 0.0652\text{ m}$$

$$\text{PE} = ½kx^2$$

$$\text{PE} = ½(4{,}600\text{ N/m})\cdot(0.0652\text{ m})^2$$

$$\text{PE} = 9.8\text{ J}$$

An increase in PE is due to the addition of weight.

5. C is correct.

Frequency = # of cycles / time

$f = 2$ cycles / 1 s

$f = 2\ \text{s}^{-1}$

$v = \lambda f$

$v = (6\ \text{m})\cdot(2\ \text{s}^{-1})$

$v = 12\ \text{m/s}$

6. B is correct.

The decibel rating for sound is based on a logarithmic scale:

Sound intensity equation:

$I\ (\text{dB}) = 10\log_{10}(I / I_0)$

where $I_0 = 10^{-12}\ \text{W/m}^2$ and is the threshold of human hearing

Example:

sound intensity decreases from $10\ \text{W/m}^2$ to $1\ \text{W/m}^2$

$I_1\ (\text{dB}) = 10\log_{10}(10\ \text{W/m}^2 / 10^{-12}\ \text{W/m}^2)$

$I_1 = 130\ \text{dB}$

$I_2\ (\text{dB}) = 10\log_{10}(1\ \text{W/m}^2 / 10^{-12}\ \text{W/m}^2)$

$I_1 = 120\ \text{dB}$

$I_1 - I_2 = 130\ \text{dB} - 120\ \text{dB}$

$I_1 - I_2 = 10\ \text{dB}$

Thus, the sound intensity decreases by 10 dB if the intensity decreases by a factor of 10.

7. A is correct.

Buoyancy force:

$F_B = \rho g V$

where V is the volume of water displaced.

When the two blocks are stacked on each other:

$F_B = \rho g V_0 = F_1 + F_2$

When the two blocks are both in the water:

$F_{B1} = \rho g V_1 = F_1$

$F_{B2} = \rho g V_2 = F_2$

continued…

Total volume displaced in both cases:

$V_0 = (1/\rho g)\cdot(F_1 + F_2)$

$V_1 + V_2 = (1/\rho g)\cdot(F_1 + F_2)$

$V_0 = V_1 + V_2$

Same water displacement, so the water level does not change.

8. C is correct.

To find # excess electrons, divide Q_1 by the charge of a single electron:

excess electrons = Q_1 / e

excess electrons = $(-1 \times 10^{-6}\,\text{C}) / (-1.60 \times 10^{-19}\,\text{C})$

excess electrons = 6.3×10^{12}

9. E is correct.

Capacitance depends on geometry (i.e., surface area) of a conductor and the permittivity of the dielectric used.

10. C is correct.

A flat plane mirror has no curvature, so the radius is considered infinite.

11. D is correct.

It seems that there is not enough information because neither the masses nor the radii of the sphere and cylinder are known. However, various parameters often cancel.

Find the acceleration of the two objects. The object with the larger acceleration will reach the bottom first.

An object rolling down an incline experiences three forces, hence three torques.

The forces are the force of gravity acting on the center of mass of the object, the normal force between the incline and the object, and the force of friction between the incline and the object.

If the origin is taken to be the center of the object, the force of gravity provides zero torque.

This can be seen by noting that the distance between the origin and the point of application of the force is zero.

$\tau_{\text{gravity}} = F_{\text{gravity}} r = mg(0) = 0$

Similarly, the normal force contributes zero torque because the direction of the force is directed through the origin (pivot point).

$\tau_{normal} = F_{normal}\,\text{r} \sin \theta = F_{normal}(R)\cdot(\sin 180°) = F_{normal}(R)\cdot(0) = 0$

Use a coordinate system where the x-axis is parallel to the incline and the y-axis is perpendicular. The object is rolling in the positive x-direction.

continued…

Dynamical equation for linear motion along the x-direction:

$$F_{\text{net}} = ma$$

$$(mg \sin \theta - f) = ma$$

Note that the normal force is only in the y-direction and thus does not directly contribute to the acceleration in the x-direction.

Dynamical equation for rotational motion:

$$\tau_{\text{net}} = I\alpha$$

$$fR = I\alpha$$

(Note that the frictional force is perpendicular to the r vector, and $\sin 90° = 1$)

where R is the radius of the object, f is the force of friction, and I is the moment of inertia

A relation is needed to couple these two dynamical equations.

The equation of constraint imposed by the restriction that objects roll without slipping:

$$\alpha = a / R$$

To find the linear acceleration, use the equation of constraint to eliminate α from the rotational equation by replacing it with a / R:

$$fR = I(a / R)$$

The force of friction is of no interest, so rearrange this last expression:

$$f = Ia / R^2$$

Substitute this into the linear dynamic equation from above in place of f:

$$mg \sin \theta - (Ia / R^2) = ma$$

Solving this for a:

$$a = mg \sin \theta / [(m + (I / R^2)]$$

$$a = g \sin \theta / [1 + (I / mR^2)]$$

For a sphere,

$$I = (2/5)mR^2$$

so:

$$a_{\text{sphere}} = g \sin \theta / (1 + 2/5)$$

$$a_{\text{sphere}} = (5/7)g \sin \theta$$

For the cylinder,

$$I = ½mR^2$$

continued...

So:

$a_{\text{cylinder}} = g \sin \theta / (1 + ½)$

$a_{\text{cylinder}} = (2/3)g \sin \theta$

Since 5/7 > 2/3, the acceleration of the sphere is greater than the acceleration of the cylinder so that the sphere will reach the bottom first.

Interestingly, neither the mass nor the size of the sphere or cylinder enters into the result.

Indeed, the mass and radius cancel.

Since neither the masses nor the radii were given in the statement of the problem, it would not be possible to solve this problem by brute force numerical calculation.

12. D is correct.

The moment of inertia:

I = mass × the square of the distance from the rotational axis

$I = mr^2$

Since,

$r_1 = r_2$

I_1 / I_2 is the ratio of the masses.

Therefore, a moment of inertia I is directly proportional to the mass m.

Doubling the mass doubles the moment of inertia, giving a ratio of 2 : 1 comparing the more massive to the less massive sphere.

13. B is correct.

power = work / time

$P = W / t$

$P = (1{,}000 \text{ J}) / (40 \text{ s})$

$P = 25 \text{ W}$

14. A is correct.

A spring mass oscillator has a frequency:

$\omega = \sqrt{(k / m_0)}$

$f = \omega / 2\pi$

$f = \sqrt{(k / m_0)} / 2\pi$

continued…

Period:

$T = 1 / f$

$T_0 = 2\pi / \sqrt{(k / m_0)}$

If mass is doubled:

$T = 2\pi / \sqrt{(k / 2\ m_0)}$

$T = 2\pi / \sqrt{(k / m_0)} \times \sqrt{2}$

$T = \sqrt{2}\ T_0$

The period increases by a factor of $\sqrt{2}$.

15. B is correct.

For a pipe or tube that is closed at one end and open at the other end:

harmonic wavelengths λ_n have values of n, which are only odd (n = 1, 3, 5, …).

After the fundamental wavelength (n = 1), the consecutive harmonic wavelength is the third harmonic (n = 3).

The harmonic wavelength can be determined from the equation:

$\lambda = 4L / n$

$\lambda = 4(1.5\ m) / 3$

$\lambda = 2\ m$

16. D is correct.

$\rho = m / V$

$V = (4/3)\pi r^3$

$\rho = m / (4/3)\pi r^3$

$\rho = (115\ kg) / (4/3)\pi \times (0.6\ m)^3$

$\rho = 127\ kg/m^3$

17. A is correct.

When a potential difference exists across a wire (for example connecting a battery to both ends), electrons flow from high to low potential.

For a battery, the electrons flow along the wire towards the positive terminal to lower their potential.

18. C is correct.

$1\ A = 1\ C/s$

19. E is correct.

For a concave spherical mirror:

$r = 2f$

If the object is located at the radius of curvature:

$r = d_o$

$d_o = 2f$

Use the mirror equation:

$1 / f = 1 / d_o + 1/ d_i$

$1 / f = 1 / 2f + 1 / d_i$

$1 / 2f = 1 / d_i$

$d_i = 2f = d_o$, the image and object are located at the same point.

Use the magnification equation:

$m = -d_i / d_o$

$m = h_i / h_o$

$-(2f) / 2f = h_i / h_o$

$-1 = h_i / h_o$

$-h_i = h_o$

If the object is located at the radius of curvature, then the image distance is equal to the object distance, and the image is inverted because the height is the negative of the object height.

20. D is correct.

$813.0 \text{ rpm} \cdot (1 \text{ min} / 60 \text{ s})\cdot(2\pi \text{ rad/rev}) = 85.14 \text{ rad/s}$

21. A is correct.

An isobaric process is a process in which the pressure remains constant throughout.

From the diagram, the volume changes, but the pressure is constant, consistent with an *isobaric process*.

22. E is correct.

$F = ma$, on Earth

$20 \text{ N} = m(4 \text{ m/s}^2)$

$m = 20 \text{ N} / (4 \text{ m/s}^2)$

$m = 5 \text{ kg}$

continued…

$W = ma$, on Moon

$W = (5\ \text{kg})\cdot(1.62\ \text{m/s}^2)$

$W = 8.1\ \text{N}$

23. C is correct.

Frequency is the number of cycles a wave makes in a specific time.

$f = \#\ \text{cycles} / \text{time}$

The figure shows the wave making 2 cycles in 4 seconds.

$f = 2\ \text{cycles} / 4\ \text{s}$

$f = 0.5\ \text{s}^{-1} = 0.5\ \text{Hz}$

24. A is correct.

$\frac{1}{2}\rho v^2 = \rho gh$

cancel ρ from both sides of the expression

$\frac{1}{2}v^2 = gh$

$v^2 = 2gh$

$v^2 = 2(9.8\ \text{m/s}^2)\cdot(0.8\ \text{m})$

$v^2 = 15.68\ \text{m}^2/\text{s}^2$

$v = 3.95\ \text{m/s} \approx 4\ \text{m/s}$

25. B is correct.

By Newton's Third Law, F_1 and F_2 form an *action-reaction* pair.

The ratio of their magnitudes equals 1.

26. D is correct.

$P = I^2R$

$I^2 = P / R$

$I^2 = 16\ \text{W} / 18\ \Omega$

$I^2 = 0.89\ \text{A}^2$

$I = 0.94\ \text{A}$

27. B is correct.

$c = \lambda f$

$f = c / \lambda$

Wavelength and frequency are inversely proportional.

28. C is correct.

velocity = frequency × wavelength

$v = f\lambda$

$\lambda = v / f$

$f = 1 / T$

$\lambda = v \times T$

$\lambda = 360 \text{ m/s} \times 4.2 \text{ s}$

$\lambda \approx 1{,}512 \text{ m}$

29. B is correct.

Given the angular velocity, the linear velocity at a given radius is calculated by:

$v = \omega r$

where v is the linear velocity and ω is the angular velocity

Convert from revolutions/minute to radians/second to arrive at an answer in m/s:

1 revolution = 2π radians and 1 min = 60 sec

angular velocity in radians/s is given by:

$\omega = (8.3 \text{ rev/min})\cdot(2\pi \text{ rad/rev})\cdot(1 \text{ min/60 s})$

$\omega = 0.87 \text{ rad/s}$

Calculate linear velocity. The problem gives the diameter, but the formula requires radius:

$v = (0.87 \text{ rad/s})\cdot(9 \text{ m})$

$v = 7.8 \text{ m/s}$

30. E is correct.

1,700 N is the magnitude of the instantaneous force a ball hits a bat if a batter hits a ball pitched with a 1,700 N instantaneous force.

31. D is correct.

In an elastic collision, KE and momentum are conserved:

KE before collision = KE after collision

$\text{Block}_1 + \text{Block}_2 = \text{Block}_1 + \text{Block}_2$

$½m_1v_1^2 + ½m_2v_2^2 = ½m_1u_1^2 + ½m_2u_2^2$

Initially,

$v_2 = 0$

$½m_1v_1^2 = ½m_1u_1^2 + ½m_2u_2^2$

Solve for u_2:

$[m_1(v_1^2 - u_1^2) / m_2]^{½} = u_2$

Momentum:

$m_1v_1 = m_1u_1 + m_2u_2$

Solve for u_2:

$m_1(v_1 - u_1) / m_2 = u_2$

Set equal and isolate m_2:

$[m_1(v_1^2 - u_1^2) / m_2]^{½} = m_1(v_1 - u_1) / m_2$

$m_2 = m_1(v_1 - u_1)^2 / (v_1^2 - u_1^2)$

Solve for m_2:

$m_2 = [(2.8 \text{ kg})\cdot(8.5 \text{ m/s} - (-1.1 \text{ m/s}))^2] / [(8.5 \text{ m/s})^2 - (-1.1 \text{ m/s})^2]$

$m_2 = [(2.8 \text{ kg})\cdot(92 \text{ m}^2/\text{s}^2)] / (71 \text{ m}^2/\text{s}^2)$

$m_2 = 3.6 \text{ kg}$

32. B is correct.

Work is defined as:

$\text{W} = Fd$

where d is the displacement in the direction of the force; in this case, it is $-h$, where the minus sign indicates that the direction of the displacement is down

The force is the force of gravity:

$F = -mg$

where again the minus sign indicates that the force is directed downward

continued…

Therefore:

$W = Fd$

$W = (-mg)\cdot(-h)$

$W = mgh$

33. A is correct.

Frequency = 1 / period

$f = 1 / T$

$f = 1.6$ kHz

$v = \lambda f$

$v = (0.25\text{ m})\cdot(1.6\text{ kHz})$

$v = (0.25\text{ m})\cdot(1{,}600\text{ s}^{-1})$

$v = 400$ m/s

34. E is correct.

60 mL = 60 g water

$Q = (mc\Delta T)_{beaker} + (mc\Delta T)_{water}$

Change in temperature is the same.

$Q = \Delta T[(m_{beaker})\cdot(c_{beaker}) + (m_{water})\cdot(c_{water})]$

$2{,}200\text{ cal} = (25\text{ °C})\cdot[(m_{beaker})(0.18\text{ cal/g·°C}) + (60\text{ g})\cdot(1\text{ cal/g·°C})]$

$2{,}200\text{ cal} / 25\text{ °C} = (m_{beaker})\cdot(0.18\text{ cal/g·°C}) + 60\text{ cal/°C}$

$88\text{ cal/°C} = (m_{beaker})\cdot(0.18\text{ cal/g·°C}) + 60\text{ cal/°C}$

$m_{beaker} = (88\text{ cal/°C} - 60\text{ cal/°C}) / (0.18\text{ cal/g·°C})$

$m_{beaker} = (28\text{ cal/°C}) / (0.18\text{ cal/g·°C})$

$m_{beaker} = 156$ g

35. C is correct.

Spherical mirror equation:

$f = -r / 2$

The negative is only included to imply convergence of the lens.

$f = r / 2$

$r = 2f$

$r = (2)\cdot(20\text{ cm})$

$r = 40$ cm

36. C is correct.

Find R_{eq} for entire circuit:

R_1:  in series:

2 Ω 1 Ω

$R_1 = 2\ \Omega + 1\ \Omega$

$R_1 = 3\ \Omega$

R_2:  in series:

5Ω 1 Ω

$R_2 = 5\ \Omega + 1\ \Omega$

$R_2 = 6\ \Omega$

R_1 and R_2 are in parallel:

$1 / R_3 = 1 / R_1 + 1 / R_2$

$1 / R_3 = 1 / 3\ \Omega + 1 / 6\ \Omega$

$1 / R_3 = 2 / 6\ \Omega + 1 / 6\ \Omega$

$1 / R_3 = 1 / 2\ \Omega$

$R_3 = 2\ \Omega$

R_3 is in series with the 4 Ω resistor:

$R_{total} = 2\ \Omega + 4\ \Omega$

$R_{total} = 6\ \Omega$

Calculate the current around the entire circuit:

$\text{V} = IR$

$12\ \text{V} = I(6\ \Omega)$

$I = 2\ \text{A}$

Calculate voltage drop across parallel section using R_3 as equivalent resistor:

$\text{V} = IR$

$\text{V} = (2\ \text{A})\cdot(2\ \Omega)$

$\text{V} = 4\ \text{V}$

*Note: voltage drop is equal across parallel branches.

Calculate current running through the branch the 2 Ω is on:

$\text{V} = IR$

continued…

$4\ \text{V} = I(3\ \Omega)$

$I = 1.33\ \text{A}$

Calculate power dissipated by the 2 Ω resistor:

$P = I^2R$

$P = (1.33\ \text{A})^2(2\ \Omega)$

$P = 3.5\ \text{W}$

37. C is correct.

An object becomes electrostatically charged when a charge imbalance exists.

Charge can only be transferred by electrons because protons are not mobile.

Thus, electron transfer creates an electrostatic charge.

38. C is correct.

gauge pressure = $P_{absolute} - P_{atm}$

$\rho gh = P_{absolute} - P_{atm}$

gauge pressure = ρgh

gauge pressure = $(10^3\ \text{kg/m}^3)\cdot(9.8\ \text{m/s}^2)\cdot(10\ \text{m})$

gauge pressure = 9.8×10^4 Pa

39. E is correct.

The distance between any two adjacent nodes is = ½λ.

$\frac{1}{2}\lambda = 75\ \text{cm}$

$\lambda = 150\ \text{cm}$

40. C is correct.

$v = \lambda f$

$\lambda = v / f$

$\text{Hz} = \text{s}^{-1}$

$\lambda = (4{,}900\ \text{m/s}) / (640\ \text{s}^{-1})$

$\lambda = 7.7\ \text{m}$

41. A is correct.

There is no force acting in the direction of motion (constant velocity), so no work is done.

$$W = Fd \cos \theta$$

Assuming no friction, Stacey performed no work because the mass experienced no acceleration due to constant velocity.

$$F = ma$$

$$F = (15 \text{ kg})\cdot(0 \text{ m/s}^2)$$

$$F = 0 \text{ N}$$

$$W = (0 \text{ N})\cdot(100 \text{ m})$$

$$W = 0 \text{ J}$$

42. E is correct.

The force delivered by the block to her hand and her hand to the block are equal by Newton's Third Law.

The time of impact is the same for the block and the hand.

Impulse is calculated by:

$$J = F\Delta t$$

Thus, if the force and time are equal for the block and the hand, they experience the same impulse.

43. E is correct.

The textbooks follow a straight path. By turning the car's steering wheel, Karen pulls her car door into the path of the textbooks, giving the illusion that the textbooks have a force acting on them.

44. D is correct.

$$d = (v_f^2 - v_i^2) / 2a$$

$$d = [(21 \text{ m/s})^2 - (5 \text{ m/s})^2] / 2(4 \text{ m/s}^2)$$

$$d = [(441 \text{ m}^2/\text{s}^2) - (25 \text{ m}^2/\text{s}^2)] / (8 \text{ m/s}^2)$$

$$d = (416 \text{ m}^2/\text{s}^2) / (8 \text{ m/s}^2)$$

$$d = 52 \text{ m}$$

45. C is correct.

$$\Delta L / L_0 = \alpha\Delta T$$

$$\Delta T = \Delta L / \alpha L_0$$

$$\Delta T = 3 \times 10^{-3} \text{ m} / (30 \times 10^{-6} \text{ K}^{-1})\cdot(10 \text{ m})$$

$$\Delta T = 10 \text{ K}$$

46. E is correct.

$F = ma$

$a = F / m$

$a = 10\ \text{N} / 2\ \text{kg}$

$a = 5\ \text{m/s}^2$

$v_f = v_i + at$

$v_f = 0\ \text{m/s} + (5\ \text{m/s})\cdot(15\ \text{s})$

$v_f = 75\ \text{m/s}$

$\text{KE} = \frac{1}{2}mv^2$

$\text{KE} = \frac{1}{2}(2\ \text{kg})\cdot(75\ \text{m/s})^2$

$\text{KE} = 5{,}625\ \text{J}$

47. D is correct.

The beat frequency is twice per second, or 2 Hz.

$f_{beat} = | f_1 - f_2 |$

$2\ \text{Hz} = | 680\ \text{Hz} - f_2 |$

$f_2 = 678\ \text{Hz or } 682\ \text{Hz}$

48. D is correct.

$F = ma$

$a = F / m$

The force is an inverse-square with respect to distance.

$F = (kqQ) / r^2$

F decreases as q moves away from Q, and F decreases asymptotically (never reaching zero).

The charges move away from each other because both are positive.

49. A is correct.

For a diverging lens, regardless of the object position, the image is:

- virtual
- upright
- reduced

50. D is correct.

The magnitude and direction of the combined vectors A and B if vector A has a length of 5 units and is directed to the North (N), while vector B has a length of 11 units and is directed to the South (S) is 6 units, S

$$5N + (-11S) = -6S$$

51. C is correct.

Newton's First Law: an object either remains at rest or continues to move at a constant velocity unless acted upon by an external force.

Newton's Second Law: $F = ma$

Newton's Third Law: when one body exerts a force on a second body, the second body simultaneously exerts a force equal in magnitude and opposite in direction on the first body.

While the conservation of momentum is related to all three of Newton's laws, Newton's Third Law is the closest to the concept that momentum can be transferred from one object to another; the total momentum does not change during this process.

52. B is correct.

Ignoring the dissipative forces due to friction, the total mechanical energy of a pendulum is conserved. Gravitational PE is constantly being converted into KE and vice versa.

The total mechanical energy (i.e., KE + PE) remains constant.

53. D is correct.

Volumetric flow rate is constant:

$$v_1A_1 = v_2A_2$$

$$v_1 / v_2 = A_2 / A_1$$

$$v_1 / v_2 = (\pi / 4\ D_2^2) / (\pi / 4\ D_1^2)$$

$$v_1 / v_2 = (8\text{ cm})^2 / (3\text{ cm})^2$$

$$v_1 / v_2 = 7$$

$$v_1 = 7v_2$$

If the diameter increases, then the speed decreases by a factor of 7.

54. E is correct.

Changing the current in a nearby wire changes the magnetic field around it and induces a voltage in the wire close to it by Faraday's Law.

Moving a magnet close to the wire or the wire close to a magnetic field changes the magnetic field around the wire and induces a voltage by Faraday's Law.

55. B is correct.

The angular speed changes according to the kinematic relation:

$$\Delta\omega = \alpha\Delta t$$

$$\Delta t = \Delta\omega / \alpha$$

$$\Delta t = (0 \text{ rad/s} - 96.0 \text{ rad/s}) / (-1.5 \text{ rad/s}^2)$$

The angular acceleration is negative because the wheel is slowing, and the initial ω is in the positive direction.

$$\Delta t = 64.0 \text{ s}$$

56. C is correct.

Isothermal process experiences a change in entropy:

$$\Delta s = Q / \mathrm{T}$$

An isothermal process has work equal to the heat added.

$$Q = \mathrm{W}$$

$$\Delta s = W / \mathrm{T}$$

$$W = \mathrm{T}\Delta s$$

$$W = (273 \text{ K})\cdot(2.6 \text{ J/K})$$

$$W = 710 \text{ J}$$

$$W = 7.1 \times 10^2 \text{ J}$$

57. A is correct.

The rock starts at the velocity Jack throws it, which is downward and therefore negative.

The velocity continues to increase negatively due to acceleration from gravity which is constant.

Thus, the velocity is linear with a downward slope.

$$v_f = v_i + at$$

$$y = \mathrm{m}x + \mathrm{b}$$

58. E is correct.

By the definition of impulse:

$$F\Delta t = m\Delta v$$

Therefore:

$$F = m\Delta v/\Delta t$$

continued…

The problem is asking to find the force on the block.

Proceed by focusing on the block, first finding Δv, and then using the relation above to find the force.

To determine Δv, calculate velocity after the collision using conservation of momentum:

$m_i v_i = (m_i + m_f)v_f$

$v_f = m_i v_i / (m_i + m_f)$

$v_f = (0.1\ \text{kg})\cdot(50\ \text{m/s}) / (0.1\ \text{kg} + 0.9\ \text{kg})$

$v_f = (5\ \text{kg}\cdot\text{m/s}) / 1\ \text{kg}$

$v_f = 5\ \text{m/s}$

$F\Delta t = m(v_f - v_i)$

$F = m(v_f - v_i) / \Delta t$

$F = (0.9\ \text{kg})(5\ \text{m/s} - 0\ \text{m/s}) / 0.01\ \text{s}$

$F = (4.5\ \text{kg m/s}) / (0.01\ \text{s})$

$F = 450\ \text{N}$

59. A is correct.

Use conservation of momentum:

$m_A v_{Ai} + m_B v_{Bi} = m_A v_{Af} + m_B v_{Bf}$

$(2\ \text{kg})\cdot(0.6\ \text{m/s}) + 0 = 0 + (2.5\ \text{kg})v_{Bf}$

$v_{Bf} = (1.2\ \text{kg}\cdot\text{m/s}) / (2.5\ \text{kg})$

$v_{Bf} = 0.48\ \text{m/s}$

60. C is correct.

The plane mirror is double the distance from an object, so ½h is required for the minimum length.

Law of reflection:

$\theta_1 = \theta_2$

$h = 2x$

$x = ½h$

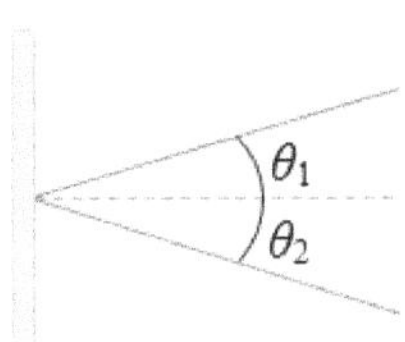

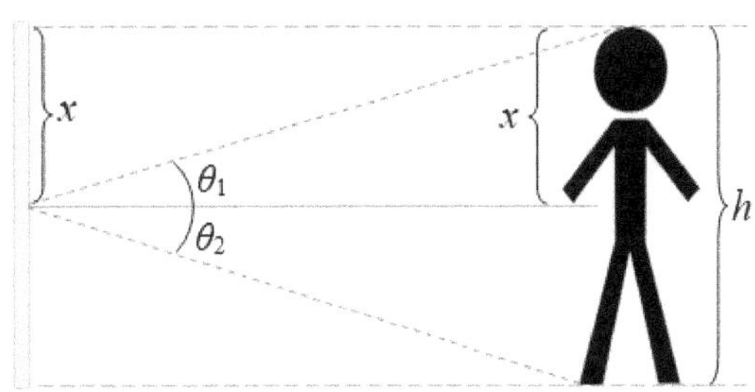

Notes for active learning

Appendix

Common Physics Equations and Conversions

Constants and Conversion Factors

1 unified atomic mass unit	$1\ u = 1.66 \times 10^{-27}$ kg
	$1\ u = 931\ MeV/c^2$
Proton mass	$m_p = 1.67 \times 10^{-27}$ kg
Neutron mass	$m_n = 1.67 \times 10^{-27}$ kg
Electron mass	$m_e = 9.11 \times 10^{-31}$ kg
Electron charge magnitude	$e = 1.60 \times 10^{-19}$ C
Avogadro's number	$N_0 = 6.02 \times 10^{23}\ mol^{-1}$
Universal gas constant	$R = 8.31$ J/(mol·K)
Boltzmann's constant	$k_B = 1.38 \times 10^{-23}$ J/K
Speed of light	$c = 3.00 \times 10^8$ m/s
Planck's constant	$h = 6.63 \times 10^{-34}$ J·s
	$h= 4.14 \times 10^{-15}$ eV·s
	$hc = 1.99 \times 10^{-25}$ J·m
	$hc = 1.24 \times 10^3$ eV·nm
Vacuum permittivity	$\varepsilon_0 = 8.85 \times 10^{-12}\ C^2/N \cdot m^2$
Coulomb's law constant	$k = 1/4\pi\varepsilon_0 = 9.0 \times 10^9\ N \cdot m^2/C^2$
Vacuum permeability	$\mu_0 = 4\pi \times 10^{-7}$ (T·m)/A
Magnetic constant	$k' = \mu_0/4\pi = 10^{-7}$ (T·m)/A
Universal gravitational constant	$G = 6.67 \times 10^{-11}\ m^3/kg \cdot s^2$
Acceleration due to gravity at Earth's surface	$g = 9.8\ m/s^2$
1 atmosphere pressure	$1\ atm = 1.0 \times 10^5\ N/m^2$
	$1\ atm = 1.0 \times 10^5$ Pa
1 electron volt	$1\ eV = 1.60 \times 10^{-19}$ J
Balmer constant	$B = 3.645 \times 10^{-7}$ m
Rydberg constant	$R = 1.097 \times 10^7\ m^{-1}$
Stefan constant	$\sigma = 5.67 \times 10^{-8}\ W/m^2K^4$

Units

Name	Symbol
meter	m
kilogram	kg
second	s
ampere	A
kelvin	K
mole	mol
hertz	Hz
newton	N
pascal	Pa
joule	J
watt	W
coulomb	C
volt	V
ohm	Ω
henry	H
farad	F
tesla	T
degree Celsius	°C
electronvolt	eV

Prefixes

Factor	Prefix	Symbol
10^{12}	tera	T
10^{9}	giga	G
10^{6}	mega	M
10^{3}	kilo	k
10^{-2}	centi	c
10^{-3}	mili	m
10^{-6}	micro	μ
10^{-9}	nano	n
10^{-12}	pico	p

Newtonian Mechanics

Translational Motion	$v = v_0 + a\Delta t$ $x = x_0 + v_0\Delta t + \frac{1}{2}a\Delta t^2$ $v^2 = v_0^2 + 2a\Delta x$ $\vec{a} = \frac{\sum \vec{F}}{m} = \frac{\vec{F}_{net}}{m}$
Rotational Motion	$\omega = \omega_0 + \alpha t$ $\theta = \theta_0 + \omega_0 t + \frac{1}{2}\alpha t^2$ $\omega^2 = \omega_0^2 + 2\alpha\Delta\theta$ $\vec{\alpha} = \frac{\sum \vec{\tau}}{I} = \frac{\vec{\tau}_{net}}{I}$
Force of Friction	$\lvert\vec{F}_f\rvert \leq \mu\lvert\vec{F}_n\rvert$
Centripetal Acceleration	$a_c = \frac{v^2}{r}$
Torque	$\tau = r_{\perp}F = rF\sin\theta$
Momentum	$\vec{p} = m\vec{v}$
Impulse	$\vec{J} = \Delta\vec{p} = \vec{F}\Delta t$
Kinetic Energy	$K = \frac{1}{2}mv^2$
Potential Energy	$\Delta U_g = mg\Delta y$
Work	$\Delta E = W = F_{\parallel}d = Fd\cos\theta$
Power	$P = \frac{\Delta E}{\Delta t} = \frac{\Delta W}{\Delta t}$
Simple Harmonic Motion	$x = A\ cos\ (\omega t)$ $x = A\ cos\ (2\pi ft)$

a = acceleration

A = amplitude

E = energy

F = force

f = frequency

h = height

I = rotational inertia

J = impulse

K = kinetic energy

k = spring constant

ℓ = length

m = mass

N = normal force

P = power

p = momentum

L= angular momentum

r = radius of distance

T = period

t = time

U = potential energy

v = velocity or speed

W = work done on system

x = position

y = height

α = angular acceleration

Center of Mass	$x_{cm} = \frac{\sum m_i x_i}{\sum m_i}$	μ = coefficient of friction
		Θ = angle
Angular Momentum	$L = I\omega$	τ = torque
Angular Impulse	$\Delta L = \tau \Delta t$	ω = angular speed
Angular Kinetic Energy	$K = \frac{1}{2} I\omega^2$	
Work	$W = F\Delta r \cos\theta$	
Power	$P = Fv \cos\theta$	
Spring Force	$\lvert\vec{F}_s\rvert = k\lvert\vec{x}\rvert$	
Spring Potential Energy	$U_s = \frac{1}{2} kx^2$	
Period of Spring Oscillator	$T_s = 2\pi\sqrt{m/k}$	
Period of Simple Pendulum	$T_p = 2\pi\sqrt{\ell/g}$	
Period	$T = \frac{2\pi}{\omega} = \frac{1}{f}$	
Gravitational Body Force	$\lvert\vec{F}_g\rvert = G\frac{m_1 m_2}{r^2}$	
Gravitational Potential Energy of Two Masses	$U_G = -\frac{Gm_1 m_2}{r}$	

Electrostatics, Magnetism, Circuits

Electric Field	$\vec{E} = \frac{\vec{F}_E}{q}$
Electric Field Strength	$\lvert\vec{E}\rvert = \frac{1}{4\pi\varepsilon_0}\frac{\lvert q\rvert}{r^2}$
Electric Field Strength	$\lvert\vec{E}\rvert = \frac{\lvert\Delta V\rvert}{\lvert\Delta r\rvert}$
Electrostatic Force Between Charged Particles	$\lvert\vec{F}_E\rvert = \frac{1}{4\pi\varepsilon_0}\frac{\lvert q_1 q_2\rvert}{r^2}$
Electric Potential Energy	$\Delta U_E = q\Delta V$
Electrostatic Potential due to a Charge	$V = \frac{1}{4\pi\varepsilon_0}\frac{q}{r}$
Capacitor Voltage	$V = \frac{Q}{C}$
Capacitance of Parallel Plate Capacitor	$C = \kappa\varepsilon_0\frac{A}{d}$
Electric Field Inside a Parallel Plate Capacitor	$E = \frac{Q}{\varepsilon_0 A}$
Capacitor Potential Energy	$U_C = \frac{1}{2}Q\Delta V = \frac{1}{2}C(\Delta V)^2$
Current	$I = \frac{\Delta Q}{\Delta t}$
Resistance	$R = \frac{\rho l}{A}$
Power	$P = I\Delta V$
Current	$I = \frac{\Delta V}{R}$
Resistors in Series	$R_s = \sum_i R_i$
Resistors in Parallel	$\frac{1}{R_p} = \sum_i \frac{1}{R_i}$

A = area
B = magnetic field
C = capacitance
d = distance
E = electric field
ϵ = emf
F = force
I = current
l = length
P = power
Q = charge
q = point charge
R = resistance
r = separation
t = time
U = potential energy
V = electric potential
v = speed
κ = dielectric constant
ρ = resistivity
θ = angle
Φ = flux

Capacitors in Parallel	$C_p = \sum_i C_i$
Capacitors in Series	$\frac{1}{C_s} = \sum_i \frac{1}{C_i}$
Magnetic Field Strength (from a long straight current-carrying wire)	$B = \frac{\mu_0 I}{2\pi r}$
Magnetic Force	$\vec{F}_M = q\vec{v} \times \vec{B}$ $\vec{F}_M = \lvert q\vec{v}\rvert \lvert\sin\theta\rvert \lvert\vec{B}\rvert$ $\vec{F}_M = I\vec{l} \times \vec{B}$ $\vec{F}_M = \lvert I\vec{l}\rvert \lvert\sin\theta\rvert \lvert\vec{B}\rvert$
Magnetic Flux	$\Phi_B = \vec{B} \cdot \vec{A}$ $\Phi_B = \lvert\vec{B}\rvert \cos\theta \lvert\vec{A}\rvert$
Electromagnetic Induction	$\epsilon = \frac{-\Delta\Phi_B}{\Delta t}$ $\epsilon = Blv$

Fluid Mechanics and Thermal Physics

Density	$\rho = \frac{m}{V}$
Pressure	$P = \frac{F}{A}$
Absolute Pressure	$P = P_0 + \rho g h$
Buoyant Force	$F_b = \rho V g$
Fluid Continuity Equation	$A_1 v_1 = A_2 v_2$
Bernoulli's Equation	$P_1 + \rho g y_1 + \frac{1}{2}\rho v_1^2 = P_2 + \rho g y_2 + \frac{1}{2}\rho v_2^2$
Heat Conduction	$\frac{Q}{\Delta t} = \frac{kA\Delta T}{d}$
Thermal Radiation	$P = e\sigma A(T^4 - T_C^4)$
Ideal Gas Law	$PV = nRT = Nk_BT$
Average Energy	$K = \frac{3}{2}k_BT$
Work	$W = -P\Delta V$
Conservation of Energy	$\Delta E = Q + W$
Linear Expansion	$\Delta l = \alpha l_o \Delta T$
Heat Engine Efficiency	$n_c = \lvert W/Q_H \rvert$
Carnot Heat Engine Efficiency	$n_c = \frac{T_H - T_C}{T_H}$
Energy of Temperature Change	$Q = mc\Delta T$
Energy of Phase Change	$Q = mL$

A = area

c = specific heat

d = thickness

e = emissivity

F = force

h = depth

k = thermal conductivity

K = kinetic energy

l = length

L = latent heat

m = mass

n = number of moles

n_c = efficiency

N = number of molecules

P = pressure or power

Q = energy transferred to system by heating

T = temperature

t = time

E = internal energy

V = volume

v = speed

W = work done on a system

y = height

σ = Stefan constant

α = coefficient of linear expansion

ρ = density

Optics

Wavelength to Frequency	$\lambda = \frac{v}{f}$	d = separation
		f = frequency or focal length
Index of Refraction	$n = \frac{c}{v}$	h = height
		L = distance
Snell's Law	$n_1 \sin\theta_1 = n_2 \sin\theta_2$	M = magnification
Thin Lens Equation	$\frac{1}{s_i} + \frac{1}{s_0} = \frac{1}{f}$	m = an integer
		n = index of refraction
Magnification Equation	$\lvert M\rvert = \left\lvert\frac{h_i}{h_o}\right\rvert = \left\lvert\frac{s_i}{s_o}\right\rvert$	R = radius of curvature
		s = distance
Double Slit Diffraction	$d \sin\theta = m\lambda$	v = speed
	$\Delta L = m\lambda$	x = position
		λ = wavelength
Critical Angle	$\sin\theta_c = \frac{n_2}{n_1}$	θ = angle
Focal Length of Spherical Mirror	$f = \frac{R}{2}$	

Acoustics

Standing Wave/ Open Pipe Harmonics	$\lambda = \frac{2L}{n}$	f = frequency
Closed Pipe Harmonics	$\lambda = \frac{4L}{n}$	L = length
Harmonic Frequencies	$f_n = nf_1$	m = mass
Speed of Sound in Ideal Gas	$v_{sound} = \sqrt{\frac{yRT}{M}}$	M = molecular mass
Speed of Wave Through Wire	$v = \sqrt{\frac{T}{m/L}}$	n = harmonic number
Doppler Effect (approaching stationary observer)	$f_{observed} = (\frac{v}{v - v_{source}})f_{source}$	R = gas constant
Doppler Effect (receding stationary observer)	$f_{observed} = (\frac{v}{v + v_{source}})f_{source}$	T = tension
Doppler Effect (observer moving towards source)	$f_{observed} = (1 + \frac{v_{observer}}{v})f_{source}$	v = velocity
Doppler Effect (observer moving away from source)	$f_{observed} = (1 - \frac{v_{observer}}{v})f_{source}$	y = adiabatic constant
		λ = wavelength

Modern Physics

Photon Energy	$E = hf$	B = Balmer constant
		c = speed of light
Photoelectric Electron Energy	$K_{max} = hf - \phi$	E = energy
Electron Wavelength	$\lambda = \frac{h}{p}$	f = frequency
		K = kinetic energy
Energy Mass Relationship	$E = mc^2$	m = mass
Rydberg Formula	$\frac{1}{\lambda} = R(\frac{1}{n_f^2} - \frac{1}{n_i^2})$	p = momentum
		R = Rydberg constant
Balmer Formula	$\lambda = B(\frac{n^2}{n^2 - 2^2})$	v = velocity
		λ = wavelength
Lorentz Factor	$\gamma = \frac{1}{\sqrt{1 - \frac{v^2}{c^2}}}$	ϕ = work function
		γ = Lorentz factor

Geometry and Trigonometry

Rectangle	$A = bh$	A = area
		C = circumference
Triangle	$A = \frac{1}{2}bh$	V = volume
		S = surface area
Circle	$A = \pi r^2$	b = base
	$C = 2\pi r$	h = height
Rectangular Solid	$V = lwh$	l = length
		w = width
Cylinder	$V = \pi r^2 l$	r = radius
	$S = 2\pi rl + 2\pi r^2$	θ = angle
Sphere	$V = \frac{4}{3}\pi r^3$	
	$S = 4\pi r^2$	
Right Triangle	$a^2 + b^2 = c^2$	
	$\sin\theta = \frac{a}{c}$	
	$\cos\theta = \frac{b}{c}$	
	$\tan\theta = \frac{a}{b}$	

Trigonometric Functions for Common Angles

θ	$\sin\theta$	$\cos\theta$	$\tan\theta$
0°	0	1	0
30°	1/2	$\sqrt{3}/2$	$\sqrt{3}/3$
37°	3/5	4/5	3/4
45°	$\sqrt{2}/2$	$\sqrt{2}/2$	1
53°	4/5	3/5	4/3
60°	$\sqrt{3}/2$	1/2	$\sqrt{3}$
90°	1	0	∞

Notes for active learning

Glossary of Physics Terms

A

Absolute humidity (or saturation value) – the maximum amount of water vapor that could be present in 1 m^3 of air at a given temperature.

Absolute magnitude – a classification scheme that compensates for the differences in the distance to stars; calculates the brightness that stars would appear to have if they were at a defined, standard distance of 10 parsecs.

Absolute scale – temperature scale set so that zero is the theoretically lowest temperature possible (this would occur when random motion of molecules has ceased).

Absolute zero – the theoretically lowest temperature possible, at which temperature the molecular motion vanishes; –273.16 °C or 0 K.

Absorptance – the ratio of the total absorbed radiation to the total incident radiation.

Acceleration – the rate of change of velocity of a moving object with respect to time; the SI units are m/s^2; by definition, this change in velocity can result from a change in speed, a change in direction, or a combination of changes in speed and direction.

Acceleration due to gravity – the rate of change in velocity produced in a body due to the Earth's attraction; denoted by the letter *g* (SI unit – m/s^2); on the surface of the Earth, its average value is 9.8 m/s^2; increases when going towards the poles from the equator; decreases with altitude and with depth inside the Earth; the value of g at the center of the Earth is zero.

Achromatic – capable of transmitting light without decomposing it into its constituent colors.

Acid rain – rainwater with a pH of less than 5.7 is acid rain. It is caused by the gases NO_2 (from car exhaust fumes) and SO_2 (from the burning of fossil fuels) dissolving in the rain. Acid rain kills fish, kills trees, and destroys buildings and lakes.

Acoustics – the science of the production, transmission, and effects of sound.

Acoustic shielding – a sound barrier that prevents the transmission of acoustic energy.

Adiabatic – any change in which there is no gain or loss of heat.

Adiabatic cooling – the decrease in temperature of an expanding gas that involves no additional heat flowing out of the gas; the cooling from the energy lost by expansion.

Adiabatic heating – the increase in temperature of the compressed gas that involves no additional heat flowing into the gas; the heating from the energy gained by compression.

Afocal lens – a lens of zero convergent power whose focal points are infinitely distant.

Air mass – a large, uniform body of air with nearly identical temperature and moisture conditions throughout.

Albedo – the fraction of the total light incident on a reflecting surface, especially a celestial body, which is reflected in all directions.

Allotropic forms – elements with several structures with different physical properties (e.g., graphite and diamond).

Alpha (α) particle – the nucleus of a helium atom (i.e., two protons and two neutrons) emitted as radiation from a decaying heavy nucleus (α-decay).

Alternating current – the flow of charge (i.e., current) traveling in one direction for one-hundredth of a second but the opposite direction for the next hundredth of a second. An electric current that first moves in one direction, then in the opposite direction with a regular frequency.

Amorphous – term that describes solids that have neither definite form nor structure.

Amp – unit of electric current; equivalent to coulomb/second.

Ampere – the unit amp; the SI unit of electric current; one ampere is the flow of one coulomb of charge per second.

Amplitude – the maximum absolute value attained by the disturbance of a wave or by any quantity that varies periodically. The height of the wave crest above the average position.

Amplitude (of an oscillation) – the maximum displacement of a body from its mean position during an oscillatory motion.

Amplitude (of waves) – the maximum displacement of particles of the medium from their mean positions during the propagation of a wave.

Angle of contact – the angle between tangents to the liquid surface and the solid surface inside the liquid; both tangents are drawn at the point of contact.

Angle of incidence – the angle of an incident (arriving) ray or particle to a surface; measured from a line perpendicular to the surface (the normal).

Angle of reflection – the angle of a reflected ray or particle from a surface; measured from a line perpendicular to the surface (the normal).

Angle of refraction – the angle between the refracted ray and the normal.

Angle of repose – the degree of inclination of a plane with the horizontal such that a body placed on the plane is on the verge of sliding but does not.

Angstrom – a unit of length; 1 = 10^{-10} m.

Angular acceleration – the rate of change of angular velocity of a body moving along a circular path; denoted by a.

Angular displacement – the angle described at the center of the circle by a moving body along a circular path. It is measured in radians.

Angular momentum (or *moment of momentum*) – the cross-product of position vector and momentum.

Angular momentum quantum number – from the quantum mechanics model of the atom, one of four descriptions of the energy state of an electron wave; describes the energy sublevels of electrons within the principal energy levels of an atom.

Angular velocity – the rate of change of angular displacement per unit of time.

Annihilation – when a particle and an antiparticle combine and release their rest energies in other particles.

Antineutrino – the antiparticle of neutrino; has zero mass and spin ½.

Archimedes principle – a body immersed in a fluid experiences an apparent loss of weight equal to the weight of the fluid displaced by the body.

Area – the amount of surface enclosed within the boundary lines.

Astronomical unit – the radius of the Earth's orbit is defined as one astronomical unit (A.U.).

Atom – the smallest unit of an element that can exist alone or in combination with other elements.

Atomic mass unit – relative mass unit (amu) of an isotope based on the standard of the ^{12}carbon isotope; one atomic mass unit (1 amu) = 1/12 the mass of a ^{12}C atom = 1.66×10^{-27} Kg.

Atomic number – the number of protons in the nucleus of an atom.

Atomic weight – weighted average of the masses of stable isotopes of an element as they occur in nature; based on the abundance of each isotope of the element and the atomic mass of the isotope compared to ^{12}carbon.

Avogadro's number – the number of ^{12}carbon atoms in exactly 12.00 g of C is 6.02×10^{23} atoms or other chemical units; the number of chemical units in one mole of a substance.

Avogadro's Law – under the same temperature and pressure conditions, equal volumes of gases contain an equal number of molecules.

Axis – the imaginary line about which a planet or other object rotates.

B

Background radiation – ionizing radiation (e.g., alpha, beta, gamma rays) from natural sources.

Balanced forces – when some forces act on a body and the resultant force is zero; see *Resultant forces*.

Balmer lines – lines in the spectrum of the hydrogen atom in the visible range; produced by the transition between a high energy level to n = 2, with n being the principal quantum number.

Balmer series – a set of four-line spectra; narrow lines of color emitted by hydrogen atom electrons as they drop from excited states to the ground state.

Bar – a unit of pressure; equal to 10^5 Pascal.

Barometer – an instrument that measures atmospheric pressure; used in weather forecasting and determining elevation above sea level.

Baryon – subatomic particle composed of three quarks.

Beat – a phenomenon of the periodic variation in sound intensity due to the superposition of waves differing slightly in frequency; rhythmic increases and decreases of volume from constructive and destructive interference between two sound waves of slightly different frequencies.

Bernoulli's theorem – states that the total energy per unit volume of a non-viscous, incompressible fluid in a streamline flow will remain constant.

Beta (β) particle – high-energy electron emitted as ionizing radiation from a decaying nucleus (β-decay); also a *beta ray*.

Big bang theory – the current model of galactic evolution in which the universe is assumed to have been created by an intense and brilliant explosion from a primeval fireball.

Binding energy – the net energy required to break a nucleus into its constituent protons and neutrons; also, the energy equivalent released when a nucleus is formed.

Biomass – the chemical energy stored in fast growing plants.

Black body – an ideal body which would absorb all incident radiation and reflect none.

Black body radiation – electromagnetic radiation emitted by an ideal material (the black body) perfectly absorbs and perfectly emits radiation.

Black hole – the remaining theoretical core of a supernova that is so dense that even light cannot escape.

Bohr model – model of the structure of the atom that attempted to correct the deficiencies of the solar system model and account for the Balmer series.

Boiling point – the temperature at which a phase change of liquid to gas occurs through boiling; the same temperature as the condensation point.

Boundary – the division between two regions of differing physical properties.

Boyle's Law – for a given mass of a gas at a constant temperature, the volume of the gas is inversely proportional to the pressure.

Brewster's Law – states that the refractive index of a material is equal to the tangent of the polarizing angle for the material.

British thermal unit (Btu) – the amount of energy or heat needed to increase the temperature of one pound of water one-degree Fahrenheit.

Brownian motion – the continuous random motion of solid microscopic particles suspended in a fluid medium due to their ongoing bombardment by atoms and molecules.

Bulk modulus of elasticity – the ratio of normal stress to the volumetric strain produced in a body.

Buoyant force – the upward force on an object immersed in a fluid.

C

Calorie – a unit of heat; 1 Calorie = 4.186 joule.

Candela – the SI unit of luminous intensity defined as the luminous intensity in each direction of a source that emits monochromatic photons of frequency 540×10^{12} Hz and has a radiant intensity in that direction of 1/683 W/sr.

Capacitance – the ratio of the charge stored per increase in potential difference.

Capacitor – an electrical device used to store charge and energy in the electrical field.

Capillarity – the rise or fall of a liquid in a tube of a fine bore.

Carnot's theorem – no engine operating between two temperatures can be more efficient than a reversible engine working between the same two temperatures.

Cathode rays – negatively charged particles (electrons) emitted from a negative terminal in an evacuated glass tube.

Celsius scale of temperature – the ice-point is taken as the lower fixed point (0 °C), and the steam-point is taken as the upper fixed point (100 °C); the interval between the ice-point and the steam-point is divided into 100 equal divisions; the unit division on this scale is 1 °C; previously called the centigrade scale; the relationship relates the temperatures on the Celsius scale and the Fahrenheit scale, $C/100 = (F - 32) / 180$; the temperature of a healthy person is 37 °C or 98.6 °F.

Center of gravity – the point through which all the weight of an object appears to act.

Centrifugal force – an apparent outward force on an object in circular motion; a consequence of the third law of motion.

Centripetal force – the radial force required to keep an object moving in a circular path.

Chain reaction – a self-sustaining reaction where some of the products can produce more reactions of the same kind (e.g., in a nuclear chain reaction, neutrons are the products that produce more nuclear reactions in a self-sustaining series).

Charles' Law – for a given mass of a gas at constant pressure, the volume is directly proportional to the temperature.

Chromatic aberration – an optical lens defect causing color fringes due to the lens bringing different colors of light to focus at different points.

Circular motion – the motion of a body along a circular path.

Closed system – the system which cannot exchange heat or matter with the surroundings.

Coefficient of areal expansion – the fractional change in surface area per degree of temperature change; see *Coefficient of thermal expansion.*

Coefficient of linear expansion – the fractional change in length per degree of temperature change; see *Coefficient of thermal expansion.*

Coefficient of thermal expansion – the fractional change in the size of an object per degree of change in temperature at constant pressure; the SI unit is K^{-1}.

Coefficient of volumetric expansion – the fractional change in volume per degree of temperature change; see *Coefficient of thermal expansion.*

Coherent source – when there is a constant phase difference between waves emitted from different parts of the source.

Combustion (or *burning*) – the combining of a substance with oxygen.

Compass – a magnet, which is free to rotate and indicate direction.

Complementary colors – two colors which, when mixed, give white. For examples: blue + yellow = white; red + cyan = white and green + magenta = white.

Compound – a substance made up of two or more elements chemically combined.

Compression – a part of a longitudinal wave in which the density of the particles of the medium is higher than the typical density.

Compressive stress – a force that tends to compress the surface as the Earth's plates move into each other.

Concave lens – a lens that spreads out light rays.

Condensation – the changing of a gas to a liquid state.

Condensation (sound) – a compression of gas molecules; a pulse of increased density and pressure that moves through the air at the speed of sound.

Condensation (water vapor) – where more vapor or gas molecules are returning to the liquid state than are evaporating.

Condensation nuclei – tiny particles such as dust, soot, or salt crystals suspended in the air on which water condenses.

Condensation point – the temperature at which a gas or vapor changes back to liquid; see *Boiling point.*

Conduction – the transfer of heat from a region of higher temperature to a region of lower temperature by increased kinetic energy moving from molecule to molecule, without the movement of the solid.

Constructive interference – the condition in which two waves are arriving at the same place at the same time and in phase add amplitudes to create a new wave.

Control rods – material inserted between fuel rods in a nuclear reactor to absorb neutrons and control the rate of the nuclear chain reaction.

Convection – transfer of heat from a region of higher temperature to a region of lower temperature by the displacement of high-energy molecules. For example, the displacement of warmer, less dense air (higher kinetic energy) by cooler, denser air (lower kinetic energy).

Conventional current – the opposite of electron current; considers an electric current to consist of a drift of positive charges that flow from the positive terminal to the negative terminal of a battery.

Convex lens – a lens that brings light rays together.

Corrosion – an undesired process where a metal is converted to one of its compounds (e.g., rusting).

Coulomb – a unit used to measure the quantity of electric charge; equivalent to the charge resulting from the transfer of 6.24 billion particles such as the electron.

Coulomb's Law – the relationship between charge, distance, and magnitude of the electrical force between two bodies; the force between two charges is directly proportional to the product of charges and inversely proportional to the square of the distance between the charges.

Covalent bond – a chemical bond formed by the sharing of a pair of electrons.

Covalent compound – chemical compound held by covalent bonds.

Crest – the point of maximum positive displacement on a transverse wave.

Critical angle – the limit to the angle of incidence when all light rays are reflected internally.

Critical mass – the mass of fissionable material needed to sustain a chain reaction.

Current – a flow of charge. Unit is Ampere (A).

Curvilinear motion – the motion of a body along a curved path.

Cycle – a complete vibration.

Cyclotron – a device used to accelerate the charged particles.

D

De-acceleration – negative acceleration when the velocity of a body decreases with time.

Decibel – unit of sound level; if P_1 & P_2 are two amounts of power, the first is said to be n decibels greater, where n = 10 log10 (P_1/P_2).

Decibel scale – a nonlinear scale of loudness based on the ratio of the intensity level of a sound to the intensity at the hearing threshold.

Density – the mass of a substance per unit volume.

Destructive interference – the condition in which two waves arriving at the same point at the same time out of phase add amplitudes that cancel to create zero total disturbance; see *constructive interference*.

Dewpoint temperature – the temperature at which condensation begins.

Dew – condensation of water vapor into droplets of liquid on surfaces.

Diffraction – the bending of light around the edge of an opaque object.

Diffuse reflection – light rays reflected in many random directions, as opposed to the parallel rays reflected from a perfectly smooth surface such as a mirror.

Diopter – unit of measure of the refractive power of a lens.

Direct current – an electrical current that always flows in one direction only (i.e., from the positive terminal to the negative terminal).

Direct proportion – when two variables increase or decrease in the same ratio (at the same rate).

Dispersion – the splitting of white light into its component colors of the spectrum.

Displacement – a vector quantity for the change in the position of an object as it moves in a particular direction; also the shortest distance between the initial position and the final position of a moving body.

Distance – a scalar quantity for the length of the path traveled by a body irrespective of its direction.

Distillation – the vaporization of a liquid by heating and then the condensation of the vapor by cooling.

Doppler effect – an apparent change in the frequency of sound or light due to the relative motion between the source of the sound or light and the observer.

E

Echo – a reflected sound distinguished from the original sound, usually arriving 0.1 s or more after the original sound.

Einstein mass-energy relation – $E = mc^2$; E is the energy released, m is the mass defect, and c is the speed of light.

Elastic potential energy – the potential energy of a body by its configuration (i.e., shape).

Elastic strain – an adjustment to stress in which materials recover their original shape after stress is released.

Electric circuit – consists of a voltage source that maintains an electrical potential, a continuous conducting path for a current to follow, and a device where the electrical potential does work; a switch in the circuit is used to complete or interrupt the conducting path.

Electric current – the flow of electric charge; the electric force field produced by an electrical charge.

Electric field line – an imaginary curve tangent to which at any given point gives the direction of the electric field at that point.

Electric field lines – a map of an electric field representing the direction of the force that a test charge would experience; the direction of an electric field shown by lines of force.

Electric generator – a mechanical device that uses wire loops rotating in a magnetic field to produce electromagnetic induction to generate electricity.

Electric potential energy – potential energy due to the position of a charge near other charges.

Electrical conductors – materials that have electrons free to move throughout the material (e.g., metals); allows electric current to flow through the material.

Electrical energy – a form of energy from electromagnetic interactions.

Electric force – a fundamental force that results from the interaction of electrical charges; it is the most powerful force in the universe.

Electrical insulators – electrical nonconductors, or materials that obstruct the flow of electric current.

Electrical nonconductors – materials with electrons that do not move easily within the material (e.g., rubber); also called *electrical insulators*.

Electrical resistance – the property of opposing or reducing electric current.

Electrode – a conductor, which dips into an electrolyte and allows the electrons to flow to and from the electrolyte.

Electrolysis – the production of a chemical change using electricity. Electrolysis can be used to split up water into hydrogen and oxygen.

Electrolyte – water solution of ionic substances that will conduct an electric current.

Electromagnet – a magnet formed by a solenoid that can be turned on and off by turning the current on and off.

Electromagnetic force – one of four fundamental forces; the force of attraction or repulsion between two charged particles.

Electromagnetic induction – the process in which current is induced in a coil whenever there is a change in the magnetic flux linked with the coil.

Electromagnetic waves – the waves due to oscillating electrical and magnetic fields and do not need any material medium for their propagation; can travel through a material medium (e.g., light waves and radio waves); travel in a vacuum with a speed 3.0×10^8 m/s.

Electron – a subatomic particle that has the smallest negative charge possible; usually found in an orbital of an atom but is gained or lost when atoms become ions.

Electron configuration – the arrangement of electrons in orbits and sub-orbits about the nucleus of an atom.

Electron current – the opposite of conventional current; considers electric current to consist of a drift of negative charges that flows from the negative terminal to the positive terminal of a battery.

Electron pair – a pair of electrons with different spin quantum numbers that may occupy an orbital.

Electron volt – the energy gained by an electron moving across a potential difference of 1 V equals 1.60×10^{-19} Joules.

Electronegativity – the comparative ability of atoms of an element to attract bonding electrons.

Electroplating – where metal is covered with a layer of another metal using electricity.

Electrostatic charge – an accumulated electric charge on an object from a surplus of electrons or a deficiency of electrons.

Element – a pure chemical substance that cannot be broken down into anything simpler by chemical or physical means; there are over 100 known elements, the fundamental materials of which matter is made.

Endothermic process – the process in which heat is absorbed.

Energy – the capacity of a body to do work; a scalar quantity; the SI unit is the Joule; there are five forms: mechanical, chemical, radiant, electrical, and nuclear.

Equilibrium – the condition of a system when neither its state of motion nor its internal energy state tends to change with time; a balanced object is in equilibrium.

Escape velocity – the minimum velocity with which an object must be thrown upward to overcome the gravitational pull and escape into space; the escape velocity depends on the mass and radius of the planet/star, but not on the mass of the body being thrown upward.

Evaporation – a process of more molecules leaving a liquid for the gaseous state than returning from the gas to the liquid; can occur at any given temperature from the surface of a liquid; takes place from the surface of the liquid; causes cooling; faster if the surface of the liquid is large, the temperature is higher, and the surrounding atmosphere does not contain a large amount of vapor of the liquid.

Exothermic process – the process in which heat is evolved.

F

Fahrenheit scale of temperature – the ice-point (lower fixed point) is taken as 32 °F, and the steam-point (upper fixed point) is taken as 212 °F; the interval between these two points is divided into 180 equal divisions; the unit division on the Fahrenheit scale is 1 °F; the relationship relates the temperatures on the Celsius scale and the Fahrenheit scale, C/100 = (°F – 32) / 180; the temperature of a healthy person is 37 °C or 98.6 °F.

Farad – the SI unit of capacitance; the capacitance of a capacitor that, if charged to 1 C, has a potential difference of 1 V.

Faraday – the electric charge required to liberate a gram equivalent of a substance; 1 Faraday = 9,6485 coulomb/mole.

Fermat's principle – an electromagnetic wave takes a path involving the least time when propagating between two points.

First Law of Motion – every object remains at rest or in a state of uniform straight-line motion unless acted on by an unbalanced force.

Fluid – matter that can flow or be poured; the individual molecules of fluid can move, rolling by another.

Focus – the point to which rays that are initially parallel to the axis of a lens or mirror converge or from which they appear to diverge.

Force (F) – a push or pull which tends to change the state of rest or uniform motion, the direction of motion or the shape and size of a body; a vector quantity; the SI unit is a Newton, denoted by N; one N is the force which when acting on a body of mass 1 kg produces an acceleration of 1 m/s^2.

Force of gravitation – the force with which two objects attract by their masses; acts even if the two objects are not connected; an action-at-a-distance force.

Fossil fuels – formed from the remains of plants and animals that lived millions of years ago.

Fracture strain – an adjustment to stress in which materials crack or break because of the stress.

Fraunhofer lines – the dark lines in the spectrum of the sun or a star.

Freefall – the motion of a body falling to Earth with no other force except the force of gravity acting on it; free-falling bodies are weightless.

Freezing – the changing of a liquid to a solid state.

Freezing point – the temperature at which a phase change of liquid to solid occurs; the same temperature as the melting point for a given substance.

Frequency – the number of oscillations completed in 1 second by an oscillating body.

Frequency (of oscillations) – the number of oscillations made by an oscillating body per second.

Frequency (of waves) – the number of waves produced per second.

Friction – the force that resists the motion of one surface relative to another with which it is in contact; caused by the humps and crests of surfaces, even those on a microscopic scale; the area of contact is small, and the consequent high pressure leads to local pressure welding of the surface; in motion, the welds are broken and remade continually.

Fuel – any substance that burns in oxygen to produce heat.

Fuel rod – long zirconium alloy tubes are containing fissionable material for use in a nuclear reactor.

Fundamental charge – the smallest common charge known; the magnitude of the charge of an electron and a proton, which is 1.60×10^{-19} coulombs.

Fundamental frequency – the lowest frequency (longest wavelength) at which a system vibrates freely and can set up standing waves in an air column or on a string.

Fundamental properties – a property that cannot be defined in more straightforward terms other than to describe how it is measured; the fundamental properties are length, mass, time, and charge.

Fuse – a safety device in an electric circuit. If the current (i.e., the flow of charge) gets too high, the wire in the fuse melts, which breaks the circuit switching off the current.

G

g – a symbol representing the acceleration of an object in free fall due to the force of gravity; its magnitude is 9.80 m/s^2.

Gamma (γ) ray – a high energy photon of short wavelength electromagnetic radiation emitted by decaying nuclei (γ-decay).

Gases – a phase of matter composed of molecules that are relatively far apart moving freely in constant, random motion and have weak cohesive forces acting between them, resulting in the characteristic indefinite shape and indefinite volume of a gas.

Graham's Law of Diffusion – the diffusion rate of a gas is inversely proportional to the square root of its density.

Gram-atomic weight – the mass in grams of one mole of an element that is numerically equal to its atomic weight.

Gram-formula weight – the mass in grams of one mole of a compound that is numerically equal to its formula weight.

Gram-molecular weight – the gram-formula weight of a molecular compound.

Gravitational constant (G) – term in the equation for Newton's Law of Gravitation; numerically, equal to the force of gravitation, which acts between two bodies with a mass of 1 kg each separated by 1 m; the value of G is 6.67×10^{-11} Nm^2/kg^2.

Gravitational potential at a point – the amount of work done against the gravitational forces to move a particle of unit mass from infinity to that point.

Gravitational potential energy (PE) – the energy possessed by a body by its height from the ground; equals *mgh.*

Gravity – the gravitational attraction at the surface of a planet or other celestial body.

Greenhouse effect – the process of increasing the temperature of the lower parts of the atmosphere through redirecting energy back toward the surface; the absorption and re-emission of infrared radiation by carbon dioxide (CO_2), water vapor and a few other gases in the atmosphere.

Ground state – the energy state of an atom with its electrons at the lowest energy state possible for that atom.

H

Half-life – time required for one-half of the unstable nuclei in a radioactive substance to decay into a new element.

Halogens – the elements in group seven in the periodic table.

Hard water – water that finds it difficult to form lather with soap.

Heat – a form of energy that makes a body hot or cold; measured by the temperature-effect, it produces in any material body; the SI unit is the Joule (J).

Heisenberg uncertainty principle – states that there is a fundamental limit to the precision with which certain pairs of physical properties of a particle (i.e., complementary variables) can be known simultaneously (e.g., one cannot measure the exact momentum and position of a subatomic particle at the same time – the more certain is one, the less certain is the other).

Hertz (Hz) – unit of frequency; equivalent to one cycle per second.

Hooke's Law – within the elastic limit, stress is directly proportional to strain.

Horsepower – unit of power; 1 hp = 746 Watts.

Humidity – the ratio of water vapor in a sample of air to the volume of the sample.

Huygens' principle – each point on a light wavefront can be regarded as a source of secondary waves, the envelope of these secondary waves determining the position of the wavefront later.

Hypothesis – a tentative explanation of a phenomenon that is compatible with the data and provides a framework for understanding and describing that phenomenon.

I

Ice-point – the melting point of ice under 1 atm pressure; equal to 0 °C or 32 °F.

Ideal gas equation – PV = nRT.

Immiscible liquids – liquids that do not mix to form a solution, e.g. oil and water.

Impulse – equal to the product of the force acting on a body and the time for which it acts; if the force is variable, the impulse is the integral of Fd_t from t_0 to t_1; the impulse of a force acting for a given time interval is equal to change in momentum produced over that interval; $J = m(v - u)$, assuming that the mass m remains constant while the velocity changes from v to u; the SI units are kg m/s.

Impulsive force – acts on a body for a short time but produces a large change in the momentum of the body.

Incandescent – matter emitting visible light because of high temperature (e.g., a light bulb, a flame from any burning source, the Sun).

Incident ray – line representing the direction of motion of incoming light approaching a boundary.

Index of refraction – the ratio of the speed of light in a vacuum to the speed of light in a material.

Indicator – a substance, which shows using a color change if a substance is acidic or basic.

Inertia – the property of matter that causes it to resist any change in its state of rest or of uniform motion; there are three kinds of inertia: the inertia of rest, the inertia of motion and the inertia of direction; the mass of a body is a measure of its inertia.

Infrasonic – sound waves at a frequency below the range of human hearing (less than 20 Hz).

Insulators – materials that are poor conductors of heat or electricity (e.g., wood or glass); materials with air pockets slow down the movement of heat because the air molecules are far apart.

Intensity – a measure of the energy carried by a wave.

Interference – the redistribution of energy due to the superposition of waves with a phase difference from coherent sources, resulting in alternate light and dark bands.

Intermolecular forces – interaction between molecules.

Internal energy – the sum of the kinetic energy and potential energy of the molecules of an object.

Inverse proportion – the relationship in which the value of one variable increases while the value of a second variable decreases at the same rate (in the same ratio).

Ion exchange – a method of removing hardness from water, and it replaces the positive ions that cause the hardness with H^+ ions.

Ion – a charged atom or group of atoms, e.g. Na+.

Ionic bond – a force of attraction between oppositely charged ions in a compound, resulting from a transfer of electrons.

Ionization – a process of forming ions from molecules.

Ionized – an atom or a particle with a net charge because it has gained or lost electrons.

Isobaric process – pressure remains constant.

Isochoric process – volume remains constant.

Isostasy – a balance or equilibrium between adjacent blocks of Earth's crust.

Isothermal process – in which temperature remains constant.

Isotope – atoms of the same element with the same atomic number (i.e., number of protons) but with a different mass number (i.e., number of neutrons).

J

Joule (J) – the unit used to measure work and energy; can also measure heat; 1 J = 1N·m.

Joule's Law of Heating – states that the heat produced when current (I) flows through a resistor (R) for a given time (t) is given by $Q = I^2Rt$.

K

Kelvin scale of temperature (K) – the lower fixed point is taken as 273.15 K (0 °C), and the steam-point (the upper fixed point) is taken as 373.15 K (100 °C); the interval between these two points is divided into 100 equal parts; each division is equal to 1 K.

Kelvin's statement of Second Law of Thermodynamics – it is impossible that, at the end of a cycle of changes, heat has been extracted from a reservoir, and an equal amount of work has been produced without producing some other effect.

Kepler's Laws of Planetary Motion – the three laws describing the motion of the planets.

Kepler's First Law – in planetary motion, each planet moves in an elliptical orbit, with the Sun located at one focus.

Kepler's Second Law – a radius vector between the Sun and a planet moves over equal areas of the ellipse during equal time intervals.

Kepler's Third Law – the square of the period of an orbit is directly proportional to the cube of the radius of the major axis of the orbit.

Kilocalorie (Kcal) – the amount of energy required to raise the temperature of 1 kg of water by 1 °C; 1 Kcal = 1,000 calories.

Kilogram – the fundamental unit of mass in the metric system of measurement.

Kinetic energy (KE) – possessed by a body due to its motion; $KE = \frac{1}{2}mv^2$, where m is mass and v is velocity.

L

Laser – a device that produces a coherent stream of light through stimulated emission of radiation.

Latent heat – energy released or absorbed by a body during a constant-temperature phase change.

Latent heat of vaporization – the heat absorbed when one gram of a substance changes from the liquid phase to the gaseous phase; also, the heat released when one gram of gas changes from the gaseous phase to the liquid phase.

Latent heat of fusion – the quantity of heat required to convert one unit mass of a substance from a solid state to a liquid state at its melting point without a change in its temperature; the SI unit is J kg^{-1}.

Latent heat of sublimation – the quantity of heat required to convert one unit of mass of a substance from a solid state to a gaseous state without a change in its temperature.

Law of Conservation of Energy – states that energy can neither be created nor destroyed but can be transformed.

Law of Conservation of Mass – mass (including single atoms) can neither be created nor destroyed in a chemical reaction.

Law of Conservation of Matter – matter can neither be created nor destroyed in a chemical reaction.

Law of Conservation of Momentum – the total momentum of a group of interacting objects remains constant in the absence of external forces.

Law of the lever – when a lever is balanced, the sum of the clockwise moments equals the sum of the anti-clockwise moments.

Lenz's Law – the induced current always flows in such a direction that it opposes the cause producing it.

Lever – a rigid body, which is free to turn about a fixed point called the fulcrum.

Light – a form of energy.

Light-year – the distance that light travels in a vacuum in one year (365.25 days); approximately 9.46×10^{15} m.

Line spectrum – an emission (of light, sound, or other radiation) spectrum consisting of separate isolated lines (discrete frequencies or energies); can be used to identify the elements in a matter of unknown composition.

Lines of force – lines drawn to make an electric field strength map, with each line originating on a positive charge and ending on a negative charge; each line represents a path on which a charge would experience a constant force; having the lines closer indicates a more energetic electric field.

Liquids – a phase of matter composed of molecules that have interactions stronger than those found in gas but not strong enough to keep the molecules near the equilibrium positions of a solid, resulting in the characteristic definite volume but the indefinite shape of a liquid.

Liter – a metric system unit of volume; usually used for liquids.

Longitudinal strain – the ratio of change in the length of a body to its initial length.

Longitudinal waves – the particles of the medium oscillate along the direction of propagation of a wave (e.g., sound waves).

Loudness – a subjective interpretation of a sound that is related to the energy of the vibrating source, related to the condition of the transmitting medium and the distance involved.

Lubricant – a substance capable of reducing friction (i.e., force that opposes the direction of motion).

Luminosity – the total amount of energy radiated into space each second from the surface of a star.

Luminous – objects that produce visible light (e.g., the Sun, stars, light bulbs, burning materials).

Lunar eclipse – when the Earth passes between the sun and the moon.

Lyman series – a group of lines in the ultraviolet region in the spectrum of hydrogen.

M

Magnetic domain – tiny physical regions in permanent magnets, approximately 0.01 to 1 mm, have magnetically aligned atoms, giving the domain an overall polarity.

Magnetic field – the region around a magnet where other magnetic objects experience its magnetic force; a model used to describe how magnetic forces on moving charges act at a distance.

Magnetic poles – the ends, or sides, of a magnet about which the force of magnetic attraction seems to be concentrated.

Magnetic quantum number – from the quantum mechanics model of the atom, one of four descriptions of the energy state of an electron wave; describes the energy of an electron orbital as the orbital is oriented in space by an external magnetic field, a kind of energy sub-sublevel.

Magnetic reversal – the changing of polarity of the Earth's magnetic field as the north magnetic pole and the south magnetic pole exchange positions.

Magnetic wave – the spread of magnetization from a small portion of a substance from an abrupt change in the magnetic field.

Magnification – the ratio of the size of the image to the size of the object.

Magnitude – the size of a measurement of a vector; scalar quantities that consist of a number and unit only.

Malleable – can be hammered into sheets (e.g., metals).

Malus Law – the intensity of the light transmitted from the analyzer varies directly as the square of the cosine of the angle between the plane of transmission of the analyzer and the polarizer.

Maser – microwave amplification by stimulated emission of radiation.

Mass (m) – the quantity of matter in a body; the SI unit is the kg; remains the same everywhere; a measure of inertia, which means resistance to a change of motion.

Mass defect – the difference between the sum of the masses of the individual nucleons of a nucleus and mass of nucleus.

Mass number – the sum of the number of protons and neutrons in a nucleus; used to identify isotopes (e.g., Uranium-238).

Matter – anything that occupies space and has mass.

Mean life – the average time during which a system (e.g., atom, nucleus) exists in a specified form.

Mechanical energy – the sum of the potential energy (PE) and the kinetic energy (KE) of a body; energy associated with the position of a body.

Mechanical waves – require a material medium for propagation (e.g., sound waves and water waves); also *elastic waves*.

Megahertz (MHz) – unit of frequency; equal to 106 Hertz.

Melting – the changing of a solid to a liquid state.

Melting point – the temperature at which a phase change of solid to liquid takes place.

Metal – matter having the physical properties of conductivity, malleability, ductility, and luster.

Meter – the fundamental metric unit of length.

MeV – a unit of energy; equal to 1.6×10^{-13} joules.

Millibar – a measure of atmospheric pressure equivalent to 1,000 dynes per cm^2.

Miscible fluids – fluids that can mix in any proportion.

Mixture – matter made of unlike parts with variable composition and separated into their parts by physical means.

Model – a mental or physical representation of something that cannot be observed directly; used as an aid to understanding.

Modulus of elasticity – the ratio of stress to the strain produced in a body.

Modulus of rigidity – the ratio of tangential stress to the shear strain produced in a body.

Mole – the amount of a substance that contains Avogadro's number of atoms, ions, molecules, or any other chemical unit; 6.02×10^{23} atoms, ions, or other chemical units.

Molecule – two or more atoms chemically combined.

Moment – a measure of the turning effect of a force. Moment of a force = force × perpendicular distance from the fulcrum.

Momentum – a measure of the quantity of motion; the product of the mass and the velocity of a body; SI units are kg·m /s.

Monochromatic light – consisting of a single wavelength.

N

Natural frequency – the frequency of oscillation of an elastic object in the absence of external forces; depends on the size, composition, and shape of the object.

Negative electric charge – one of two types of electric charge; repels other negative charges and attracts positive charges.

Negative ion – atom or particle that has a surplus or imbalance of electrons and a negative charge.

Net force – the resulting force after vector forces are summed; if a net force is zero, the vector forces have canceled, and there is no unbalanced force.

Neutralization – the reaction between an acid and a base to give salt and water.

Newton (N) – a unit of force defined as $kg{\cdot}m/s^2$; 1 Newton is needed to accelerate a 1 kg mass by 1 m/s^2.

Newton's First Law of Motion – a body continues in a state of rest or uniform motion in a straight line unless an external (unbalanced) force acts upon it.

Newton's Law of Gravitation – the gravitational force of attraction acting between two particles is directly proportional to the product of their masses and inversely proportional to the square of the distance between them; the force of attraction acts along the line joining the two particles; real bodies having spherical symmetry act as point masses with their mass assumed to be concentrated at their center of mass.

Newton's Second Law of Motion – the rate of change of momentum is equal to the force applied; the force acting on a body is directly proportional to the product of its mass and acceleration produced by force in the body.

Newton's Third Law of Motion – for every action, there is an equal and opposite reaction; the action and the reaction act on two different bodies simultaneously.

Noise – sounds made up of groups of waves of random frequency and intensity.

Non-uniform acceleration – when the velocity of a body increases by unequal amounts in equal intervals of time.

Non-uniform speed – when a body travels unequal distances in equal intervals of time.

Non-uniform velocity – when a body covers unequal distances in equal intervals of time in a direction, or when it covers equal distances in equal intervals but changes its direction.

Normal (*N*) – a line perpendicular to the surface of a boundary.

Nuclear energy – the form of energy from reactions involving the nucleus.

Nuclear fission – the splitting of a heavy nucleus into more stable, lighter nuclei with an accompanying release of energy.

Nuclear force – one of four fundamental forces; a strong force of attraction that operates over short distances between subatomic particles; overcomes the electric repulsion of protons in a nucleus and binds the nucleus.

Nuclear fusion – a nuclear reaction of low mass nuclei fusing to form a more stable and more massive nucleus with an accompanying release of energy.

Nuclear reactor – a steel vessel in which a controlled chain reaction of fissionable materials releases energy.

Nucleons – a collective name for protons and neutrons in the nucleus of an atom.

Nucleus – the central, positively charged, dense portion of an atom; contains protons and neutrons.

O

Octet rule – during bonding, atoms tend to reach an electron arrangement with eight electrons in the outermost shell.

Ohm – unit of resistance; 1 ohm = 1volt/ampere.

Ohm's Law – the current flowing through a conductor is directly proportional to the potential difference across the ends of the conductor. At a constant temperature, the voltage across a conductor is proportional to the current flowing through it. Voltage = current × resistance (V = IR).

Open system – a system across whose boundaries matter and energy can pass.

Optical fiber – a long, thin thread of fused silica; used to transmit light; based on total internal reflection.

Orbital – the region of space around the nucleus of an atom where an electron is likely to be found.

Origin – the point on a graph where the x and the y variables have a value of zero at the same time.

Oscillatory motion – the to and fro motion (periodic) of a body about its mean position; also, *vibratory motion.*

Oxidation – the addition of oxygen or the losing of electrons.

P

Pascal – a unit of pressure equal to the pressure resulting from a force of 1 N acting uniformly over an area of 1 m^2.

Pascal's Law – states that the pressure exerted on a liquid is transmitted equally in all directions.

Paschen series – a group of lines in the infrared region in the spectrum of hydrogen.

Pauli exclusion principle – no two electrons in an atom can have the same four quantum numbers; a maximum of two electrons can occupy a given orbital.

Peltier effect – the evolution or absorption of heat at the junction of two dissimilar metals carrying current.

Period (of a wave) – the time for a wave to travel through a distance equal to its wavelength; denoted by T; period of a wave = 1/frequency of the wave.

Period (of an oscillation) – the time to complete one oscillation; does not depend upon the mass of the bob and amplitude of oscillation; directly proportional to the square root of the length and inversely proportional to the square root of the acceleration due to gravity.

Periodic wave – a wave in which the particles of the medium oscillate continuously about their mean positions regularly at fixed intervals of time.

Periodic motion – a motion that repeats at regular time intervals.

Permeability – the ability to transmit fluids through openings, tiny passageways, or gaps.

pH scale – a scale from 0 to 14. If the pH of a solution is 7 it is neutral; if the pH of a solution is less than 7 it is acidic; if the pH of a solution is greater than 7 it is basic.

Permanent hardness – hardness in water that cannot be removed by boiling. It is caused by calcium sulfate.

Phase – when particles in a wave are in the same state of vibration (i.e., in the same position and the direction of motion).

Phase change – the action of a substance changing from one state of matter to another; always absorbs or releases internal potential energy that is not associated with a temperature change.

Photons – quanta of energy in the light wave; the particle associated with light.

Photoelectric effect – the emission of electrons in some materials when the light of a suitable frequency falls on them.

Physical change – a change of the state of a substance but not in the identity of the substance.

Pitch (sound) – depends on the frequency of the wave.

Planck's constant – proportionality constant in the ratio of the energy of vibrating molecules to their frequency of vibration; a value of 6.63×10^{-34} J·s.

Plasma – a phase of matter; a hot highly ionized gas consisting of electrons and atoms that have been stripped of their electrons because of high kinetic energies.

Plasticity – the property of a solid whereby it undergoes a permanent change in shape or size when subjected to a stress.

Plastic strain – an adjustment to stress in which materials become molded or bent out of shape under stress and do not return to their original shape after the stress is released.

Polarized light – light whose constituent transverse waves are vibrating in the same plane.

Polaroid – a film that transmits only polarized light.

Polaroid or polarizer – a device that produces polarized light.

Positive electric charge – one of the two types of electric charge; repels positive and attracts negative charges.

Positive ion – atom or particle with a net positive charge due to an electron or electrons being torn away.

Positron – an elementary particle having the same mass as an electron but an equal and positive charge.

Potential difference (or *voltage*) – the force, which moves the electrons around the circuit. Unit is Volt (V).

Potential energy (PE) – possessed by a body by its position or configuration; see the *Gravitational potential energy* and *Elastic potential energy*.

Power – scalar quantity for the rate of doing work; the SI unit is Watt; 1 W = 1 J/s. The rate at which energy is converted from one form to another. Unit is Watts (W). Power = voltage × current (P = VI).

Pressure – a measure of force per unit area (e.g., kilograms per square meter (kg/m^2). Unit is Pascal (Pa).

Primary coil – part of a transformer; a coil of wire connected to a source of alternating current.

Primary colors – three colors (red, yellow, and blue) combined in various proportions to produce any other color. When red, green, and blue are combined, it results in white. For example, red + green + blue = white.

Principal quantum number – from the quantum mechanics model of the atom, one of four descriptions of the energy state of an electron wave; describes the main energy level of an electron regarding its most probable distance from the nucleus.

Principle of calorimetry – states that if two bodies of different temperature are in thermal contact, and no heat is allowed to go out or enter the system; heat lost by the body with higher temperature is equal to the heat gained by the body of lower temperature (i.e., the heat lost = the heat gained).

Products – chemicals produced in a chemical reaction.

Progressive wave – a wave that transfers energy from one part of a medium to another.

Projectile – an object thrown into space horizontally or at an acute angle and under the action of gravity; the path followed by a projectile is its trajectory; the horizontal distance traveled by a projectile is its range; the time is taken from the moment it is thrown until the moment it hits the ground is its time of flight.

Proof – a measure of ethanol concentration of an alcoholic beverage; double the concentration by volume (e.g., 50% by volume is 100 proof).

Properties – qualities or attributes that, taken together, are usually unique to an object (e.g., color, texture, size).

Proportionality constant – a value applied to a proportionality statement that transforms the statement into an equation.

Pulse – a wave of short duration confined to a small portion of the medium at any given time; also a *wave pulse*.

Q

Quanta – fixed amounts; usually referring to fixed amounts of energy absorbed or emitted by matter.

Quantum limit – the shortest wavelength; present in a continuous x-ray spectrum.

Quantum mechanics – model of the atom based on the wave nature of subatomic particles and the mechanics of electron waves; also *wave mechanics*.

Quantum numbers – numbers that describe the energy states of an electron; in the Bohr model of the atom, the orbit quantum numbers could be any whole number (e.g., 1, 2, 3, etc.); in the quantum mechanics model of the atom, four quantum numbers are used to describe the energy state of an electron wave (n, m, l, and s).

Quark – one of the hypothetical fundamental particles; has a charge with magnitudes of one-third or two-thirds of the charge on an electron.

R

Rad – a measure of radiation received by a material (radiation-absorbed dose).

Radiant energy – the form of energy that can travel through space (e.g., visible light and other parts of the electromagnetic spectrum).

Radiation – the emission and propagation of waves transmitting energy through space or some medium. Heat transfer through invisible rays, which travel outwards from the hot object without a medium.

Radioactive decay – the natural, spontaneous disintegration or decomposition of a nucleus.

Radioactive decay constant – a specific constant for a isotope that is the ratio of the rate of nuclear disintegration per unit of time to the total number of radioactive nuclei.

Radioactive decay series – series of decay reactions that begins with one radioactive nucleus that decays to a second nucleus that decays to a third nucleus and so on, until a stable nucleus is reached.

Radioactive Decay Law – the rate of disintegration of a radioactive substance is directly proportional to the number of undecayed nuclei.

Radioactivity – spontaneous emission of particles or energy from an atomic nucleus as it disintegrates.

Rarefaction – a part of a longitudinal wave where the density of the particles of the medium is less than the typical density.

Real image – an image generated by a lens or mirror that can be projected onto a screen.

Reactants – chemicals that react together in a chemical reaction.

Rectifier – converts alternating current to direct current.

Rectilinear motion – the motion of a body in a straight line.

Reduction – the removal of oxygen or the gaining of electrons.

Reflected ray – a line representing the direction of motion of light reflected from a boundary.

Reflection – the bouncing back of light from a surface.

Refraction – the bending of a light wave, a sound wave, or another wave from its straight-line path as it travels from one medium to another.

Refractive index – the ratio of the speed of light in a vacuum to that in the medium.

Relative density – (i.e., *specific gravity*) is the ratio of the density (mass of a unit volume) of a substance to the density of given reference material. *Specific gravity* usually means relative density concerning water. The term *relative density* is more common in modern scientific usage.

Relative humidity – the percentage of the amount of water vapor present in a specific volume of the air to the amount of water vapor needed to saturate it.

Resistance (R) – the opposition of a conductor to current (i.e., the flow of charge). A good conductor has a low resistance, and a bad conductor has a high resistance.

Resolving power – the ability of an optical instrument to produce separable images of different points of an object.

Resonance – when the frequency of an external force matches the natural frequency of the body.

Restoring force – the force which tends to bring an oscillating body back to its mean position whenever it is displaced from the mean position.

Resultant force – a single force, which acts on a body to produce the same effect on it as done by other forces collectively; see *balanced forces*.

Reverberation – apparent increase in the volume of sound caused by reflections from the boundary surfaces, usually arriving within 0.1 seconds after the original sound.

Rigid body – an idealized extended body whose size and shape are fixed and unaltered when forces are applied.

S

Salt – when a metal replaces the hydrogen of an acid.

Saturated air – air in which an equilibrium exists between evaporation and condensation; the relative humidity is 100 percent.

Saturated solution – the apparent limit to dissolving a given solid in a specified amount of water at a given temperature; a state of equilibrium exists between dissolving solute and solute coming out of solution.

Scalar quantity – a physical quantity described entirely by its magnitude.

Scientific law – a relationship between quantities; usually described by an equation in the physical sciences; describes a broader range of phenomena and is more important than a scientific principle.

Scientific principle – a relationship between quantities within a specific range of observations and behavior.

Second – the standard unit of time in the metric and English systems of measurement.

Second Law of Motion – the acceleration of an object is directly proportional to the net force acting on that object and inversely proportional to the mass of the object.

Secondary coil – part of a transformer; a coil of wire in which the voltage of the original alternating current in the primary coil can be stepped up or down by electromagnetic induction.

Secondary colors – formed when two primary colors (i.e., red, green, blue) are mixed. The three secondary colors are yellow, magenta, and cyan. For example, red + green = yellow; red + blue = magenta; and blue + green = cyan.

Second's pendulum – a simple pendulum whose period on the surface of the Earth is 2 seconds.

Semiconductors – elements whose electrical conductivity is intermediate between that of a conductor and an insulator.

Shear strain – the ratio of the relative displacements of one plane to its distance from the fixed plane.

Shear stress – the restoring force developed per unit area when deforming force acts tangentially to the surface of a body, producing a change in the shape of the body without any volume change.

Siemens – the derived SI unit of electrical conductance; equal to the conductance of an element with a resistance of 1 ohm; also written as ohm^{-1}.

Simple harmonic motion – the vibratory motion occurs when the restoring force is proportional to the displacement from the mean position and is directed opposite to the displacement.

Simple pendulum – a heavy point mass (a small metallic ball) suspended by a light inextensible string from frictionless rigid support; a simple machine based on the effect of gravity.

Snell's Law – states that the ratio of *sin* i to *sin* r is a constant and is equal to the refractive index of the second medium concerning the first.

Solar eclipse – happens when the moon passes between the sun and the Earth.

Solenoid – a cylindrical coil of wire that becomes electromagnetic when current passes through it.

Solids – a phase of matter with molecules that remain close to fixed equilibrium positions due to strong interactions between the molecules, resulting in the characteristic definite shape and definite volume of a solid.

Solution – a mixture of a solute (usually a solid) and a solvent (usually a liquid).

Sonic boom – sound waves that pile up into a shock wave when a source is traveling at or faster than the speed of sound.

Sound – a form of energy.

Specific gravity – see *relative density*.

Specific heat – the amount of heat energy required to increase the temperature of 1 g of a substance by 1 °C; each substance has its specific heat value.

Speed – a scalar quantity for the distance traveled by a body per unit of time; if a body covers the distance in time, its speed is given by distance/time; SI units are m/s.

Spin quantum number – from the quantum mechanics model of the atom, one of four descriptions of the energy state of an electron wave; describes the spin orientation of an electron relative to an external magnetic field.

Stable equilibrium – a body is in stable equilibrium if, when slightly moved, its center of gravity rises.

Standing waves – the condition where two waves of equal frequency traveling in opposite directions meet and form stationary regions of maximum displacement due to constructive interference and stationary regions of zero displacement due to destructive interference.

State of motion – when a body changes its position concerning a fixed point in its surroundings; the states of rest and motion are relative to the frame of reference.

State of rest – when a body does not change its position concerning a fixed point in its surrounding; the states of rest and motion are relative to the frame of reference.

Steam-point – the temperature of steam over pure boiling water under 1 atm pressure; taken as the upper fixed point (100 °C or 212 °F) for temperature scales.

Stefan-Boltzmann Law – the amount of energy radiated per second per unit area of a perfectly black body is directly proportional to the fourth power of the absolute temperature of the surface of the body.

Sublimation – the changing of a solid directly to a gas. (Iodine is an example of a substance that sublimes).

Superconductors – some materials in which, under certain conditions, the electrical resistance approaches zero.

Super-cooled – water in the liquid phase when the temperature is below the freezing point.

Supersaturated – containing more than the average saturation amount of a solute at a given temperature.

Surface tension – the property of a liquid due to which its surface behaves like a stretched membrane.

Suspension – a mixture of a liquid and a finely divided insoluble solid.

T

Temperature – a measure of the hotness or coldness of a body; according to the molecular model, measures the average kinetic energy of the molecules; heat flows from a body at a higher temperature to a body at a lower temperature.

Temporary hardness – hardness in water that can be removed by boiling. It is caused by calcium hydrogen carbonate.

Tensional stress – the opposite of compressional stress; when one part of a plate moves away from another part that does not move.

Tesla (T) – SI unit of magnetic flux density; the magnetic flux density of a magnetic flux of 1 Wb through an area of 1 m^2.

Thermal capacity – the quantity of heat required to raise the temperature of the body by one degree (1 K or 1 °C).

Thermal equilibrium – when two bodies in contact are at the same temperature, and there is no heat flow between them; also, the average temperature of the bodies in thermal equilibrium.

Thermal expansion – the increase in the size of an object when heated.

Thermometer – a device used for the measurement of temperature; the mercury thermometer is commonly used.

Third Law of Motion – when two objects interact, the force exerted on one object is equal in size and opposite in direction to the force exerted on the other object; forces always occur in matched pairs that are equal and opposite.

Titration – the process of adding one solution from a burette to a measured amount of another solution to find out exactly how much of each is required to react.

Total internal reflection – a condition where all light is reflected from a boundary between materials; occurs when light travels from a denser to a rarer medium, and the angle of incidence is greater than the critical angle.

Transformation of energy – converting one form of energy into another (e.g., when a body falls, its potential energy is converted to kinetic energy).

Transverse wave – a wave in which the particles of the medium oscillate in a direction perpendicular to the direction of propagation of the wave (e.g., water waves, light waves, radio waves).

Trough – the point of maximum negative displacement on a transverse wave.

U

Ultrasonic – sound waves too high in frequency (above 20,000 Hz) to be heard by the human ear.

Unbalanced forces – when some forces act on a body and the resultant force is not zero.

Uniform acceleration – when the velocity of a body increases by equal amounts in equal intervals of time.

Uniform circular motion – the motion of an object in a circular path with uniform speed; accelerated motion.

Uniform speed – when a body travels equal distances in equal intervals of time.

Uniform velocity – when a body travels along a straight line in a direction with equal distance in equal time intervals.

Universal Law of Gravitation – every object is attracted to every other object with force directly proportional to the product of their masses and inversely proportional to the square of the distance between the centers of the two masses.

Unpolarized light – light consisting of transverse waves vibrating in all possible random directions.

Unstable equilibrium – a body is in unstable equilibrium if, when slightly moved, its center of gravity falls.

V

Van der Waals force – general term for weak attractive intermolecular forces.

Valency – the number of electrons an atom wants to gain, lose, or share to have a full outer shell.

Vapor – the gaseous state of a substance that is generally in a liquid state.

Vector quantity – a quantity that needs magnitude and direction to describe it.

Velocity (v) – distance traveled by a body in a direction per unit time; the displacement of the body per unit time; a vector quantity; the SI units are m/s.

Vibration – a back and forth motion that repeats itself.

Virtual image – an image formed when the reflected or refracted light rays appear to meet; this image cannot be projected on a screen.

Volt (V) – a unit of potential difference equivalent to joules/coulomb.

Voltage drop – the difference in electric potential across a resistor or other part of a circuit that consumes power.

Volume – the amount of space an object occupies.

W

Watt (W) – SI unit for power; equivalent to joule/s.

Wave – a disturbance or oscillation that moves through a medium.

Wavelength (λ) – the distance between the two nearest points on a wave in the same phase; the distance between two adjacent crests or two adjacent troughs.

Wave (mechanical) – a periodic disturbance produced in a material medium due to the vibratory motion of the particles of the medium.

Wave mechanics – alternate name for quantum mechanics derived from the wavelike properties of subatomic particles.

Wave motion – the movement of a disturbance from one part of a medium to another involving the transfer of energy but not the transfer of matter.

Wave period – the time required for two successive crests (or successive regions) of the wave to pass a given point.

Wave velocity – the distance traveled by a wave in one second; depends on the medium through which it passes.

Weight – the force with which a body is attracted towards the center of the Earth; the SI unit is N; the gravitational units are kg·wt and g·wt; the weight of a body is given by *mg*. Weight = mass × acceleration due to gravity.

Weightlessness – the state when the apparent weight of a body becomes zero; objects while falling freely under the action of gravity are seemingly weightless.

Wien's Displacement Law – states that for a black body, the product of the wavelength corresponding to its maximum radiance and its absolute temperature is constant.

Work – work is done when a force acting on a body displaces it; work = force × displacement (W = Fd) in the direction of the force; work is a scalar quantity; the SI unit is Joule.

Y

Young's modulus of elasticity – the ratio of normal stress to the longitudinal strain produced in a body.

Z

Zeeman effect – the splitting of the spectral lines in a spectrum when the source is exposed to a magnetic field.

Zeroth Law of Thermodynamics – states that if body A is in thermal equilibrium with body B, and B is also in thermal equilibrium with C, then A is necessarily in thermal equilibrium with C.

www.ingramcontent.com/pod-product-compliance
Lightning Source LLC
LaVergne TN
LVHW082014200926
848481LV00010B/2858